T0321395

Fundamentals of Laboratory Animal Science

Fundamentals of Laboratory Animal Science

Edited by

Enqi Liu

Research Institute of Atherosclerotic Disease,
Xi'an Jiaotong University Cardiovascular
Research Center, Xi'an, Shaanxi, China
and
Laboratory Animal Center, Xi'an Jiaotong University
Health Science Center, Xi'an, Shaanxi, China

Jianglin Fan

Department of Molecular Pathology, Faculty of
Medicine, Graduate School of Medical Sciences,
University of Yamanashi, Yamanashi, Japan

CRC Press
Taylor & Francis Group
Boca Raton London New York

CRC Press is an imprint of the
Taylor & Francis Group, an **informa** business

CRC Press
Taylor & Francis Group
6000 Broken Sound Parkway NW, Suite 300
Boca Raton, FL 33487-2742

© 2018 by Taylor & Francis Group, LLC
CRC Press is an imprint of Taylor & Francis Group, an Informa business

No claim to original U.S. Government works

Printed on acid-free paper

International Standard Book Number-13: 978-1-4987-4351-8 (Hardback)

Library of Congress Cataloging-in-Publication Data

Names: Liu, Enqi, editor. | Fan, Jianglin, editor.
Title: Fundamentals of laboratory animal science / [edited by] Enqi Liu and Jianglin Fan.
Description: Boca Raton : CRC Press, 2017. | Includes bibliographical references and index.
Identifiers: LCCN 2017006542 | ISBN 9781498743518 (hardback : alk. paper)
Subjects: | MESH: Laboratory Animal Science | Animals, Laboratory | Animal Experimentation | Models, Animal
Classification: LCC SF996.5 | NLM QY 50 | DDC 616.02/73--dc23
LC record available at https://lccn.loc.gov/2017006542

Visit the Taylor & Francis Web site at
http://www.taylorandfrancis.com

and the CRC Press Web site at
http://www.crcpress.com

Contents

Preface

Laboratory animals have become increasingly important for biomedical research, and it is estimated that approximately 70% of biomedical research is associated with the use of experimental animals. For just over a century the Nobel Prize has been awarded each year in recognition of the world's greatest biomedical advances. Of the 106 Nobel Prizes awarded for physiology or medicine, 95 were directly dependent on animal study. Animal experiments not only expand our knowledge of science, but also greatly improve human and animal health. Since 1900, animal research achievements have helped extend the longevity of Americans by approximately 25 years. However, there are those who disapprove of animal research because they believe such experiments to be expensive and harmful to animals. Therefore, it is incumbent on researchers to find a way to effectively implement animal study based on the 3Rs (reduction, replacement, and refinement) principle and animal welfare.

The curriculum of Laboratory Animal Science is provided in the majority of medical schools and agriculture universities in the world. However, a suitable and practical textbook that contains state-of-the-art techniques is still lacking. In 2004, 2008, and 2014, with the assistance of the Chinese Ministry of Health, we successfully published two books in Chinese titled *Medical Laboratory Animal Science* and *Animal Models of Human Diseases*, which have been used as the standard textbooks for graduate school students in China and have become best sellers in medical schools in China. Recently, after consultation with professionals who perform biomedical studies using laboratory animal models, we feel that it is necessary to publish a textbook of *Laboratory Animal Science* in English for readers worldwide. For this purpose, we decided to compile the current textbook, which will be useful for teaching and training in this field. This book focuses on theoretical instruction, experimental operation, and skills training, and systematically covers both basic information of laboratory animals and techniques for animal experiments.

This book is divided into eight chapters. Chapter 1 provides an overview of laboratory animal science, summarizing some basic concepts and categories of laboratory animals and animal experiments. In addition, it introduces a brief history, applications in biomedicine, and management of laboratory animals. The welfare of laboratory animals has attracted increasing attention in recent years. Chapter 2 describes in detail the definition of animal welfare and the measures taken to ensure protection of experimental animals. Chapter 3 discusses the standardization of laboratory animals and stresses its importance for animal experimentation. This chapter outlines several key elements that are closely related to animal experimentation such as animal genetics, health, facilities, nutrition, and the like. Chapter 4 introduces the most commonly used laboratory animals, such as mice, rats, hamsters, guinea pigs, rabbits, dogs, pigs, rhesus monkeys, chickens, and fish and mainly discusses and compares animal anatomy, biology, husbandry, and applications in biomedical research. Chapter 5 discusses models of human disease and introduces how to select and create an animal model for studying specific human diseases. Information about the newest genetically modified animals is also included in this chapter. Chapter 6 provides a summary of the technologies used in animal experimentation. All of these skills ensure research goes smoothly while greatly reducing animal pain and distress. Chapter 7 discusses how to design a suitable animal experiment. It introduces the content and procedure of animal experimental design. Chapter 8 in which readers will learn the whole process and protocol of an animal experiment, summarizes the organization and management of animal experiments. We believe that this book will become not only a standard textbook for undergraduate and postgraduate students, but also a life tool on the bookshelf for veterinarians, researchers, animal-care staff, and other professionals who are involved in animal experiments.

Acknowledgments

We would like to thank Mr. John Sulzycki, a senior editor at CRC Press, Taylor & Francis Group, for his help toward publishing this book. We also thank our staff Mr. Lu Li and Ms. Hong Zhu for their help in taking and preparing photographs for this book. This work could not be accomplished without support from the Natural Science Foundation of Shaanxi Province (2014PT013 and 2017JZ028), 2015 Graduate Teaching Research and Teaching Reform Project of Xi'an Jiaotong University, and Grants-in-Aid for Scientific Research from the Ministry of Education, Culture, Sports and Technology, Japan (22390068, 25670190, and 15H04718). Finally, we would like to thank our spouses, Yali and Jing, for their love, support, and also tolerance of our absence.

Enqi Liu
Jianglin Fan

Editors

Enqi Liu, PhD is a professor and director of the Research Institute of Atherosclerotic Disease, Xi'an Jiaotong University Cardiovascular Research Center and Laboratory Animal Center, Xi'an Jiaotong University Health Science Center. Dr. Liu earned his bachelor's degree in animal genetics and master's degree in veterinary medicine from China Agricultural University, and his PhD from Saga University in Japan.

Currently, Dr. Liu also holds the chair of the professor at Xi'an Jiaotong University Health Science Center, Xi'an. His research interests include laboratory animal science, experimental pathology, atherosclerosis, lipid metabolism, and translational research. He has published more than 70 articles in peer-reviewed journals. He has also published several professional books including *Laboratory Animal Genetics and Breeding* (2002), *Medical Laboratory Animal Science* (2004, 2008), and *Animal Models of Human Diseases* (2014).

Jianglin Fan, MD, PhD, FAHA is a professor and chairman of molecular pathology at University of Yamanashi School of Medicine, Yamanashi, Japan. He also holds the chair of adjunct professor at Xi'an Medical University and Xi'an Jiaotong University. Dr. Fan earned his MD from the Yanbian University, School of Medicine and his PhD from Saga Medical School. He completed a postdoctoral fellowship at Gladstone Institute at the University of California, San Francisco. Dr. Fan's research focuses on atherosclerosis, including elucidation of the molecular mechanisms of atherosclerosis, lipid and lipoprotein metabolism, and inflammation. One of his major achievements was developing novel transgenic and knockout rabbit models for the study of cardiovascular disease. His work has led to more than 150 publications in prestigious peer-reviewed journals.

Contributors

Changqing Gao
Department of Laboratory Animal
Central South University
Changsha, Hunan, China

Dongmei Tan
Laboratory Animal Center
Chongqing Medical University
Chongqing, China

Yi Tan
Laboratory Animal Center
Chongqing Medical University
Chongqing, China

Qi Yu
Institute of Basic and Translational
    Medicine
Xi'an Medical University
Xi'an, Shaanxi, China

Sihai Zhao
Laboratory Animal Center
Xi'an Jiaotong University Health
    Science Center
Xi'an, Shaanxi, China

CHAPTER 1

Laboratory Animals and Biomedical Research

Enqi Liu and Jianglin Fan

CONTENTS

1.1 INTRODUCTION OF LABORATORY ANIMAL SCIENCE

1.1.1 Laboratory Animal Science

High-quality laboratory animals and accurate animal experimental results in biomedical research were initially called for in the 1950s. Thus, an independent specialty on laboratory animals and animal experimentation came into being. The content of laboratory animal science includes animal heredity, breeding, quality control, disease prevention, and animal welfare. The science of animal experimentation refers to animal experiments to obtain novel, scientific experimental data under an animal welfare guarantee.

Some of the major tasks of laboratory animal science are to provide laboratory animals, obtain accurate, reproducible data, and collect scientific information for biomedical research.

Animal experimentation is mainly applied in the fields of medicine, biology, veterinary medicine, and agriculture. In terms of animal quantities used in experimentation, most of the animals are used in the medical field, including teaching and training, medical research, and function detection and security inspection for medicines, biological products, and food. Thus, the science of medical laboratory animals and medical animal experimentation is referred to as medical laboratory animal science.

As a new and independent discipline, laboratory animal science has been rapidly developed since the rediscovering of Mendelism. In 1944, the American Academy of Science put the issue of laboratory animal standardization on the agenda for the first time, which is usually regarded as the starting point of modern laboratory animal science. In 1966, "laboratory animal science" first appeared in scientific literatures, marking the birth of this new discipline.

Laboratory animal science is a relatively young science. However, it has built an entire theory system and has derived subdisciplines such as laboratory animal breeding, laboratory animal microbiology, laboratory animal environmental ecology, laboratory animal nutrition, laboratory animal medicine, comparative medicine, and laboratory animal husbandry.

Laboratory animal science has great significance. First, it is an important means to study biomedicine and can directly affect the establishment, implementation, and reliability of the biomedical projects. Second, the improvement and development of laboratory animal science has brought various projects into a new field of vision and promoted the development of biomedicine.

1.1.2 Laboratory Animals

1.1.2.1 Species of Laboratory Animals

There are a great variety of animals in nature. Currently, there are more than 1.5 million animal species in the world. Based on natural classification, morphological characters, internal structure, biogenesis, and kinship, animals can be classified into *phylum, class, order, family, genus,* and *species.* Additionally, animals can be subdivided into *subphylum, subclass, suborder, subfamily, subgenus,* and *subspecies.*

A species is often defined as a group of living organisms consisting of similar individuals capable of exchanging genes or interbreeding, which is one of the basic units of biological classification and a taxonomic rank. Species is the basic unit of biological classification and its formation is the result of natural selection. For example, C57BL/6 inbred strain mice, which are the most commonly used experimental tool in modern biomedical research, belong to phylum Vertebrata, class Mammalia, subclass Eutheria, order Rodentia, suborder Myomorpha, family Muridae, genus *Mus,* and species *Mus musculus.* At present, the laboratory mice used worldwide in biomedical research are mainly derived from four subspecies: *Mus musculus domesticus, Mus musculus musculus, Mus musculus molossinus*, and *Mus musculus castaneus.*

Only a few animals are used for scientific research and animal experimental study in nature. Except a small number of invertebrates, the majority of laboratory animals are mammalian animals that belong to phylum Vertebrata. Among them, the quantity of rodents accounts for more than 80% of all vertebrates, and mice account for more than 70% of the rodents.

1.1.2.2 Laboratory Animals and Animals for Research

To distinguish animals used in scientific research from other animals, all animals in nature can be classified as laboratory animals, economical animals, or wild animals based on the existing state.

In a narrow sense, laboratory animals are often defined as animals specially bred for biomedical research. To meet the requirements of biomedical research, teaching, therapeutics, assessment, diagnosis, biological product manufacturing, etc., laboratory animals are artificially bred and reproduced into standardized species or strains. According to the definition, laboratory animals should have the following three features:

- From the perspective of genetic control, laboratory animals must come from a clear source, be bred artificially, and have defined genetic background. Therefore, laboratory animals belong to genetically defined animals. In accordance with genetic backgrounds, they can be divided into isogenic animals and heterogeneic animals. Isogenic animals refer to inbred strain and F1 hybrid animals. Heterogeneic animals are mainly outbred strain and F2 hybrid animals. In addition, inbred strain animals include common inbred strain, congenic inbred strain, coisogenic inbred strain, recombinant inbred strain, separate inbred strain animals, and others.

- From the perspective of microbial control, the microorganisms and parasites carried by laboratory animals should be controlled strictly. To guarantee the accuracy, sensitivity, and repeatability of animal experimentation, microbiological and parasitological control not only exclude animal diseases, but also asymptomatic infection and nonpathogenic pathogens, which may interfere with experimental results. Accordingly, laboratory animals in some countries are classified arbitrarily into four levels: conventional (CV), clean (CL), specific-pathogen free (SPF), germ free (GF), and gnotobiotic animals (GA). Microorganisms and parasites carried by SPF and GF animals are artificially monitored and animals themselves must be depolluted by artificial cesarean section or embryo transfer.
- From the perspective of application, the ultimate goal of all laboratory animals is to be used for scientific study. Currently, laboratory animals serve as frontier sentries in the fields of biomedicine, pharmacy, chemical industry, agriculture, husbandry, environmental protection, commodity inspection, foreign trade, military industry, and aerospace. Rather than humans, laboratory animals are used to perform scientific research and verify many scientific truths. On account of the complexity of biological phenomena, no other methods can completely replace laboratory animals so far.

According to the narrow concept of laboratory animals, mice, rats, hamsters, guinea pigs, rabbits, and dogs have completely become qualified laboratory animals after being artificial bred for many years. However, the work of laboratory animalization on other mammals, birds, fish, and nonhuman primates is in progress. Thus, they are not actually laboratory animals in a strict sense.

Economic animals are also known as domestic animals. They are directionally domesticated, bred, and produced for economic traits (meat, milk, eggs, fur, wool, etc.) as indicators of artificial selection to meet the needs of human social life. Many economic animals, such as pigs, horses, cattle, sheep, goats, chickens, ducks, geese, pigeons, and fish, are also applied in biomedical research. Some of these economic animals have become very close to meeting the strict standards of being laboratory animals. However, compared with the "standard" animals like mice and rats, their quality is not perfect.

Wild animals refer to animals living in natural conditions. For the purpose of biomedical research, humans sometimes capture these animals from nature to conduct animal experimental studies rather than to perform artificial breeding or feeding. Except in rare cases, wild animals such as frogs, toads, and salamanders, which are extensively used in biomedicine teaching, wild fish, invertebrates, birds, and nonhuman primates, which are also used in scientific research, are generally not used in artificial breeding.

Compared with the narrow concept of laboratory animals, there is a so-called broad concept, that is, animals for research or laboratory animals. It refers to animals from wild animals, economic animals, or other artificial breeding animals, selected artificially and bred directionally for biomedical research.

The scope of animals for research is much more than that of laboratory animals. Distinguishing animals for research from laboratory animals is important but also has some limitations.

In the aspect of animal experimental reproducibility, there is big difference between animals for research and laboratory animals. The reproducibility of study means that different workers who perform the same experiments can obtain the same results using the same strain animals, albeit at different locations and times. Reproducibility demands that the accuracy of animal experimentation reaches that of a chemical reaction, which requires the laboratory animals to also possess the "purity" of chemical reagents. To achieve good reproducibility and comparability for scientific research, it is necessary to strictly conduct genetic control, microbiological control, nutritional control, and environmental control. As wild animals live in the wild and survive through natural selection in accordance with the principle of "survival of the fittest," it is difficult for wild animals and economic animals to meet these strict requirements, which can only be achieved by laboratory animals. However, economic animals, such as domestic animals, are under artificial selection and their stock selection is based on economic traits and maximum benefit. Although laboratory animals are also under artificial selection, their strain selection is based on the requirements of biomedical research. In the process of strain selection, laboratory animals are bred into a stock or strain to serve as ordinary "reagents" for scientific research. The diseases occurring in laboratory animals, which are similar to human diseases, can also be reserved with genetic methods to build "disease model" strains as special "reagents" for biomedical research or human substitutes.

Obviously, from the narrow concept of laboratory animals, it is difficult to establish popular models for biomedical research. For example, most zebra fish, *Caenorhabditis elegans*, and drosophila, which bring epoch-making discovery to genetics, do not belong to outbred strain, inbred strain, or SPF animals. The narrow concept of laboratory animals has been developed on the basis of rodents such as mice and rats, and achieved great success. Nevertheless, this concept does not necessarily apply to other animals. For example, inbred strain laboratory animals mean animals from continuous sibmating over 20 generations and the inbreeding coefficients are greater than 0.99. In addition to rodents, most animals cannot tolerate such high inbreeding to produce inbred strain animals. Thus, chickens, quail, and other birds can be considered as inbred strain animals as long as the inbreeding coefficient is greater than 0.5.

For convenience, laboratory animals mentioned in this textbook refer to animals for research unless otherwise specified. Animals for research include vertebrates and invertebrates such as *Drosophila* and *C. elegans*. This textbook is mainly about vertebrates for research, which are commonly used in biomedical science. Mammals for research account for the vast majority of vertebrates and they are the main laboratory animal groups that should be provided protection by Animal Welfare Act or animal welfare board/organization.

1.1.2.3 Standardization of Animals for Research

Standardization of animals for research refers to standardization of production conditions (environment, facilities, and nutrition), quality (microbiological, parasitological, and genetic quality control), and animal experimental conditions. In order to reach "standardization," it is necessary to carry out genetic control, microbiological and parasitological control, nutritional control, and environmental and facility control.

Genetic, microbiological and parasitological, nutritional, environmental, and facility control will be introduced in Chapter 3.

1.1.2.4 Genetically Modified Laboratory Animals

Gordon et al. (1980) injected exogenous DNA into the male pronucleus of a mouse zygote by means of microinjection technology and successfully obtained transgenic animals, which carried exogenous genes and exhibited stable transgenic hereditary characteristics. Meanwhile, embryonic stem (ES) cells carrying exogenous genes were injected into normal blastocysts and developed into chimera mice. Soon afterwards, genetically modified (GM) mice whose endogenous genes were knocked out (gene inactivation) and knock-in mice in which either of two genes associated with certain physiological phenomena were replaced by the other gene (gene functional change) were successfully produced. Subsequently, the cloned animals were produced by nuclear transfer technology, mutant model animals were induced by N-ethyl-N-nitrosourea (ENU), and knock-down animals were obtained using RNA interference (RNAi) technology. In recent years, new knock-out technologies, such as zinc-finger nucleases (ZFNs), transcription activator-like effector nucleases (TALENs), and clustered regulatory interspaced short palindromic repeat (CRISPR)/Cas-based RNA-guided DNA endonucleases are being introduced following the increasing maturity of animal genetic engineering. The biggest advantage of these technologies is that they are independent of ES cells and enable a broad range of genetic modifications by inducing DNA double-strand breaks that stimulate error-prone nonhomologous end joining or homology-directed repair at specific genomic locations. Therefore, a large number of GM animals are generated through the application of these advanced technologies, which greatly increases the resource abundance of laboratory animals.

From 1990 through 1999, the number of animals for scientific research in Great Britain decreased from 3,207,094 to 2,656,753, a decrease of 17%. However, the use of GM animals increased from 48,255 to 511,607, a more than 10 times increase. It has increased by 14% in 2000 from 1991 and reached 581,740, accounting for 21.4% of Britain's total use of animals in 2000. Figure 1.1 shows that the use of GM animals increased considerably over the period from 215,300 in 1995 to 2.03 million animals in 2013 (+1.82 million or +845%), with the breeding of GM animals and fundamental biological research being the main primary purposes accounting for the rise. A total of 1.94 million GM animals were used in biomedical study in 2014, and nearly all were mice (91% or 1.76 million), zebrafish (8% or 154,000), or rats (1% or 20,100 procedures). We retrieved five journals related to cardiovascular diseases, including *Circulation, Circulation Research, Arteriosclerosis, Thrombosis, Vascular Biology, Cardiovascular Research*, and *Atherosclerosis*, and found that 184 articles published were related to animal experimentation for the study of human atherosclerosis in

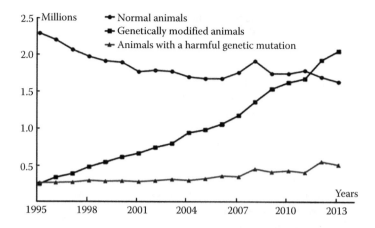

Figure 1.1 Procedures by genetic status of animal in Great Britain, 1995–2013. All EU countries count the total number of procedures for each species rather than the total number of animals, these figures differ by 1%–2% due to reuse and longer studies taking place over several years (with the same animal).

2012. Among them, GM mice were used in 162 articles, while articles that involved using normal mice, rabbits, rats, nonhuman primates, and hamsters were only 9, 6, 5, 1, and 1, respectively. It has been proven that GM animal models play an irreplaceable role in atherosclerosis study, which is the foundation of modern cardiovascular pathology.

Currently, almost all pathogeneses of human disease can be simulated by GM animal models. The utilization of GM animals has greatly shortened the time, reduced the research cost, and improved the results reliability for biomedical research. Therefore, GM animal models are the core model for current biomedical research, and GM mice are predominant in the application of GM animal models.

The generation and application of GM animals are dealt with in Chapter 5.

1.1.3 Animal Experimentation

1.1.3.1 Implications of Animal Experimentation

Large evidence demonstrated that many of the landmark achievements in the field of biomedical research have come from laboratory animals. According to the statistics from the Congress Office of Technology Assessment and National Association for Biomedical Research, laboratory animals have been used for more than 70% of the biomedical research projects funded by the United States government. Since the beginning of the twentieth century, these prizes have charted the world's greatest medical advances. Of the 107 Nobel Prizes awarded for physiology or medicine, 95 were dependent on research using animals. This includes every prize awarded for the past 30 years.

From the viewpoints of science, economy, and ethics, viewpoints, it is necessary to conduct animal experimentation for development of biomedicines. The type of animal experimentation indicated in this textbook refers to animals in biomedical research and the experimental processes that may bring pain or injury to the animals. Experimental study *in vitro*, which involves animal organs, tissues, and cells, is outside the scope of animal experimentation.

The life cycles of most laboratory animals are very short. For example, mice can only survive 2–3 years and their reproductive cycles are only 70 days. Laboratory animals used for experiments are convenient for observing and studying the whole breeding process and life course of an animal in a very short period. The developmental process of human disease is very complicated. Humans cannot be used for experimentation study to explore the pathogenesis and prevention mechanism of diseases. However, animal diseases and biological phenomena can be analogized to humans. Laboratory animals are prone to infection from some diseases, which are similar to human diseases. Study on laboratory animals can mimic human diseases and provide prevention and therapy strategies, both for humans and other animals. Therefore, animal experimentation is beneficial for both humans and animals.

Tephromyelitis (commonly known as poliomyelitis) is an ancient and terrible disease. After the World War II, there was an epidemic in Europe and America. From 1948 to 1952, 11,000 patients died of poliomyelitis, and paralysis and limb atrophy caused by poliomyelitis occurred in 200,000 people in the United States. Currently, it is believed that children will no longer be infected by poliomyelitis after injection of the poliomyelitis vaccine. In addition to poliomyelitis, children can also be protected from typhoid, tetanus, diphtheria, whooping cough, and smallpox infection after injection of the corresponding vaccines. Most healthy adults also benefit from vaccines. However, few people know that the vaccines have come from numerous animal experimentations.

In addition, animal experimental study is in favor of improving animal welfare. For example, canine distemper is an infectious dog disease that is caused by viruses and spreads easily. In the past, 80% of puppies died of canine distemper infection in Great Britain. From the 1920s to 1930s, scientists conducted animal experimentation to study a canine distemper vaccine, but they were met with fierce opposition from organizations that opposed animal use for research. These organizations even proposed dog protection bills to the British House of Commons and tried to make using dogs in biomedical research illegal. Fortunately, biomedical research organizations persuaded the government to accept animal experimentation. Scientists finally developed a canine distemper vaccine and 200,000 puppies survive in Great Britain every year now. In addition, feline immunodeficiency virus (FIV) and leukemia virus (FeLV) are the main causes of cat death, and about 15% of cats were infected by FIV or FeLV. Since the successful development of vaccines by animal experimentation, cats have been protected from FIV and FeLV.

1.1.3.2 Principles of Animal Experimentation

Currently, although most laboratory animals are far removed from humans in evolutionary history, they have been of great value in animal experimentation. As tissues, organs, physiological, and metabolic characteristics of laboratory animals are similar to those of humans and other animals, the basic principles of physiology, anatomy, medicine, and surgery can be applied to all animals, including humans.

The discoveries of the genome and post-genome era have proven that any animal (even invertebrates) could become a valuable model for studying human diseases. For example, *C. elegans* is less than 1 mm in length, lives in the soil and is not similar with humans. However, it contains a 40% genome sequence homology with humans and shares 49% of the protein amino acids with humans. Through the experimental study of *C. elegans*, the key genes governing organ development and programmed cell death were discovered, and it was proven that these genes also existed in higher animals (including humans). The discoveries from *C. elegans* have opened the door to explore human cell differentiation and evolution, and have profound influences on studying the pathogenesis of many human diseases.

In the history of biological evolution, mice parted ways with humans 75 million years ago. Since the genome sequence was completed in December 2002, it has been estimated that there were 27,000–30,500 protein-coding genes in mice, and 99% genes have "matched" sequences in the human genome sequence. Studying every mouse gene exerts the most profound effect on understanding human self and human diseases.

The reactions to drugs or some chemicals are quite different between laboratory animals and humans. Animal genomes can be engineered by modern biological technology. "Humanized" animals reduce the difference with humans in certain aspects and become more appropriate models. For example, GM technology can change the drug metabolism of rodent animals and narrow the difference with humans.

In the field of biomedical research, animal experimentation has been one of the most effective methods. Routine animal experimentation being replaced by other methods is known as alternative methods.

Laboratory animals are not available for all biomedical research. Due to the restriction of ethical and moral laws and regulations, not every animal can be used in biomedical research.

1.1.3.3 *Laboratory Animals and Animal Models of Human Disease*

In the field of biomedical research, laboratory animals are both carriers of experimental study and models of human diseases. They can also be used to produce biological products and perform biological detection. Laboratory animals are "heroes" in military, aerospace, and many other fields. In the future, laboratory animals may not only turn into donors for human organ transplants, but may also become the "core models" of functional genomics and disease genomics.

Most laboratory animal models are created for studying the occurrence, development, and treatment of human diseases. Although this textbook focuses on the generation and application of animal models of human diseases, not all laboratory animals used in biomedical research can be considered as human disease models. However, it is difficult to distinguish animal models of human diseases from laboratory animals for biomedical research.

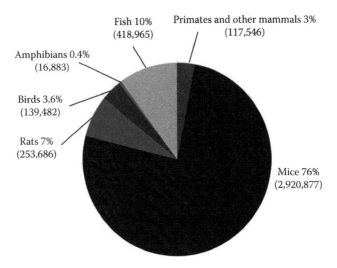

Figure 1.2 Distribution of laboratory animal species used in Great Britain, 2014.

In 2014, a total of 3.87 million animals were used in Great Britain (Figure 1.2). Of those, 1.94 million (50%) were GM animals that were not used in further procedures. The remaining 1.93 million animals (50%) were used for experimental purposes, and of these, 1.04 million (27% of the total 3.87 million animals) were used for basic research; 508,000 (13% of the total 3.87 million) were for regulatory use; 358,000 (9% of the total 3.87 million) were used for translational/applied research; and 20,000 (0.5% of the total 3.87 million) for other purposes. Of those 1.04 million animals used for basic research purposes, 89% (924,000 animals) were for the study of oncology or specified organ systems, 7% (78,000) were for the study of animal biology (including ethology/animal behavior), and 4% (38,000) were for other purposes. Of all the 1.04 million animals used for basic research purposes, the three most common purposes were targeted at the nervous system (23% or 235,000), the immune system (22% or 224,000), or for oncology (13% or 133,000).

Based on the above analysis and the British statistical results in 2014, if most GM animals and harmful genetic defect animals can be used to study human disease models, it can be simply presumed that about half of the laboratory animals for biomedical research served as animal models to study human diseases.

In China, 16 million animals have been used every year, and about 7 million animals were used for teaching, drugs, and chemical tests. Unlike other countries, there is a great usage of hamsters, accounting for about 30% (nearly 5 million) of the laboratory animal use. Hamsters are mainly used as biological materials to produce vaccines. Less than one-third of the laboratory animals are actually used as animal models of human disease for biomedical research.

The abovementioned statistic results simply reflect the relationship between laboratory animals for animal models of human disease and for biomedical research.

The concept, creation principles, and selection of animal models of human diseases for biomedical research will be introduced in Chapter 5.

1.1.3.4 Results of Animal Experimentation

In animal experimentation, animal reactions (animal experimental results) can be summarized with the following formula:

$$R = (A + B + C) \times (D + E) + F$$

where R represents the reaction of animals, that is, experimental processing results; A, the animal (not plant) response; and B, the different species causing different effects on experimental results. By means of gene manipulation, animal species can be transformed to influence animal experimental results; C represents the individual reaction, genetic quality control can reduce the difference between individuals; D, the influence on experimental results caused by the interaction among environmental factors, species, and individual animals; E, the stress reaction. More accurate results can be obtained through improving animal welfare and relieving animal pain or distress; F represents environmental error.

Normal animal experimental technology will be introduced in Chapter 6. Animal experimental design will be introduced in Chapter 7. The basic knowledge and skills of animal experimentation can be mastered through learning, which can lay the foundation to work in biomedical research.

1.2 HISTORY AND APPLICATION OF LABORATORY ANIMALS

1.2.1 History of Laboratory Animals

The most commonly used laboratory mice originated in Africa, while the laboratory rats originated in Central Asia. According to historical records, the history of mice bred by humans is at least 3000 years. In 1100 BC, breeding of spot mice was recorded in ancient Chinese books. In the eighteenth century, fancy mice in China and Japan spread to Europe and mated with local species. Their offspring is the main source of modern laboratory mice.

The use of mouse in scientific research has a long history. As early as 1664, Robert Hooke used mice to study the characteristics of the air. Books recording mouse color inheritance were published in Europe in the eighteenth century. In general, animal use for scientific research was far less systematic and continuous before the twentieth century.

In 1902, William Castle from Harvard University had started to breed and use mice for biomedical research. At that time, mice specially bred for research use mainly came from a farm created by Abbie Lathrop in Massachusetts. Lathrop not only sold mice to scientific research institutions, but also performed production and breeding research. She also successfully generated spontaneous tumor mouse strains.

Many of the most common inbred strains used in modern biomedical research belong to strains of mice bred by William Castle and Abbie Lathrop.

William Castle and his student Clarence Little realized the value of mice in studying Mendelism and began producing mice strains using a breeding system. Little utilized consecutive inbreeding (sibmating) and created the world's first inbred strain mice—DBA in 1909. DBA represents the coat color gene, that is, the abbreviation of dilute, brown, and non-agouti. Over the next 10 years, Clarence Little and other scientists created several famous inbred strains of mice such as A, C57BL, C3H, CBA, BALB/c, 101, 129, and AKR. In fact, the most widely used inbred mouse strains were basically created from 1920 to 1930. C57BL/6 mice generated by Clarence Little in 1921 were considered as the world recognized "standard" of laboratory inbred animals. The most commonly used inbreeding rats were also bred in this period. Most of them, such as F344, M520, Z61, A732, were bred by Maynie R. Curtis and Wilhelmina F. Dunning in the Columbia University Institute for Cancer Research. By 1930, there were 12 inbred strains of rats. Since 1906, George Rommel of the United States Department of Agriculture had started making use of guinea pigs in inbreeding studies; some of those inbred strains are still widely used.

According to the latest statistics of the International Committee on Standardized Genetic Nomenclature for Mice, there are 478 inbred mouse strains and 234 inbred rat strains (Festing: University of Leicester, UK, 1998) in the world. The inbred mouse strains TA1, 615, LIBP/1, NJS, and T739 were created in China. At present, some new inbreed strain mice and rats have been produced in the world and some strains have been gradually eliminated.

Although there are many mouse and rat strains, the commonly used strains in biomedical research only include several mouse strains such as A, AKR, BALB/c, CBA, C3H, C57BL/6, DBA/2, 129, and SJL. C57BL/6 mice have been utilized for genomic sequence analysis. Commonly used rat strains include five or six strains such as F344, LEW, LOW, SHR, and SD. Inbred strain animals that lack in special biological characteristics are rarely used in scientific research. Information about the inbred strain mice and rats in the world is available at http://www.informatics.jax.org/external/festing.

In addition to mice and rats, inbred strain hamsters, guinea pigs, rabbits, and chickens have also been successfully produced and used for biomedical research.

In 1918, China began to breed mice for research and also imported mice, rats, rabbits, and golden hamsters from abroad. Kunming mice, which were introduced from India albino mice to Kunming in 1947, have been distributed across China and have become the most widely used outbred strain mice. Up to now, to provide more choices for biomedical research, scientific and technical workers have done much work on developing Chinese characteristic laboratory animals such as the Mongolian gerbil, Chinese hamster, tree shrews, Himalayan marmot, pika, reed vole (*Microtus fortis*), green swordtail (*Xiphophorus helleri*), and miniature pigs.

1.2.2 Animal Experimentation and Biomedical Research

Laboratory animals, as human substitutes, have been developed for biomedicine. In the field of modern biomedical research, the conditions to conduct experimentation can be summarized as involving four basic elements, AEIR: A represents animals; E, equipment; I, information; and R, reagents. Animals are always the most important element.

Western medicine has mainly originated in ancient Greece. In 400 BC, *Corpus Hippocraticum*, the first medical handbook, recorded using animals to conduct experimentation. Since Claudius Galen (130–201 AD), physicist and physician of ancient Rome, used pigs, dogs, and monkeys in experimentation of medical physiology, making use of animals in medical research prevailed for several centuries. Since Christianity developed and became dominant in Rome, experimental research almost completely disappeared. Not until the Renaissance was medical research carried out, although most research was about anatomy at that time. During this period, French philosopher René Descartes (1596–1650) explained life by pure mechanical principles. He believed that having a soul or not was the difference between animals and humans. He thought animals had no soul, so they did not have consciousness. He also thought humans could think and feel pain, but animals behaved like machines without feelings. Rene theoretically supported making use of animals in research. By the eighteenth century, people gradually realized that medical experimentation was very important to improve human lifestyle and health. Thus, studies on experimental medicine began. Laboratory medicine has developed since only 300 years to modern medicine.

From the following examples, we can clearly find the contributions of experimental medicine to modern medicine. In 1878, German scientist Robert Koch (1843–1910) observed *Mycobacterium tuberculosis* and then found the relationship between bacteria and disease by studying diseases in cattle and sheep. During the end of the nineteenth century and the beginning of the twentieth century, dogs were used by Ivan Petrovich Pavlov (1849–1936), a Russian physiologist, to study digestive physiology and higher neural activity, and he put forward the concept of conditioned reflex and started the physiological research on higher neural activity. In the end of the nineteenth century, German bacteriologist Friedrich Loeffler (1852–1915) used guinea pigs and other animals to study *Corynebacterium diphtheria*. He found that bacterial toxin rather than the bacteria itself was the real cause of animal death. This helped the discovery of immune therapy, which could prevent diphtheria, and opened a new era for antibiotic therapy. In 1914, Katsusaburō Yamagiwa and Kōichi Ichikawa coated asphalt on the ears of rabbits and induced skin cancer. Further research found that 3,4-benzopyrene within asphalt was a chemical carcinogen and proved the carcinogenic effects of chemicals. French physiologist Charles Ricet (1850–1935) found from animal experimentation that the essence of allergies was antigen–antibody reactions and made large contributions to the study of allergic diseases. In 1975, Georges Köhler (1946–1995) and César Milstein (1927–2002), scientists at the University of Cambridge, successfully produced monoclonal antibodies by the hybridoma technique and brought revolutionary change to antigen identification, infectious disease diagnosis, and cancer research and treatment. This was a major breakthrough for modern biomedical research. The main materials were BALB/c mice and their passaged myeloma cells. The understanding and utilization of inbred strain mice has led to the technique of fusing myeloma cells with immunized spleen cells of BALB/c mice and monoclonal antibody technique.

There are 29.1 million diabetes patients in the United States, 415 million people have diabetes worldwide, accounting for 8.3% of the total adult population. Patients with type 1 insulin-dependent diabetes mellitus will quickly die without insulin. At the beginning of the twentieth century, Frederick Banting, Charles Best, and many Canadian scientists repeatedly used dogs and rabbits to conduct experiments. They separated, purified, and identified insulin to determine its function. Currently, more than 30 million diabetes patients in the world have been treated and have a prolonged life span through insulin injections. Patients with high blood pressure (more than 270 million cases in China alone) also benefit from animal experimental study. Scientists have invented many drugs to treat and prevent high blood pressure, stroke, and heart disease caused by high blood pressure through animal experimentation. The success of canine heart surgery and transplantation experiments made it possible to perform similar surgeries for humans. At present, radiation therapy and chemical therapy (drugs), which can both kill cancer cells, have been developed by many experiments using chicken, rats, mice, and rabbits. In the past, nausea and vomiting responses were serious side effects of cancer treatment. However, a new drug was discovered to prevent nausea and vomiting using ferret research. Without animal experimentation, it is difficult to find a good way to treat cancer.

Toxicology testing became important in the twentieth century. In the nineteenth century, laws regulating drugs were more relaxed. For example, in the United States, the government could only ban a drug after a company had been prosecuted for selling products that harmed customers. However, in response to the Elixir Sulfanilamide disaster of 1937 in which the eponymous drug killed more than 100 users, the U.S. congress passed laws that required safety testing of drugs on animals before they could be marketed. Other countries enacted similar legislation. In reaction to the Thalidomide tragedy, further laws requiring safety testing on pregnant animals before a drug could be sold were passed in the 1960s.

People still have a limited understanding of human diseases and a lot of animal experimentation is still in progress. With the increase in human longevity, many people may be subject to geriatric diseases like Parkinson's disease or cancer. The further study of acquired immune deficiency syndrome (AIDS), incurable diseases, and genetic diseases is very important. Insulin injection is not a perfect method to treat diabetes. Scientists have been conducting animal experimentation in order to find ideal ways that may help cure diabetes such as slow releasing insulin technology or tissue transplants.

1.3 USE OF LABORATORY ANIMALS CURRENTLY

According to rough estimates, there were 30 million vertebrates used in scientific research in 1960. By 1970, the number multiplied by several times and reached 100–200 million.

Great Britain is one of the countries with the most accurate statistics on laboratory animals. In 1940, only about 1 million animals were used in scientific research, but the number increased to 3.5 million in 1960. In 1971, it reached 5.6 million and remained at about 5.5 million a few years later. The number has significantly decreased since 1980 and dropped to 2.62 million in 2001, less than half of the use in 1970 (Figure 1.3). In recent years, the number of laboratory animals in Great Britain has increased. It has risen to 4.11 million in 2012, an increase of 57% from 2001. The growing number of laboratory animals brought a lot of pressure to the coalition government of Prime Minister David Cameron (2010–2016).

According to records on animal species for scientific research, more than 70% of laboratory animals are rodents. In 2000, the top five widely used laboratory animals in Great Britain were mice, rats, fish, birds, and guinea pigs, accounting for 59.2%, 19.7%, 9.0%, 4.6%, and 2.6% of the total number, respectively. In 2001, rodents accounted for 82% of animals for research in Great Britain; fish, amphibians, reptiles, and birds, 14%; small mammals (except rodents), 2.3%; large mammals, 1.3%; dogs and cats, 0.4%; and nonhuman primates, 0.1%. In 2014, the most popularly used animals were mice, fish, rats, birds, others, and specially protected species, for 60%, 14%, 12%, 7%, 6%, and 1% of the total number (Figure 1.2), respectively.

In Great Britain, since the beginning of the twentieth century to the early 1970s, the number of laboratory animals for research sharply increased and remained in a stable period until the mid- to late-1970s. The number gradually decreased after 1980 but has plateaued since 1985 (Figure 1.3).

From 1945 to 1971, the number of laboratory animals for scientific research in Great Britain elevated from 1.18 million to 5.61 million, an increase of 4.75 fold, and then declined year by year. It decreased to 3.11 million in 1986, 45% less than in 1971. The number increased slightly after 1987. The reason is that a law (*Animals [Scientific Procedures] Act*) on laboratory animal statistics was revised by the British

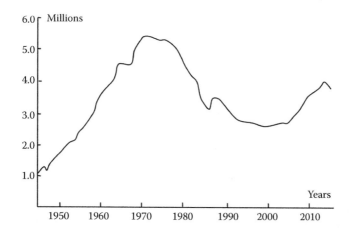

Figure 1.3 The graph of total laboratory animals used in Great Britain, 1945–2014.

government in 1986, and animals for production, treatment, and natural products, which were not counted before, were included. Therefore, the increase in statistical data does not reflect an increase in animals. In 2001, the number decreased to 2.62 million, which was the lowest. Then, it elevated year by year and reached 4.11 million in 2012, an increase of 57% from 2001. When comparing 2014 with 2013, the total of 3.87 million animals represents a decrease of 6% (Figure 1.3).

From a global perspective, the number of animals for biomedical research markedly increased in the early to mid-last century, and then has gradually fell and become relatively stable. However, the general trend is that the number of higher animals has dropped, while the number of lower animals used has increased. Traditional animal numbers have dropped, while GM animal numbers have increased. In addition, animal species have become more abundant.

Over the past 25 years, the total number of animals used in research has fallen by nearly 40% in the United States. Compared with in 1975, the use of animals in research also greatly decreased in Canada in 1992. Mouse and rat use in Japan was 11 million in 1975 and dropped to 7 million in 1989. In 2006, 1530 facilities in China used approximately 16 million laboratory animals, but the number was unknown before 2006. From 1975 to 1992, fish use has increased nearly 4 times in Canada from 1975, and mammals have been replaced by fish in many research fields (e.g., toxicology). In 2000, fish use for animal science research in Britain has increased from 122,438 to 243,019, an increase of 98% from 1999. Currently, fish substitute for rats and have become Great Britain's second most used laboratory species (Figure 1.2). The use of birds (including fertilized eggs) and ferrets has also substantially increased, but the use of primates, livestock, and dogs has been reduced.

The biological characteristics and application of laboratory animals commonly used in biomedical research will be introduced in Chapter 4.

The application areas of animal experimentation in biomedical research include improving the understanding of complex life systems, studying the cause of diseases and bettering human and animal life; looking for new diagnosis and treatment methods; producing biological products that are useful for both humans and animals (e.g., vaccines, insulin); product safety testing such as medicines, vaccines, household goods, cosmetics, food additives, etc.

1.4 MANAGEMENT OF LABORATORY ANIMALS

1.4.1 Controversy of Animal Experimentation

Animal experimentation is always an attractive topic for politicians and is also the focus of the animal welfare arguments. Currently, the real battle in developed countries exists between scientists who try to reduce the suffering and pain of animals in scientific research and extreme animal advocates such as "animal rights organizations" and "animal liberation fronts" who attempt to clamp down on animal experimentation, animal sourced foods, and even pets. In recent years, these organizations have threatened scientists who conduct animal experimentation, demolished experimental equipment through violence, destroyed experimental data, and released or stolen laboratory animals. They ignore the revolutionary progress brought by animal experimentation to both humans and animals and stubbornly believe that experimental results obtained by "mad scientists" are not reliable because animal experimentation lacks justice and impartiality.

There are also moderate organizations and individuals that oppose animal experimentation worldwide. They believe animal experimentation should be terminated because it brings extra suffering and pain to animals. People who oppose animal experimentation ignore that many people and animals are suffering from diseases that may shorten their life span. Animal experimentation may reduce suffering and pain from diseases for humans and animals, and prolong their life span. Only few research projects referred to as animal experimentation may cause obvious suffering and pain, but animals are given pain medication or anesthesia in advance. Animal pain always leads to stress, which severely interferes with the experimental results. Pain or discomfort may be caused by experimental factors, but it can be overcome by reasonable experimental design. Modern science and technology, and the latest achievements in laboratory animal medicine, may provide a new approach to further reduce and avoid pain or discomfort for animals.

Some people think that scientists have enough medical knowledge, and that there is no need to perform animal experimentation. They believe animal experimentation can be replaced by knowledge or technology such as cell culture or computer models. The fact is that we know little about diseases such as cancer, heart disease, and sudden infant death syndrome. More than 50 million people die from these diseases each year worldwide. Animal models can describe how diseases affect the human body because both human and animal bodies are very complex. Other nonanimal models only partly reflect the characteristics of diseases.

Therefore, the use of laboratory animals is helpful for the study of biomedical research, development of safe and effective drugs, and guarantees the health of human and animals. Meanwhile, it is important to emphasize animal welfare in animal experimental study.

1.4.2 Animal Welfare and Alternative Methods of Animal Experimentation

1.4.2.1 Laboratory Animal Welfare

The status of animals in philosophy, religion, and culture must be considered following the development of life ethics. Animals are no longer "automatons" without feelings. As a life form, animals also have basic living requirements and high-level psychological requirements. Animals have feelings, and are rich in emotions and abilities such as affection and love, memory, concentration and curiosity, imitation and reasoning. While coexisting with humans in an interdependent ecosystem, animals should be cared for, respected, and have their interests considered for equally, which is known as animal welfare.

Animal utilization and animal welfare are the two aspects of the unity of opposites. In terms of biomedical research, excessive requirements of animal welfare may bring a heavy economic burden to scientific research. Moreover, many animal studies have to be terminated due to different restrictions. On the contrary, ignoring unnecessary pain or discomfort of animals would go against human ethics. Therefore, real animal welfare is not only just protecting animals, but paying attention to animal welfare, merging animal welfare and animal utilization, and opposing those who perform animal experimentation by extreme methods or ways.

In animal experimental study, animal welfare has put more emphasis on animal health and comfortable living conditions. Animal welfare will worsen when the basic physiological requirements of the animal cannot be satisfied.

1.4.2.2 3Rs Theory

In the early 1950s, the number of laboratory animals for biological research markedly increased, which has drawn social public attention to animal protection and laboratory animals. The comprehensive and systematic 3Rs theory was first put forward by Zoologists William Russell and microbiologists Rex Burch in 1959. The principles of the 3Rs have subsequently become embedded in national and international legislation regulating the use of animals in scientific procedures.

3Rs is the abbreviation for Reduction, Replacement, and Refinement. More specifically, 3Rs refers to using a smaller number of animals in scientific research to acquire the same amount of experimental data, or using a certain number of animals to gain more experimental data; using other ways rather than conducting animal experimentation to achieve the same results; reducing or avoiding pain or discomfort by improving and perfecting experimental procedures to improve animal welfare.

Biomedical researchers should regard the 3Rs as a branch of biomedicine and further study 3Rs technology. The study and application of 3Rs is an opportunity rather than a threat. 3Rs is an effective approach in biomedicine and other fields. It is helpful to conduct scientific research and obtain more accurate and reliable results.

Laboratory animal welfare and the principles of the 3Rs will be specifically introduced in Chapter 2.

1.4.3 Management of Laboratory Animals

The purpose of laboratory animal management is to humanely treat animals during teaching, biomedical research, testing and other activities, guarantee animal welfare, and improve the level of biomedical research.

Great Britain is the first country that made laws to protect animals in scientific research and promulgated the world's first law of animal experimentation in 1822. The *Cruel Treatment of Cattle Act*, was one of the first pieces of animal welfare legislation. It was the first time to consider animal abuse as a crime. Although this Act was applied only to large domestic animals such as cattle, sheep, pigs, and horses, while dogs, cats, rodents, and birds were excluded, it was still a milestone in the history of animal protection and welfare.

The *Animal Welfare Act* was signed into law in 1966 by the United States, and it is the only Federal law in the United States that regulates the treatment of animals in research, exhibition, transport, and by dealers. The Act was amended 8 times. Animals covered under this Act include any live or dead cat, dog, hamster, rabbit, nonhuman primate, guinea pig, and any other warm-blooded animal determined by the Secretary of Agriculture for research, pet use, or exhibition. Excluded from the Act are birds, rats, mice, farm animals, and all cold-blooded animals.

Other laws, policies, and guidelines may include additional species coverage or specifications for animal care and use, but all refer to the *Animal Welfare Act* as the minimum acceptable standard. The *Health Research Extension Act of 1985* provides the legislative mandate for the Public Health Service (PHS) *Policy on Humane Care and Use of Laboratory Animals*. It directs the Secretary of Health and Human Services to establish guidelines for the proper care and treatment of animals used in research, and for the organization and operation of animal care committees. The Office of Laboratory Animal Welfare (OLAW) implements the PHS Policy. OLAW's responsibility for laboratory animal welfare extends beyond National Institutes of Health (NIH) to all PHS-supported activities involving animals. In Great Britain, the use of animals in scientific procedures is regulated by the *Animals (Scientific Procedures) Act 1986* (it has recently been revised to transpose *European Directive 2010/6 3/EU* on the protection of animals used for scientific purposes), an animal protection measure that requires licensing and oversight of all places, projects, and personnel involved in such work. The *Laboratory Animal Regulations* was issued by the government of China in 1988. Laboratory animal regulations or acts formulated by state/province or city governments were also included. Those laws or regulations are to be obeyed unconditionally by relevant animal producers, institutions, and researchers who perform laboratory animal production or animal experimentation. The *Guide for the Care and Use of Laboratory Animals* was edited and published by the Institute of Laboratory Animal Resources (ILAR) of National Research Council. It is imperative to abide by the *Guide for the Care and Use of Laboratory Animals* for facilities or individuals who plan to apply for relevant grants from the NIH to conduct biomedical research involving vertebrates. Many countries also reference the *Guide for the Care and Use of Laboratory Animals* to manage the work of laboratory animals.

After many years of discussion, *Directive 2010/63/EU* revising *Directive 86/609/EEC* on the protection of animals used for scientific purposes was adopted in 2010. The Directive is firmly based on the principle of the 3Rs, to replace, reduce, and refine the use of animals for scientific purposes. The scope is now wider and includes fetuses of mammalian species in their last trimester of development and cephalopods, as well as animals used for the purposes of basic research, higher education, and training.

From a legislative perspective, there is no specialized, complete law for animal protection or animal welfare in biomedical research in China. The current separate rules such as the *Law of the Protection of Wildlife* and *Animal Epidemic Prevention Law* do not refer to laboratory animals. Thus, the legal system of laboratory animal management in China should be perfected.

1.4.4 Protocol of Animal Experimentation

In accordance with the *Animal Welfare Act* and PHS policy, the Institutional Animal Care and Use Committee (IACUC) is required to oversee the responsible use of animals in university research and instructional activities as described in the PHS *Policy on Humane Care and Use of Laboratory Animals* and the *Guide for the Care and Use of Laboratory Animals*. The IACUC reviews protocols, reviews the animal care and use program, and monitors university animal facilities to ensure compliance with standards and regulatory requirements.

As mandated by the PHS policy, membership of the IACUC will comprise at least five members including the following: one veterinarian with training or experience in laboratory animal science and medicine; one practicing scientist experienced in research with animals; one member whose primary concerns are in a nonscientific area (e.g., ethicist, lawyer); one member who is not affiliated with the institution other than as a member of the IACUC. In addition, IACUC should submit annual summary and evaluation reports to the OLAW.

All animal experimentation should be reviewed and approved by IACUC before study performance. Studies submitted to the IACUC for review must comprise a completed application form, including appropriate signatures. In general, projects of conventional animal experimentation, which use common laboratory animals, can be easily approved by IACUC. However, if the project requires wild animals, higher animals, special GM animals, or nonhuman primates, IACUC will discuss carefully and may submit it to the higher IACUC, or even to national authorities for approval. In Great Britain, all living vertebrates and octopi used in scientific research are protected by the *Animals (Scientific Procedures) Act 1986*, but invertebrates (e.g., *Drosophila*, nematodes, etc.) are not covered. The utilization of wild animals for biomedical research is forbidden in Britain unless the project is approved by the Home Office. Therefore, the government bans animal experimentation using Pongidae.

The study on animal experimentation reflects the complexity of life activities. From the perspective of scientific research, accurate research conclusions are based on the similarities between human model animals and humans. However, these kinds of animals are usually in the public eye and some people vehemently oppose using these animals for experimental research.

The organization and management of animal experimentation will be reviewed and summarized in Chapter 8.

1.4.5 Laboratory Animal Science Organization

Many countries have established laboratory animal science associations, societies, or other scientific organizations, and some of them have become outstanding and influential international organizations. The American Association for Laboratory Animal Science (AALAS) (http://www.aalas.org) is an association of professionals that advances responsible laboratory animal care and uses it to benefit people and animals. AALAS provides educational materials to all laboratory animal care professionals and researchers, and administers certification programs for laboratory animal technicians (LATs) and managers. The Federation of European Laboratory Animal Science Associations (FELASA) (http://www.felasa.org) represents common interests in the furtherance of all aspects of laboratory animal science in Europe and beyond. FELASA focuses on exchanging Laboratory Animal Science information, optimizing experimental conditions, ensuring animals to be treated humanely and properly and promoting the development of European animal experimentation. The Association for Assessment and Accreditation of Laboratory Animal Care (AAALAC) (http://aaalac.org/) is a private, nonprofit organization that mainly focuses on promoting the humane care of animals by willingness to accept certification. Currently, over 650 companies, universities, hospitals, and research institutes, including famous institutes and organizations such as NIH and the American Red Cross, have passed the AAALAC certification. International Council for Laboratory Animal Science (ICLAS) (http://www.iclas.org/), which was established by the United Nations Educational, Scientific and Cultural Organization (UNESCO) in 1956, puts emphasis on promoting the health of humans and animals by utilizing and managing laboratory animals humanely in scientific research. The Fund for The Replacement of Animals in Medical Experiments (FRAME) (http://www.frame.org.uk) is committed to solve the problems of animal experimentation using the 3Rs. Chinese Association for Laboratory Animal Science (CALAS) was established in April 1987. CALAS, an academic corporate social group comprising laboratory animal science workers, has made great contributions to the prosperity and development of Chinese laboratory animal science.

The Jackson Laboratory (JAX) (http://www.jax.org/) is an independent, nonprofit biomedical research institution dedicated to contributing to a future of better health care based on the unique genetic makeup of each individual. The Laboratory's mission is to discover precise genomic solutions for disease and empower the global biomedical community in the shared quest to improve human health. JAX is a National Cancer Institute-designated Cancer Center and has NIH centers of excellence in aging and systems genetics. JAX is also the world's source for more than 7000 strains of GM mice, is home of the Mouse Genome Informatics database, and is an international hub for scientific courses, conferences, training, and education. JAX provided 3 million JAX mice annually for more than 20,000 investigators in at least 50 countries for research and drug discovery.

1.4.6 Laboratory Animal Technical Training

Laboratory animal technical training and qualification recognition are divided into animal breeding and animal experimentation technology in many countries.

It is required by the European Convention for the Protection of Vertebrate Animals used for Experimental and other Scientific Purposes and *Directive 2010/63/ EU* that scientists who conduct animal experimentation be formally trained and educated to gain the professional knowledge of laboratory animal science.

According to the requirements of the Council of Europe and European Union, FELASA has designed detailed training schemes of laboratory animal science.

Based on research field, training and qualifications can be classified into four classes: A, B, C, and D. Class A is applied to persons who take care of animals. In line with master's degree and working time, class A can be subdivided into four grades: grade one, grade two, grade three, and grade four. Class B is applied to persons who carry out animal experiments; class C, to persons who are responsible for directing animal experiments; class D, to laboratory animal science specialists who design and conduct animal experimentation, have mastered enough knowledge to improve the level of animal experimentation, and use animals humanely and scientifically. Class D requires more than 80 class hours of specialized training in laboratory animal science.

AALAS has classified qualifications of LATs into three categories: assistant laboratory animal technician (ALAT), laboratory animal technician (LAT), and laboratory animal technologist (LATG). The qualification certificates can be obtained after full-time theoretical study, technical training, and passing of the examination.

Based on the *Laboratory Animal Regulations* and *Regulation on the Quality Management of Laboratory Animal*, the Chinese government has also demanded that professionals who work in laboratory animal production and animal experimental research must be trained by the provincial or municipal Laboratory Animal Management Committee, pass the exams, and obtain qualification certificates before engaging in the corresponding work. The education and training of laboratory animal science has two purposes. First, the relevant personnel accept the following views: animal experimentation is necessary for biomedical research and cannot be completely replaced by other methods; compared to the pain suffered by animals, the benefits brought by animal experimentation are very valuable; designs and implementation of animal experimentation must protect animal welfare as much as possible. Second, study of the basic theories and knowledge of laboratory animal science and mastering the basic skills of animal breeding and animal experimentation is a must.

Contents of this textbook were selected and compiled according to the above two basic beliefs. Students can master the basic theoretical knowledge and basic skills of animal experimentation after systematic learning and training. Moreover, students can achieve B and C level of FELASA and LATG level of AALAS, and obtain qualification certificates with animal experimental study.

BIBLIOGRAPHY

Alehouse P, Coghlan A, Copley J. Animal experiments—Where do you draw the line? Let the people speak. *New Scientist* 1999;162:26–31.

Festing MFW. Origins and characteristics of inbred strains of mice, 11th listing. *Mouse Genome* 1993;91:393–450.

Flecknell PA, Avril WP. *Pain Management in Animals*. W B Saunders, Philadelphia, 2000.

Gordon JW, Scangos GA, Plotkin DJ et al. Genetic transformation of mouse embryos by microinjection of purified DNA. *Proceedings of the National Academy of Sciences USA* 1980;77:7380–4.

He Z, Li G, Li G et al. *Laboratory Animal Welfare and Animal Experimental Science*. Science Press, Beijing, China, 2011.

Home Office. *Annual Statistics of Scientific Procedures on Living Animals Great Britain 2013*. The Stationery Office, UK, 2014.

Home Office. *Annual Statistics of Scientific Procedures on Living Animals Great Britain 2014*. The Stationery Office, UK, 2015.

Kong Q, Qin C. Laboratory animal science in china: Current status and potential for the adoption of three R alternatives. *Alternative to Laboratory Animals* 2010;38:53–69.

Liu E. *Laboratory Animal Genetics and Breeding*. Gansu Nationalities Publishing House, Gansu, China, 2002.

Liu E. *Animal Models of Human Diseases*, Second Edition. People's Health Publishing House, Beijing, China, 2014.

Liu E, Yin H, Gu W. *Medical Laboratory Animals*. Science Press, Beijing, China, 2008.

National Research Council. *Guide for the Care and Use of Laboratory Animal*, Eighth Edition. National Academy Press, Washington, 2011.

National Research Council Laboratory Animal Management. *Rodents*. National Academy Press, Washington, 1996.

Reardon S. A mouse's house may ruin experiments. *Nature* 2016;530:264.

Russell WMS, Burch RL. *The Principles of Humane Experimental Technique*. Methuen, London, UK, 1959.

Singer P. *Animal Liberation: A New Ethics for Our Treatment of Animals*. Random House, New York, 1975.

Van Zutphen LFM, Baumans V, Beynen AC. *Principles of Laboratory Animal Science*, Second Edition. Elsevier Science Publishers, Amsterdam, Netherlands, 2001.

White JK, Gerdin AK, Karp NA et al. Genome-wide generation and systematic phenotyping of knockout mice reveals new roles for many genes. *Cell* 2013;154:452–64.

Zhao S, Liu E, Chu Y et al. Numbers of publications related to laboratory animals. *Scandinavian Journal of Laboratory Animal Science* 2007;34:81–6.

CHAPTER 2

Welfare of Laboratory Animals

Enqi Liu and Jianglin Fan

CONTENTS

2.1 THE CONCEPT OF ANIMAL WELFARE

2.1.1 Origin of Animal Welfare

For a long time, animals have been people's private property, goods, or materials used in scientific research, existing just as an appendage for humans.

Systematic concern for the well-being of other animals may have arisen in the Indus Valley Civilization, as religious ancestors were believed to return in animal form, and therefore arose the need for animals to be treated with respect.

In the early nineteenth century, some enlightened Europeans demonstrated insight sympathy to animals, and solved the ethical issues in legal practice. In 1809, a British lord introduced a bill to ban cruelty to animals, which was ridiculed at that time. Although the proposal was adopted in the House of Lords, it was rejected in the House of Commons. However, with the passage of the time and the progress of society, people's thinking on the interests of animals matured. In 1822, humanist Richard Martin moved a motion to prohibit cruelty to animals, which got through both Houses in the United Kingdom. The bill, called the *Cruel Treatment of Cattle Act* (nicknamed *Martin's Act*), was one of the first pieces of animal welfare legislation. It was the first time that animal abuse was considered a crime. Although this Act was applied only to large domestic animals, such as cattle, sheep, pigs, horses, etc., and dogs, cats, rodents, and birds were excluded, it was still a milestone in the history of animal protection and welfare. People's attitude toward animals began to subtly change. The *Cruel Treatment of Cattle Act* affected not only the British people, but also people in other countries. This Act was repealed by the *Cruelty to Animals Act 1849.*

In 1850, France adopted the anti-animal cruelty law. Meanwhile, Ireland, Germany, Austria, Belgium, the Netherlands, and some other countries also adopted anti-animal cruelty laws one after another. Following the creation of the Royal Society for the Prevention of Cruelty to Animals in the United Kingdom in 1824 (given Royal status in 1840), Henry Bergh founded the American Society for the Prevention of Cruelty to Animals (ASPCA) in 1866 in New York City on the belief that animals are entitled to kind and respectful treatment at the hands of humans, and must be protected under the law. In that year, the first anticruelty law was passed since the founding of ASPCA, and the organization was granted the right to enforce anticruelty laws.

If from the early nineteenth century to its end, people were concerned only with domestic animal welfare, then since the end of nineteenth century, with more and more animals used for biomedical experiments, laboratory animal welfare became people's focus, and the relevant laws and regulations of each country were promulgated one after another. The federal government of the United States promulgated the *Animal Welfare Act* in 1966; The Act was amended 8 times (1970, 1976, 1985, 1990, 2002, 2007, 2008, and 2013) and is enforced by the US Department of Agriculture (USDA), the Animal and Plant Health Inspection Service (APHIS), and animal care agencies. *Health Research Extension Act* and the United States Government Principles for the Utilization and Care of Vertebrate Animals Used in Testing, Research and Training were promulgated in 1985, and *Policy on Humane Care and Use of Laboratory Animals* (the latest revision in 2015) and *Guide for the Care and Use of Laboratory Animals* (the latest version: 8th Edition, 2011) have been adopted since then. The main content of the relevant laboratory animal welfare was included in the relevant animal welfare laws, regulations, and guidelines. Some countries enacted a single law, designed specifically to protect laboratory animal welfare in biomedical research. For example, in 1876, the *Cruelty to Animals Act* 1876 was an Act passed by the Parliament of the United Kingdom, which set limits on the practice of, and instituted a licensing system for animal experiments. The Act was replaced 110 years later by the *Animals (Scientific Procedures) Act* in 1986. In 1986, the Council of European Convention adopted *Directive 86/609/EEC* on the protection of animals used for experimental and other scientific purposes. The directive was repealed by *EU Directive 2010/63/EU*, which is the European Union (EU) legislation "on the protection of animals used for scientific purposes" and is one of the most stringent ethical and welfare standards worldwide.

In Western countries, the implementation of these regulations has guaranteed the welfare of laboratory animals.

2.1.2 The Connotation of Animal Welfare

The basic starting point is to make animals live in a healthy, happy state. For animals to be healthy, happy, and comfortable, a series of actions are offered to provide the animals with appropriate external conditions. The so-called "healthy, happy state" refers to the animals' psychologically pleasant feelings including no diseases, no abnormal behavior, no psychological depression, or stress, etc. Relying on scientific development, we can already measure and evaluate the feelings of the animals, such as whether the animals are injured or ill, or are feeling pain. The animals' behaviors, such as frustration, depression, or panic, can be objectively evaluated.

Animal welfare generally emphasizes ensuring animal health and happiness by improving external conditions. When the external conditions do not allow the health and happiness of the animal, it marks deterioration in animal welfare.

Meeting the needs of animals is the primary principle in guaranteeing animal welfare. The main needs of animals include the following: sustain life, maintain health, and keep comfortable. These three aspects determine the quality of life for animals. Artificially altering or restricting these needs can cause abnormal animal behaviors and physiology, thereby affecting animal health and the reliability of scientific experiments. Relieving animals of pain and helping them to have the following "Five Freedoms" are the basic principles of protecting animal welfare:

1. **Freedom from hunger or thirst** by ready access to fresh water and a diet to maintain full health and vigor.
2. **Freedom from discomfort** by providing an appropriate environment including shelter and a comfortable resting area.
3. **Freedom from pain, injury, or disease** by prevention or rapid diagnosis and treatment.
4. **Freedom to express (most) normal behavior** by providing sufficient space, proper facilities, and company of the animal's own kind.
5. **Freedom from fear and distress** by ensuring conditions and treatment that avoid mental suffering.

The Five Freedoms were formalized in 1979 press statement by the UK Farm Animal Welfare Council, which are also the five criteria of safeguarding animal welfare agreed by the international professional community. Currently, the basic animal welfare principles are increasingly recognized by the people, and gradually protected by the laws or regulations.

Ensuring animal welfare in biomedical research not only meets the social needs of animal protection, but is also of great significance in improving the quality of biomedical science. Therefore, it is necessary for biomedical researchers to understand the stress, pain, and distress of laboratory animals, and also to know how to take the right normative measures to avoid or mitigate these adverse reactions during an animal experiment.

2.2 STRESS, PAIN, AND DISTRESS

Humans have a moral obligation to make laboratory animals stay in the most comfortable environment. As animals and humans have the same neural structures and physiological functions, pain and distress can also occur in animals. From an ethical and scientific point of view, we need to study and assess the behavior and physiological abnormalities of the animals caused by stress, pain, and distress.

2.2.1 Stress

2.2.1.1 Concept of Stress

Physiological or biological stress refers to the body's adaptive and coping–responding process when the individuals face or perceive (cognition, evaluation) threats or challenges to the body caused by environmental changes (stressors). In 1936, Canadian psychologist Hans Selye introduced stress to the biomedical field for the first time. He thought that stress was a nonspecific systemic adaptation syndrome when the body adaptively reacted to the environment, that is, different types of stimuli, such as cold, heat, hypoxia, noise, congestion, and long-term stress, will make animals have a set of similar symptoms.

According to the stressful event, the body responds to stress by activating the sympathetic nervous system, which results in the fight-or-flight response. However, the body cannot keep this state for long periods of time, and the parasympathetic system returns the body's physiological conditions to normal (homeostasis). In animals, stress typically describes a negative condition or a positive condition that can have an impact on an animal's mental and physical well-being.

Homeostasis is a concept central to the idea of stress. In biology, most biochemical processes strive to maintain equilibrium (homeostasis), a steady state that exists more as an ideal and less as an achievable condition. Environmental factors, internal and external stimuli continually disrupt homeostasis; an organism's present condition is a state of constant flux moving about a homeostatic point that is that organism's optimal condition for living. Factors causing an organism's condition to diverge too far from homeostasis can be experienced as stress. A life-threatening situation, such as major physical trauma or prolonged starvation, can greatly disrupt homeostasis. On the other hand, an organism's attempt at restoring conditions back to or near homeostasis, often consuming energy and natural resources, can also be interpreted as stress.

Following Hans Selye, focusing on the problem of stress, many scientist's studies are no longer confined to physiological stress, but focus more on that stress-causing stimuli that are not only limited to biological ones. Biological, psychological, and social stimuli are also included, emphasizing the biological, psychological, and social stimulation of environmental threats and challenges to the survival of the body, and the reactions include physiological, psychological, or behavioral aspects.

Therefore, the modern stress theory divides the stress process into three parts: stress input, intermediary mechanism, and stress response and coping.

2.2.1.2 Stress Input (Stimulus)

Stress input, also known as stressors, refers to the environment or stimuli that may cause psychological and (or) physiological reactions after the body performs cognition evaluation. It can be divided into the following three categories:

- The individual's internal environment: the production of all the necessary materials and balance disorders within the body, such as diseases, nutrition deficiencies, endocrine disorders, changes of different enzymes and blood components in the body, etc. All this can be either part of the stress response or stressors.

- The external physical environment: the environmental conditions of the body, such as temperature, humidity, light, noise, air and air speed, air cleanliness, diet and water, drugs, pathogenic microorganisms, and parasites, which are all stressors.
- The psychosocial environment: the living conditions, living in crowded areas, relationship between individuals, conflicts between populations, relationship with humans, etc.

2.2.1.3 *Mediation Mechanism (Process)*

The mediation mechanism of stress, known as the physiological basis of stress, refers to the internal process of the body's transformation of the input message (stressors, environmental requirements) into output information (stress reaction), and it is the intermediate link of the stress reaction including psychological and physiological mediation mechanisms.

2.2.1.3.1 *Psychological Mediation Mechanism*

The psychological mediation mechanism refers to the individual's own cognitive evaluation process of the impact of the situation on themselves. Perception and cognitive evaluation are the keys to determine whether the individual should take defense and resistance against the environmental stimuli, which involves the level of the individual's information-processing ability. This depends on the climate, diet, drugs, biological relationships, and the external conditions of specific environment, and is also influenced by the individual's heredity and past experience, as well as other internal factors. Therefore, each individual will have different cognitive evaluations of the same stressor and take different coping responses, which depend on the individual's capacity of recognition and coping response.

Typically, the individual's cognitive evaluation of stressors is divided into two categories: positive stress (eustress) and negative stress (distress). The former may moderately increase the cortical arousal level to mobilize positive emotional reactions, which contributes to the proper evaluation of input information and the play of coping response capabilities. The latter causes excessive arousal (anxiety), evokes excessive emotional arousal (excitement) or low emotion (depression), reduces cognitive ability, and results in an unclear sense of self, which interferes with making the right judgment and choosing a positive coping response.

2.2.1.3.2 *Physiological Mediation Mechanism*

The physiological mediation mechanism investigates how to translate this cognition into physiological reactions when the information of the stressor is cognized and evaluated. Past studies divided it into nervous system, endocrine system, and immune system, while recent studies are more inclined to think of it as a whole, collectively referred to as the "stress system."

The neurological functioning regions of the physiological mediation mechanism are the dopamine system, amygdala and hippocampal complex, and pro-opiomelanocortin neurons in the neocortex limbic system. In the above brain areas, the amygdala is considered a key area. The amygdala is a special structure in the neocortex limbic system that exerts mediatory environmental changes on the somatic and visceral motor systems. The central nucleus of the amygdala is the main outgoing section that projects onto the brainstem autonomic center. Electrical stimulation of the amygdala in conscious animals may evoke full fear and anxiety. The perception of this experience is very real, and is associated with previous experience of special stimulation and events. The emotional experience is accompanied by increased heart rate, elevated blood pressure, dilated pupils, pallor, and other biological responses and facial expression changes. Current brain research has found that the amygdala secretes many neuropeptides, most of which are involved in stress response.

The main nerve pathways of physiological mediation mechanisms: the hypothalamic–pituitary–adrenal (HPA) axis is one of the major pathways for different external environmental stress responses. The hypothalamus is stimulated by inputs and then proceeds to secrete corticotropin-releasing hormones. This hormone is transported to its target, the pituitary gland, to which it binds, causing the pituitary gland to secrete its own messenger, adrenocorticotropic hormone, systemically into the body's blood stream. When adrenocorticotropic hormone reaches and binds its target, the adrenal gland, the adrenal gland in turn releases the final key messenger in the cascade, cortisol. Cortisol, once released, has widespread effects in the body. During an alarming situation in which a threat is detected and signaled to the hypothalamus from primary sensory and limbic structures, cortisol is one way the brain instructs the body to attempt to regain homeostasis—by redistributing energy (glucose) to areas of the body that need it most, that is, toward critical organs (the heart, the brain) and away from digestive and reproductive organs during a potentially harmful situation in an attempt to overcome the challenge at hand.

After enough cortisol has been secreted to best restore homeostasis, and the body's stressor is no longer present or the threat is no longer perceived, the heightened levels of cortisol in the body's blood stream eventually circulate to the pituitary gland and hypothalamus to which cortisol can bind and inhibit, essentially turning off the HPA-axis' stress-response cascade via feedback inhibition. This prevents additional cortisol from being released. This is biologically identified as a normal, healthy stress mechanism in response to a situation or stressor—a biological coping mechanism for a threat to homeostasis.

It is when the body's HPA axis cannot overcome a challenge and/or is chronically exposed to a threat that this system becomes overtaxed and can be harmful to the animal body and brain. A second major effect of cortisol is to suppress the body's immune system during a stressful situation, again, for the purpose of redistributing metabolic resources primarily to fight or flight organs. While not a major risk to the body if only for a short period of time, if under chronic stress, the body becomes exceptionally vulnerable to immune system attacks. This is a biologically negative consequence of exposure to a severe stressor and can be interpreted as stress in and of itself—a detrimental inability of biological mechanisms to effectively adapt to the changes in homeostasis.

In addition, under stress conditions, the HPA axis coordinates with the parasympathetic nervous system and other neuroendocrine responses.

2.2.1.4 Stress Response

After the individual's cognition, evaluation and perception of the threats of stressors, behavior, and psychological and physiological changes arising from psychological and physiological-mediated mechanisms are referred to as stress response. When stressors act on the body, the central nervous system (CNS) accepts and integrates stress information, and transfers it to the hypothalamus, which releases large amounts of catecholamine through the sympatheticoadreno-medullary system, thereby increasing blood supply to the heart, brain, and skeletal muscle. Meanwhile, the neurohormone secreted by the hypothalamus can excite the HPA axis, which widely affects the function of several body systems. However, stronger and persistent stress responses can cause disorder and imbalance in the body's physiological function, and in severe cases, can lead to pathological changes.

There is likely a connection between stress and illness. Stress can lead to anxiety, fear, anger, depression, and other poor performances in animals. When the pups and mother are completely separated after birth, animals exhibit significant sad calls. Twenty-four hours after separation, plasma glucocorticoid levels increase 10 times compared with the baseline value. When mice are placed in a new environment, their plasma corticosteroids also increase significantly, but the reaction to electric shock increases it by only 50%; the response of plasma catecholamine to the new environment is relatively small, but the intensity of the reaction to the shock is sensitive. Noise can cause animal plasma norepinephrine to immediately increase, but plasma corticosterone rises slowly.

Social confrontation can elevate the animal plasma corticosterone levels of conflicting mice regardless of a victory or failure. The elevated corticosterone levels and duration of the losers exceed those of the winners, and the norepinephrine levels in winner mice significantly increases. Repeated use of stressors can cause expected reactions. For example, regular daily exposure of rats to cold environments for 10 min for three consecutive months can cause expected reactions, that is, the elevated corticosterone before exposure exceeds that during the exposure. After two rats are separately housed for 30–60 days, when any one rat is moved into the cage of another, conflict between the settler and the intruder is formed. In the conflict, the settler is the attacker and eventually the final winner, while the intruder is on the defensive side and the final loser.

2.2.2 Pain and Distress

Although the feeling of pain is very common in humans, to clearly identify its presence or absence in animals is much more difficult because animals cannot report their subjective experience. However, several species of animals exhibit very similar responses to the behavior of humans subjected to painful stimulus. It may not be anthropomorphism to say that these animals can feel pain. For dogs, cats, some ungulates, birds, and mice, their voices and actions are not identical to humans, but very similar. Lower vertebrates, such as fish and lizards, are usually silent animals, thus their reaction to this stimulus is limited to movement behavior. In this case, pain perception seems to exist.

2.2.2.1 Discomfort and Distress

Discomfort is essential to an animal's life and it also regulates the different functions of the body. The discomfort, needs, and dysfunction that animals experience has biological meaning, and this experience is how animals regulate several functions to remove these states. For an individual, hunger, thirst, loss of spouse, and fear are all discomforts, and these states induce eating, drinking, courtship, defense or evasion, etc. When these states persist, particularly when animals do not have or lose the ability to correct the uncomfortable state, discomfort may transform into pathological behaviors that occur in cases of stress such as stereotyped or destructive behaviors. In the absence of available behaviors to achieve the desired objective, animals may also show no impulsive behavior, but become negative, passive, indifferent, and fatigued.

2.2.2.2 Pain

Pain is a kind of unpleasant feeling or distress caused by emotional perceptions, along with existing or potential body injury. Pain is caused by the body's strong internal and external stimuli, often subjective feelings and symptoms, including several psychological factors such as physical, mental, and emotional experience. Physiological responses of pain include pain perception and pain response. Pain response can be local or systemic. The body has different perceptions and responses under different circumstances, and different physical and mental activity states.

The main function of pain is to have the body swiftly make adaptive defense responses to a certain degree of pain through pain sensation and pain reaction, which has protective effects. Cats and rats would receive an electric shock when they made an attempt to eat and drink. One or two such tests would stop them from going to eat and drink, and manifest escape and defensive reactions. Fear is also a conditional response to pain. Antelope will hastily flee when first seeing a cheetah, and this fear and escape reaction is part of the animals' survival mechanism forming after millions of years of evolution.

The pain signal travels from the periphery to the spinal cord along an Aδ or C fiber. As the Aδ fiber is thicker than the C fiber, and is thinly sheathed in an electrically insulating material, it carries its signal faster than the unmyelinated C fiber. Pain evoked by the (faster) Aδ fibers is described as sharp and is felt first. This is followed by a duller pain, often described as burning, carried by the C fibers. These first-order neurons enter the spinal cord via Lissauer's tract.

Aδ and C fibers synapse on second-order neurons in the substantia gelatinosa. These second-order fibers then cross the cord via the anterior white commissure and ascend in the spinothalamic tract. Before reaching the brain, the spinothalamic tract splits into the lateral neospinothalamic tract and the medial paleospinothalamic tract.

Second-order neospinothalamic tract neurons carry information from Aδ fibers and terminate at the ventral posterolateral nucleus of the thalamus, where they synapse on third-order neurons. Paleospinothalamic neurons carry information from C fibers and terminate throughout the brain stem, one-tenth of them in the thalamus, and the rest in the medulla, pons, and periaqueductal gray matter.

Spinal cord fibers dedicated to carrying Aδ fiber pain signals, and others that carry both Aδ and C fiber pain signals up the spinal cord to the thalamus in the brain have been identified. Other spinal cord fibers, known as wide dynamic range neurons, respond not only to Aδ and C fibers, but also to the large Aβ fibers that carry touch, pressure, and vibration signals. Pain-related activity in the thalamus spreads to the insular cortex and anterior cingulate cortex; pain that is distinctly located also activates the primary and secondary somatosensory cortices.

Pain originates from peripheral regions, but its feeling is processed in the CNS. Until now, several theories concerning pain principles have been put forward, and in-depth understanding of its nature has been gradually increasing. In 1965, Ronald Melzack and Patrick Wall proposed "the gate control theory of pain," which was revised with new facts and opinions in 1982. Before the information is transmitted to the brain, the pain messages encounter "nerve gates" that control whether these signals are allowed to pass through into the brain. In some cases, the signals are passed along more readily and pain is experienced more intensely. In other instances, pain messages are minimized or even prevented from reaching the brain at all. This gating mechanism takes place in the dorsal horn of the body's spinal cord. Both small nerve fibers (C and Aδ fibers, pain fibers) and large nerve fibers (Aβ fibers, normal fibers for touch, pressure, and other skin senses) carry information to two areas of the dorsal horn. These two areas are either the transmission cells that carry information up the spinal cord to the brain or the inhibitory interneurons that halt or impede the transmission of sensory information. Pain fibers impede the inhibitory interneurons, allowing pain information to travel up to the brain. Large fiber activity, however, excites the inhibitory neurons, which diminishes the transmission of pain information. When there is more large fiber activity compared with pain fiber activity, people tend to experience less pain.

Melzack and Wall suggested that this process explains why animals tend to rub injuries after they happen. When humans bang their shin on a chair or table, for example, they might stop to rub the injured spot for a few moments. The increase in normal touch sensory information helps inhibit the pain fiber activity, therefore reducing the perception of pain.

Gate control theory is also often used to explain why massage and touch can be helpful pain management strategies during childbirth. As touch increases large fiber activity, it has an inhibitory effect on pain signals. Therefore, massage and touch is necessary and useful for some laboratory animals.

In the process of pain, many endogenous pain-causing substances in the body are also involved in nerve impulse conduction. They are neurotransmitters, such as

5-hydroxyptamine, a physical pain-causing substance from platelets in peripheral tissues, of which low concentrations can cause pain. It is mainly involved in vascular injury pain and traumatic hunger suppressing role in the form of neurotransmitters. The neuronal descending conduction fiber inhibits and reduces the pain impulse before the synapse, which is an important downstream pain regulation mechanism. Bradykinin, a strong pain-causing substance, exhibits a strong effect on pain induction in the surrounding tissue, of which very small amount can cause pain and it is involved in all types of pain. Furthermore, there are also endogenous pain-causing substances such as prostaglandins, enkephalins, endorphins, substance P, and potassium. The endogenous biochemical mechanism of pain is a complex and dynamic process, and in the pain-reaction process, wider biochemical issues are involved including changes in a variety of enzymes, the endocrine system, and immune system.

2.2.2.3 Pain Cognition

The nervous system consists of the brain, spinal cord, and a complex network of neurons. This system is responsible for sending, receiving, and interpreting information from all parts of the body. The nervous system monitors and coordinates internal organ function, and responds to changes in the external environment. This system can be divided into two parts: the central nervous system and the peripheral nervous system. The CNS has a function of processing pain perception and response activities.

There are three phases in animal pain processing:

- The first is the feeling distinction phase, in which animals receive a large amount of incoming information of malignant stimulation such as acupuncture, thermal stimulation, or electrical shock. This incoming information is not only fast, but has accurate positioning. Its function is related to fast pain and ability to process pain characteristics such as space, time, and strength.
- Next is the emotional activation phase (defensive phase). Distinction of pain sensations makes the body, or some part of the body, produce rapid contractile responses and defensive reactions to pain stimuli. This is similar to the reflex response, some of which still remains in the brain after contact with the brain is cut off, and is separated from the subjective feeling, forming a simple conditioned reflex. This pain experience used as an objective indicator in animal experiments is not credible because it does not include pain sensation. At this phase, the animals are very sensitive to pain-inducing factors and they also form a conditioned reflex to the reflection characteristics.
- The third phase is evaluation-understanding phase (recovery phase). The cognitive function of animals lies in the analysis of somatosensory information. The coordination of other sensory impulses arouses memory impression to become ready to make further reactions, and then selectively act on the processed feeling or activation mechanisms, which forces the reaction onto the slow conduction pathway. In this case, the animal may flee, escape, rest, or heal, while other functions are temporarily stopped, and the animals' movement or gestures can be suppressed. This may be considered as a biological adaptive response to ensure that the animals keep quiet and wait for the body to recover or the situation to change.

2.2.2.4 Pain Response

When noxious stimuli act on the body, besides producing pain, they also cause pain responses. The former is a subjective feeling, the latter mainly manifests as changes in body's physiological functions. As animals cannot express through language when they feel pain, it can be assumed that the same subjective experience of pain is manifested in their body when responding to noxious stimuli.

The pain reaction depends on the effects of the physiological and psychological mechanisms, such as feeling distinction, emotional activation, and cognition evaluation, on movement mechanisms. Pain responses caused by noxious stimulation involve the entire body, leading to pathological changes of each system including abnormal mental emotion. The reaction mechanism includes two aspects: one is the distribution of pain impulse conduction, when the impulse comes in, it can make other sensory nerves conduct to the CNS at the same time, thereby causing tension, ischemia, hypoxia, and the release of endogenous pain substances in the blood vessels, internal organs, and the body; the other is a projection reaction of pain, including stress reflective behaviors, such as escaping behavior and emitting help signals, caused by the noxious stimuli, changes in heartbeat, breathing, mental emotion, and biochemical bodily fluids, such as increases in the levels of catecholamines, blood sugar, thyroid hormone, and 5-hydroxytryptamine, increase of oxygen consumption, decrease of intravenous immunoglobulin, and the decline of phagocyte function.

Pain reactions mainly include three types: local reactions, reflex reactions, and behavior reactions, among which local reactions are the simplest, while behavioral are very complex.

Local reactions can be done without the involvement of CNS. Limited to a simple reaction, the stimulated part reacts to the noxious stimuli such as some degree of redness, swelling, or vascular dilation.

Reflex response refers to the body's regular response activities after being stimulated with the involvement of the CNS. Reflex responses can be divided into two types: somatic reflex response and visceral reflex response. Whichever part is noxiously stimulated, somatic reflex responses can occur. These are mainly skeletal muscle contractions that result in the animals' escape from further harm that the noxious stimuli will do to the body. The response degree is related with the spatial bounds of stimulating effects of noxious stimuli. If noxious stimuli are continuous, the induced skeletal muscle contraction is also continuous and far sites are generally involved. When the strong noxious stimulation of the body surface has strong and sudden effects, it also induces a series of visceral reflex responses, often along with somatic reflex responses. Thus, within the body, broad and general mobilization occurs including increased heart rate, peripheral vasoconstriction, elevated blood pressure, dilated pupils, and increased secretion from the sweat glands and adrenal medulla. The physiological significance of all this is to get animals on the defensive, evasive, or advantageous position to attack as much as possible.

Behavior response refers to the escape, fleeing, resistance, attack, or other overall reactions that the body makes to noxious stimuli. It is a combination of a series of somatic and visceral reflex reactions. Behavior reactions at the onset of pain are shrinkage and protection responses, and at the same time, voice reaction occurs in some animal populations, which is easily observed as the rapid obvious response. On evaluation, the understanding phase (recovery phase), a variety of different behavior responses may appear in animals, of which escape is the most common form of behavior. However, animals sometimes have to change their movement patterns for the injured limb to recover such as limping. Animals will also retreat to a safer place and avoid all activities other than necessary ones. Voice response also has some hint significance. For example, acute pain will result in screaming and roaring, while chronic pain will lead to moaning and sighing. Voice responses can evoke other animals, especially attention and sympathy from its own kind.

The changes in behavior, physiology, and pathology brought about by pain response will largely interfere with the observed results of animal experiments. Therefore, we must carefully and conscientiously administer analgesics and anxiolytics, as well as, be familiar with the application methods of these drugs because the impact of drug administration on the experimental results is very complex.

2.2.2.5 Encouragement and Suppression of Pain

A notable feature that pain is different from the body's other stimuli perceptions is a strong emotional feeling. Inevitable pain is felt after the body is injured. The pain intensity is proportional to the extent of damage. However, evidence suggests that, at least in the process of a strong pain, a sense of pain and pain response depends not only on the amount of noxious stimuli, but animal experiments confirmed that past experience, especially early life experience of individuals, plays a major role in pain production and development. If dogs that have been reared in isolated cages from birth to maturity, due to the loss of environmental stimuli under normal conditions, cannot normally respond to noxious stimuli after they grow up, they may repeatedly touch a lighted match and go to it each time after reflective retreat. They also can endure acupuncture with no behavioral or emotional changes. In contrast, dogs raised in the natural environment are able to detect potential dangers quickly, and researchers are often unable to approach them with a flame or needle. Due to lack of appropriate experience obtained in a normal environment in infancy, they cannot distinguish stimulus with damaging threats from a variety of stimuli, nor can they respond to selective reactions accordingly.

Ivan Pavlov's classical conditioned reflex experiments demonstrated the importance of situational meaning in pain responses. Under normal circumstances, electric shock of a dog's paw pads can cause a strong response in the animal, but if food is given after shocking, after repetition and repetition, electric shocks encourage animals to salivate, wag tails, and run toward the food without any pain reaction signs. Noxious stimuli in such a situation lose the original meaning and become a signal for food, as long as the electric shock is applied to the same paw pads. Application of high temperature as a conditioned stimulus will obtain similar results.

2.2.2.6 Pain Evaluation Criteria

Currently, there have been many reports on the classification criteria for pain and distress of animals, in which all researchers have tried to grade pain and distress, distinguish pain and distress signals, and evaluate the possible scope of responses in digital form. However, different pain evaluation criteria are used in different experimental approaches. Therefore, all the current evaluation methods are too subjective, resulting in dissatisfaction.

Although the evaluation of pain is a little difficult, stress, pain, and distress in animals can cause marked changes in the endocrine system and nervous system, thereby affecting the results of animal experiments, which cannot be ignored. Thus, in animal experiments, all means to reduce the stress, pain, and distress of animals should be used.

2.3 ANIMAL ETHICS

2.3.1 Attitude toward Laboratory Animals

Since Peter Singer (1975) published his famous book *Animal Liberation*, there have been many articles and comments on ethical issues about animal experiments. Singer holds the interests of all beings capable of suffering to be worthy of equal consideration and that giving lesser consideration to beings based on their species is no more justified than discrimination based on skin color. In Singer' viewpoint, it is a kind of unfair discrimination to accept animals as substitutes of humans in those experiments that cannot allow the use of humans. This discrimination was what he termed "speciesism." This speciesism, racism, and sexism belong to amoral discrimination.

Traditional ethicists opposed Singer's view subsequently, as they believe only humans have a direct sense of responsibility and obligation, and have the ability to perform these. All human potentials ought to be treated equally. The animal, however, has no moral consciousness and should not be considered on the same moral level as humans. This criticism of Singer sparked Tom Regan to write the book entitled *The Case for Animal Rights*, in which he retorted that just as we have an obligation to protect children, the elderly and patients with mental disorders, humans have a direct obligation to protect higher vertebrates rather than harm their lives. All animals have their own values, higher nonhuman vertebrate animals should enjoy the same moral respect as humans.

A public opinion survey demonstrated that the vast majority of people support animal experiments in Western society. Therefore, we do not discuss the issue of animal experiments at the moral level or necessity of animal experiments, but instead focus on discussion of the issues and regulations from the mainstream concept of using animals in scientific research with a moral limit. Each researcher needs to evaluate the reasonableness of the experiment ethics when using live animals in biomedical research. The focus of our discussion is on how to conduct such an evaluation and what errors are likely to be committed.

2.3.2 The Ethical Issues of Using Animals in Scientific Research

In scientific research, animals are merely used as a way to solve the problem, but no attention is paid to their death. In general, laboratory animals are described in the section of "Materials and Methods" in research articles. Laboratory animals are just objects in the experimental process, a living instrument in study projects, whose value lies only in that they can improve the reproducibility and effectiveness of scientific experiments. In the field of ethics, the stubborn anthropocentric idea has been seared on how to assess the value of animals. Currently, it is not surprising that animals have only the equivalent value of instruments. However, due to decades of progress, the anthropocentric theory has received more and more criticism.

In many professional ethics publications, many discussions concern the relationship between the ethical issues of animal use and human ethics. According to the general public views of Western civilization, only humans have moral and ethical consciousness. They believe that only humans have self-awareness and the ability to be responsible for their own actions, as well as the ability to adjust their decisions and make a commitment. However, more and more objections insist that although the above features may be the typical characteristics of ethics as a subject, ethics should be concerned about a wider range of objects surrounding the moral and ethical character itself. These objects may not be ethical protagonists such as humans requiring the same moral and ethical care. A simple example is, the features and capabilities of some populations have reached a very low level. For example, patients with mental disorders are the objects of moral and ethical concerns, and when these people are used as objects of experimental research, they may have little or no maintenance ability of their rights. Due to the absolute vulnerability of this population, they will accept a special form of protection, which does not provide adequate training and preparation for these people to independently and consciously make decisions to voluntarily participate in scientific research as subjects. However, when the sick and elderly are chosen as subjects, if they find that the situation they are in is the same as that of patients with mental disorders, the experiment will be subject to harsh criticism and investigation.

Any serious study of moral and ethical status of animals must carefully analyze the similarities and differences between humans and animals. For example, because animals do not have the characteristics of moral and ethical character, we accept the view that the animal is in a nonethical class and we must have enough reason to explain that it is justified not to give moral ethical concern to the human populations lacking moral ethical character. This does not make sense to apply this to humans, and based on the same inference, concern should be shown for animal welfare and health as well. Problems regarding animal moral and ethical status may also be explained in terms of the declaration of people's rights. The most basic point in human rights is that people are equal regardless of race, gender, or cultural background.

Based on the view of respecting fundamental human rights, injustice toward men and women, and different colors of different ethical groups is not reasonable. Moreover, in the debate about animal ethical status, the focus is to raise the apparent characteristic differences between humans and animals to the height of moral and ethical relationships, which are then treated as valid reasons for unequal treatment. There is a very relevant question of whether universal human rights also apply to animals. With this given specific right, out of respect for the demand for this right, the principle of equal treatment of animals and humans is beyond reproach. "In principle" means that the equal treatment is that we do not consider the possibility of vetoing the ethical requirements of the basic shared right. Rights are generally conferred as a whole, and the intrinsic value of a whole is shown only through the whole. Many people believe that recognition of the intrinsic value of animals is the necessary logical premise for discussing animal rights. Still others, though recognizing the intrinsic value of animals, oppose the idea of promotion of rights to animals. They believe that as long as we recognize the human responsibility for animals, it is sufficient; there is no need to recognize the rights of animals.

Biocentrism emphasizes that all life forms have intrinsic value. Animal centrism puts more emphasis on the intrinsic value of animals, this means that animals should not be used as only tools, but should be respected and protected. To clarify this point, we must look for evidence benefitting the view that the organism has a considerable degree of self-awareness. Regardless of admitting intrinsic value or not, the animal is the subject of species-specific importance, which is very susceptible to human-specific damage. It is an incorrect view that the intrinsic value of the animal, like all values, is "given" by people. Opposite to the instrumental value of the animal, the intrinsic inherent value of the animal is not derived from human interests or purposes. Many people recognize the inherent intrinsic value of the animal, rather than emphasize the special nature of the animal, that is, the inherent intrinsic value of the animal is a necessary prerequisite for us to evaluate the animal. It is a fallacy if we say that the inherent intrinsic value of the animal is the result of human evaluation of the animal.

Recognizing the inherent intrinsic value of the animal is admitting that we have a direct moral and ethical responsibility for the animal, which is completely different from the traditional direct responsibility. The number of people who recognize the inherent intrinsic value of animals in their value system is increasing. This trend is reflected in the reform and revision of the different animal protection laws.

Recognition of inherent intrinsic value of animals can be observed in research. Here are some examples in this regard:

- While the scientific nature of experiments is a necessary precondition, before performing the animal experiments, proceeding to ethics assessment is also very important. For example, if the experimental method is considered ethically unacceptable, the experiment must be prohibited.
- When alternative methods can be used, it is not necessary to do experiments with animals, even if alternative methods are very expensive.
- If there is no viable alternative method, and if there is a conflict of interest between humans and animals, the interests of all parties must be weighed. In this case, if the

animal experiments are still unavoidable, then researchers must be soberly aware that the destruction of intrinsic inherent value of animals cannot be justified by moral ethics.

- In the case that animal experiments are considered acceptable, the species-specific behavior of laboratory animals cannot be disrupted before, during, or after the experiments.
- Researchers engaged in animal experiments have a moral and ethical responsibility for finding alternative ways to achieve the purpose of their scientific research.

Recognition of inherent intrinsic value of animals is interpreted as the moral ethical fair principle by some ethicists. It means that animal protection cannot rely on human sympathy and compassion for animals, but should rely on humans' direct moral ethical responsibility for animals, and their respect for the intrinsic value of animals. The equal relationship between humans and animals is often mistaken as treating animals and humans in the same way. Although animals and humans have certain similar characteristics and needs, they also differ greatly. The principle of equality requires them to accept the same treatment as people in some similar cases. For example, animals can feel pain and distress in the same way as humans. Animals should receive different treatments in different situations.

For a long time, there has been the view that equal and just treatment of animals continues to improve, and is becoming closer to human treatment. However, animals themselves do not require the lifestyle of humans, they only require the lifestyle of their species. Each animal is a perfect species; not merely transitional life forms between unicellular organisms and humans. If we always evaluate animals on what characteristics are like that of humans, and assess animal intrinsic value based on their suitability to humans, it is a continuous anthropocentric way of thinking. "Animal rights" advocators would say the same, "animals have the right to be treated as such," for they oppose the utilitarian use of animals for human purposes. This means that they believe that those important behaviors only for the interests of humans are unacceptable.

2.3.3 Qualification of Animal Experiments

Recognizing the intrinsic value of animals makes people accept the view that between humans and animals, at least some aspects of moral ethics should be considered equal. The result of adhering to this view is that the principles of medical ethics required in the experiments where humans are subjects are emulated to animal experiments. Requirements in the standard of animal experiments are required to reach the standard of the experiments where human subjects are used such as in the scientific value standards of evaluating experiments, object selection criteria, and investment and benefit analysis standards. However, it should be recognized that in animal experiments, the objects' (animals') active cooperation and response cannot be obtained. The best way is to appoint Institutional Animal Care and Use Committee (IACUC) or an ethical committee to ensure that the interests of animals are fully taken into account. The Committee's assessment of animal experiments is carried out based on ethical reasoning in the form of ethics dialogue.

2.3.3.1 Dialogue Mechanism in Ethical Issues

Recognition of the intrinsic value of animals is a starting point, which means that researchers must leave space for and consider the benefit of animals. When a conflict of interest between humans and animals appears, ethical protagonists (the party that is able to make moral and ethical decisions) is the fair ethics responsible party. When faced with whether the decision is appropriate, two aspects must be carefully examined. First, a closer look must be taken at the process of making judgments to know who should play a role in the decision-making process; and second, the validity of the arguments should be examined such as the validity of the cited argument.

The important point of normative ethics is that each individual is the protagonist of ethics, who can freely choose his own behavioral pattern from different behaviors. This choice should be based on reasonable consideration of the relevant facts, values, and general principles, and this choice should be made in a fair and nonbiased attitude. This is not explained casually by saying "this is my choice." On the contrary, any additional ethical protagonist in the same situation, should have an identical choice and judgment. This is often referred to in normative ethics as unified trend (general versatility).

If effective ethical considerations and choice are not only for oneself, but also for other individuals in the same situation, out of respect for the autonomous right of others, the cited arguments in the decision-making process should be provided for other ethical protagonists to comment on and investigate. This is what we often say in this book: a person must find justified reasons and causes for his own behavior. This ethical obligation is not responsible for answering to people of high status, authorities, or organizations. The obligation of ethics is based on the recognition of fundamental equality of all ethical protagonists, regardless of their social status. As far as the moral and ethical issues are involved, any ethical protagonist can freely participate in the discussions and decision making.

Although the concept of ethical autonomy is a prerequisite, it has a strong social element in normalizing the definition and role in ethics. On the one hand, to determine an ethical issue can be a very personal and individual process, and this self-consciousness, view of life, and moral and ethical training (virtues) play a very important role in making decisions. On the other hand, unification brings normative ethics into a social process, in which rational elements in each individual's consideration are called into question when exchanging views. Therefore, establishing a dialogue mechanism concerning ethical issues is a vital component of normative ethics.

Since the ethical issues of using animals in experimental research were raised, researchers cannot be exempted from such a view-exchanging dialogue. As long as the researchers are not trained and cannot apply ethics to their own research work, participating in such view-exchanging dialogues will require a readjustment process of themselves. The subjective bias in scientific research can easily lead to the refusal or avoidance of ethical issues. In natural sciences, this bias is manifested in the reduction of the nature of the living animal into a material object, emphasizing the nonmoral and nonethical values, as well as the machine value of animals. Here, science represents the objective, rational, and universal. In contrast, ethics represents the subjective, irrational, and accidental. In other words, moral and ethics emphasize only the emotional values.

From this viewpoint, science and ethics are opposite, completely different, incompatible things. However, ethical consensus on the basis of reasoning can consider that the two sides have the possibility of agreeing with each other, and the dialogue mechanism with debate and support may be a good stopgap measure. If the ethical views of scientific researchers is excluded, the public may be led to believe that it is just a moralization, resulting in refusal to accept all animal experiments and being disgusted to listen to argumentative views. As a result, replacing the dialogue mechanism is a process of confrontation. Negotiations are full of rhetoric and sophisticated discussion to get the most out of the transaction. The results can only be that with dissatisfaction, the two sides compromise with each other, but the gap between the two sides has not shortened. The two sides should take a different way to resolve conflicts, to accept each other from a adapting measure rather fundamental strategy, and to put the conflict-concerning content into the discussion. It may be realistic to consider all purposes necessary, and then analyze the premise of their different points of view and their key differences. This process requires joint efforts to achieve consensus in a right and wrong aspect, without abetting and intimidation.

One method for such an ethical dialogue between scientific research personnel and nonscientific personnel is to establish an ethics committee such as IACUC. The members of this committee need to have some experience in guiding ethical dialogue. To learn from each other and to achieve understanding, consistent, mutual exchanges on discussion procedure, and ethical standards in several local committees are very important. Dialogue stemming from the nature of the ethics will generate results that should be acceptable to the public.

2.3.3.2 Ethical Reasoning Model

In the early stages of ethics dialogue, arguments are most frequent. If a dialogue is about the permissibility of an animal experiment, all the arguments are related with objects. Then, if the nature of the problem in the dialogue is figured out, a way to resolve the problem can be found even if there are conflicts between both sides.

The debate on whether a plan of animal experiment is feasible from an ethical point of view involves not only the issue of just agreeing or disagreeing with a random vote, but also the evaluation of a well-planned action. A person may be interested in finding out the reasons for his behavior, which may be the driving force from the research personnel, and may also be the target he wants to achieve. The former reason is called motivation, while the latter reason is called the target. In the dialogue, these reasons should be very clear and it should be discussed whether the research is important enough to warrant animal experiments.

Why should researchers give enough reason to carry out animal experiments? Why are conflicts resolved through debate? The decision to allow or disallow an animal experiment is an ethical process of the animal experiment. Finding reasons for the decision means that there are defensible and sufficient reasons to support and justify the decision. Adequate reasons tend to allow the animal experiment, whereas inadequate reasons tend to disallow the animal experiment. For example, an experiment may cause severe discomfort and pain to the laboratory animals or the experiment is of very low scientific value. In the dialogue, these criticisms cannot simply be discarded. Reasons for doing animal experiments should be described in great detail to be prepared for relevant opposing views. In short, the purpose of the dialogue is to achieve normative specifications for conduct. Debate can maximize sufficient reasons for specifications for fair conduct.

Ethical reasoning can also evaluate the relevant quality of the object. Good arguments should be based on support. The argument for the conclusion supporting this argument is tenable and associated.

To obtain good reasons for specific conduct, the following points should be considered:

- Have the necessary scientific quality standards of animal experiments been established?
- Are animals used in the experiments expected to be in pain or harmed? Should the assessment of harmful effects include degree of harm, duration, length, and frequency?
- Are there any replaced, reduced, or refined methods to animal experiments?
- What is the extent of the importance of animal experiments? Is it very important or only somewhat important? Or is it actually only of possible importance?
- Will the importance of the animal experiment be able to compensate for the harmful effects on animals?

These issues require detailed examination, and in the process of consideration, the balance of the two sides should be kept. First, intuitive moral and ethical judgment about the controversial issues in the experiment must be clearly articulated. Then, observing ethical rules, related issues can be thought about. Finally, we must also strive to establish the relation between these rules and the intuitive moral and ethical judgments, and obtain the final conclusion through favorable reasoning based on this relation. Here we use the example of whether acute toxicity experiments are ethically acceptable to illustrate the process. For example, a private company plans to test the toxicity of an insecticide, otherwise it will not be in the market under national law. In such experiments, rats are used and it is assumed that the experiment will cause discomfort and pain to the rats.

First, it should be clear what the intuitive moral and ethical judgments here are. Before a pesticide is used by consumers, detection of its toxicity is right. On the other hand, it is wrong because it can cause distress for rats. In the case that such intuitive ethical judgments are incompatible with each other, they are not enough for a final correct decision. Then a critical examination of our intuitive ethical judgment must be done. In the dialogue conducted in the IACUA, examination should be done on these intuitive judgments, reasoning clues, and their applicability with the related ethics laws in this case. For example, here, there are rules of promoting the health of others; the rule of respecting independent rights and protecting them from pain;

the rule of do no harm; the rule of well-doing; and the rule of being fair. These rules are used to make critical examinations of intuitive judgment by establishing links between them.

The process of reasoning has three consecutive steps. First, the intuitive judgment should be examined from a viewpoint of ethics laws or rules, and then compared with appropriate ethical facts to be critically evaluated. For example, making rats suffer as an intuitive judgment. Is it correctly used here? If there is a better reason to explain why making rats suffer is necessary, then the above intuitive judgment is misused. In the second step, contrary to the intuitive judgment and the information related to the issue, critical examination is made on ethical rules. Ethical rules should not be separated from intuitive judgment and experience because they are derived from judgment and experience. The topic of discussion is likely to be extended. For example, the rule of whether fair or not means that animals and human beings should be treated equally. It can be surmised that as long as rats have the ability to perceive pain, rats should not be considered for experiment. The latter view emphasizes the equal treatment of animals and people, which is not a fair or correct interpretation of the rule of equality. In the third step, the process of establishing a relationship between ethical rules and intuitive judgment continues until an impartial decision is reached. The critical inspection of intuitive judgment and the interactive examination of ethical rules are carried out until some degree of consensus is reached between the acceptable intuitive judgment and ethical rules. In the above example, the intuitive judgment, making rats suffer is wrong, is valid; however, we should also admit that it is unacceptable for this judgment to be more important than alleviating human from distress because of the validity of the rules of promoting health for others. It cannot be inferred, however, that the obligation resulting from this rule is greater than the obligation to protect rats against distress. Therefore, the final conclusion is that we have a responsibility to allow for toxicity testing using rats. The determination of this conclusion should also be based on how important the results of toxicity tests are for the expected improvement of human health. If this test is not performed, the official approval of pesticides going on the market will only be delayed a few months. For example, a few months later, an official will approve the sale of pesticides in the market according to other additional information available. Here, the intuitive ethical judgments tend to deny pesticide experiments using animals. The recognized intuitive ethical judgment and authoritative ethical laws can form a sufficient reason to make a negative decision: under the current circumstances, toxicity testing is considered to be ethically impermissible. Generally speaking, the reality is more complex. For example, if the pesticide is far superior to similar existing products, or its going on the market is critical for the economic survival of the company, the ethical rule of well-doing should also be considered, increasing the possibility of redetermination of the already recognized intuitive moral and ethical judgments.

2.3.4 New Ethical Issues Resulting from Advances in Biotechnology

Transgenic and gene-targeting technology is a biotechnology method that currently exerts a great impact on biomedical research. However, the implementation of this technology challenges the reduction of the "3Rs"(replacement, refinement, and reduction). To obtain 3–4 "founder" animals by genetic engineering, a large number of animals, probably more than 100–150, will be used to generate them. Moreover, the phenotypes of genetically modified (GM) animal "founders" are also different, and more than one passage is often needed for their maintenance. Therefore, the facilities for breeding GM animals have markedly increased recently, resulting in a sharp increase in cost. At the same time, the generation of GM animals also puts forward a new ethics question. The focus is on the deformation of the animal nature (wild type or traditional inbred type) into a new animal (GM animal). Different from the previous debate on the balance between animal welfare and scientific purposes, the debate here is precisely on the generation of GM animals.

Compared with traditional gene targeting, animal numbers may be decreased when creating GM animals using the up-to-date technology of the CRISPR-Cas9 system.

The 3Rs principles used to seek a specific balance between animal distress and research objectives are not applicable in the case of GM animals.

- When a large number of animals need to be sacrificed to establish a successful GM animal founder, although reducing the number of animals used remains an ethics-related issue, which demands a decision, it was ignored.
- When oocytes and embryo culture are used in these techniques, the mention of *in vitro* methods to replace animal experiments is inappropriate. The GM changes that ethics are concerned with are behavior at the cellular level or whole animal body level. The original intention of replacement, however, remains on the strains of new GM animals used for the establishment or follow-up of traditional animal experiments (poison mechanics, serum research, diagnostic testing, and basic research). The selection is not related with replacement.
- It is also not related with refinement because operation on oocytes or embryos will not cause any pain. Researchers using this biotechnology also believe that these technologies refine animal experiments. Their reason is that compared with cruel, direct mutation-inducing techniques, such as cancer-inducing methods with radiation and chemicals, GM technology is more in line with ethical requirements. The contrary situation is the low success rate of microinjection and the lower expected results of gene knockout. Biotechnology experts consider the use of tiny tools as refinement, but critics argue that ensuring a high level of controllability and predictability is refinement. The two definitions have validity, but when they are not distinguished, the result will be a muddled mess.
- The rule of do no harm assumes that an individual is hurt (some scholars believe that animal experiments have such problems), or an individual's behavior may cause harm. Whether the introduction or knockout/in of a gene in the mouse genome is in line with the rule of do no harm is difficult to answer, using the traditional definition of ethical rules.
- The rule of well-being also includes the implementation of euthanasia when distress is unbearable. Whether GM technology can be considered as well-being is more important than any other discussion on the benefit (for humans) of using GM animals for animal experiments. In other words, the question of whether changing the genetic background of an animal is correct from ethical and moral perspectives must be examined based on the social norms of civilization rather than simply concluded in the scientific community.

The inherent intrinsic value of animals has been recognized in the debate on the use of animals in Western countries. One of the results of this recognition is that every change of animal nature, such as instincts of mice, must be justified in the ethical sense. The introduction of genes from another species into the genome of mice belongs to this change.

GM animal models have been created for biomedical experiments, and as in all other animal experiments, the same ethical evaluation must be performed. New moral ethical issues concerning the GM animal creation itself, rather than their applications in the experiments, is a complex problem that requires a more extensive social dialogue. Through such a dialogue, we will develop new ethical norms to evaluate the acceptability of the application of these biotechnologies.

When animals are used for research work, the researchers clearly understand that the animal experiment has become a scientific process studying living life. In animal experiments, the instrumental value of animals is a key value in experimental success. With the recognition that vertebrates have their own values, animal experiments are no longer seen as valid if they think from an ethical point of view.

In addition, every individual has the responsibility to put animal experiment 3Rs into practice. Researchers will have to consider whether their findings can compensate the applied methods. To solve this problem, the ability to conduct ethical reasoning and to check consistency with other people in ethical debates is required. This "ethics dialogue" should be guided by the ethics committee. Ethical reasoning in terms of both content and form must meet certain criteria in order for researchers to have a good sense of ethics and conduct their research with a clear conscience. With a respect for the nature of public concern for morals and ethics, and with an attitude for learning, soliciting mass opinions, and consciousness of prescient worries, we may make tradeoffs between the interests of animals and interests of humans. We should oppose the neglect of the intrinsic value of animals, be accepted by civil society and contribute to social stability. For those researchers engaged in animal experiments, obtaining the recognition of the ethics committee and the support from public, who have neither direct interest nor knowledge in the related study, will be enriching.

2.4 PRINCIPLES OF THE 3Rs

2.4.1 Formation of the 3Rs in Animal Experiments

In 1959, in their book *The Principles of Humane Experimental Technique,* William Russell and Rex Burch put forward the intact 3Rs principles in animal experiments in scientific studies for the first time. These are the replacement of animal experiments using other methods, the reduction of the number of animals used and the refinement of the experimental process to reduce animal distress. The scientific studies referred to here actually include three aspects: (1) acquisition of new knowledge, (2) knowledge imparted in biomedical teaching, and (3) the use of animals to detect chemicals, drugs, and equipment safety, and effectiveness. With the progress and development of *in vitro* technology, the space and conditions for implementation of the 3R principles are provided.

One of the results of the debate and discussion on the ethical issues in animal experiments is that each person engaged in research has the duty and responsibility to find alternative methods to replace animal experiments. However, in the present case, not conducting any animal experiments is unrealistic. Therefore, the ethical evaluation mechanisms of animal experiments were set up to try to find a relative balance between human interests and the interests of animals. Today, more respect is given to the intrinsic value of animals, and researchers are facing the dilemma that animal experiments cannot be done, but have to be done. The proposal of the 3Rs principles in line with ethical laws may be inevitable.

Although it is only in recent years that the concept of alternative methods to animal experiments has been widely accepted, the 3Rs principles in animal experiments were practiced since 1960. The refinement of anesthesia for animal experiments is very important. In 1846, Ether Dome was the first to use anesthesia on patients, and shortly after that, the same anesthetics are applied in animal experiments. Since 1876, in the United Kingdom, in accordance with the Royal Society for the Prevention of Cruelty to Animals, in experiments causing pain to animals, the use of anesthesia has been mandatory. Tissue culture is another alternative with a long history. In 1885, Wilhelm Roux successfully kept chicken embryo cells viable in warm salt solution. The successful *in vitro* cell growth in glassware was first reported by Ross Harrison in 1907. Since antibiotics were added to the medium, cell culture technology has developed rapidly and has now become a very standard technology. Today, alternative methods are being applied in biomedicine, veterinary science, and all aspects of teaching. The power to develop alternative methods is largely for ethical reasons, but other factors also play a role. For example, the use of laboratory animals is very expensive, time-consuming, and difficult to standardize. The alternative methods used to replace the whole animal are generally less complex systems, with easy-to-control experimental conditions to achieve standardization. However, the simplification of the alternative methods is only an advantage when the mechanism is studied at the level of organs/tissues/cells. The reactions in the simplified system are different from those in the overall system, which is a demerit of the simplicity of alternative methods. This is roughly because difference between species limits the application of the data obtained in animals to humans.

- Replacement refers to no use of live vertebrate animals in experiments and use of some alternative methods in other scientific research conditions to achieve a certain purpose. Replacement can be further divided into relative and absolute. The former refers to the application of cultured *in vitro* cells, tissues or organs of vertebrates, while the latter refers to not using animals, such as the use of cultured cells or tissue of humans and vertebrates, or computer models. Only after other alternative technologies have been tried and failed, can scientists use animals for experiments.
- Reduction means that if laboratory animals must be used in a research program, and no reliable substitute method is available, scientists should consider using the

least number of animals required to achieve the research target. The ethical and economical objectives of reducing the amount of animals used is to minimize the number of animals experiencing distress, pain, and anxiety, and to avoid unnecessary waste of animals, medicines, laboratory supplies, and other resources. To ensure the scientific validity of the experimental results, there are three basic ways to reduce the amount of animals used: (1) animals used should be shared as much as possible in different laboratory research projects; (2) high-quality laboratory animals should be used, such as SPF or GM animals, for "quality replaces number"; and (3) reasonable experimental design is required to control biological sources of variability in the experiments.

• Refinement refers to minimizing experimental damage to the animal's body, and the relief of pain and stress felt by the animals through improving animal facilities, feeding management, experimental conditions, and operating techniques. The principle of refinement is not only in line with ethical requirements, but from a technical and scientific perspective, many documents should also be consulted before starting. The refinement of the animal experimental process is very valuable to the scientific results and repeatability of animal experiments.

The proposal of the 3Rs principle has had a positive impact on the amendments and supplements of laws and regulations related to animal experiments in some Western countries. The 3Rs principle is also widely adopted in the ethical evaluation and arguments for biomedical research programs and experimental procedures. Although researchers have the right to conduct research according to their own unique ways, they can only enjoy academic freedom within the framework of animal welfare regulations and use animals in the most refined way. In Western countries, ethical evaluation of application and implementation is generally practiced, and the system of issuing a permit has become an important component in applying the 3Rs principles to scientific research.

Currently, the concept and theory of the 3Rs principles have become more accepted by researchers worldwide, and some related funds, publications and organizations, as well as personnel holding international or regional conferences engaged in 3Rs research have emerged. If the emphasis on the 3Rs principles was mainly out of consideration for animal welfare, the change in recent years has made people realize that the application of the 3Rs principles is not only for animal protection or ethical responsibility, but also for the development of biomedical science requirements.

2.4.2 Application and Development of the 3Rs Theory

2.4.2.1 *In Vitro Technologies*

In vitro technology is the most important class of alternative animal experiment methods. It should be noted that not every *in vitro* method can be viewed as an alternative method. In some research fields, *in vitro* methods have always been inseparable from the study itself. In this case, the *in vitro* method does not affect the animal experiment, and is therefore not considered a substitute for animal experiments.

In vitro methods include the study of organelles, cells, tissues, and organs. Tissue culture is a broad term for the technology of the survival of *in vitro* cells, tissues, organs, or part of organ in nutrient medium for at least 24 h, that is, a created environment similar to the normal *in vivo* physiological conditions for cells or part of an organ. Tissue culture can be divided into two broad categories: organ culture and cell culture. Organ culture also includes the culture of part of a tissue or organ as well as culture of the entire organ. The purpose of organ culture is to maintain the structural and functional relationship between cells and tissues in the related organs. However, the culture of a complete three-dimensional organ prevents nutrients from reaching each cell and promptly removing metabolic waste. Thus, the cultured organ only has a limited time to live, and this technique requires obtaining fresh material from animals or humans each time.

Today, modern tissue culture technologies can reconstruct an organ *in vitro*. For example, different skin tissues are separated, and the large number of skin cells obtained from the cell culture under appropriate conditions is seeded on a suitable substrate to reconstruct the skin with a three-dimensional structure. Skin reconstruction technology has expanded our knowledge of the biological processes of the skin, and it is also used to replace animal experiments for screening skin corrosion and radiation. Cell culture is different from organ culture. The junctions between cells are destroyed by enzymatic or mechanical methods. When dispersed cells obtained from tissue are cultured in culture medium, it is referred to as a primary culture. When cell culture is performed for two generations, it is a continuous or subculture.

The types of cells and culture technologies are divided into either monolayer culture (adherent culture) or suspension culture. The survival time of primary cells is limited, but in some cases, spontaneous or induced transformed cells can continue splitting and live forever, resulting in the formation of cell lines. If some cells are given appropriate physical or chemical stimuli, they can be differentiated. In the differentiation process, the cell's original characteristics are more or less restored. If the cell is stem cell, it can differentiate its biological properties, the same as in *in vivo* development. Cells that slowly proliferate can be saved in liquid nitrogen for later use. Cell lines are long and homogeneous, making them less dependent on the cell donor animals. The application of molecular biology technologies for cell modification can also improve cell lines. GM cells have new biological characteristics such as having human receptor proteins. These cells can become research models of receptor-binding assays for drug-screening experiments.

Cell culture is also often used for vaccine production. For example, monkey kidney cells are usually used to culture and produce the polio vaccine. In the Netherlands, as technology continues to improve, fewer monkeys are used in the production of this vaccine (Table 2.1).

2.4.2.2 *Using Lower Life Forms*

In experiments, lower life forms, such as bacteria, fungi, insects, and mollusks, are sometimes used to reduce the number of vertebrates. The best example of this

Table 2.1 Effects of Dutch Vaccine Technology Advances on the Number of Rhesus Monkeys Used

Year	Technology	No. of Monkeys Used
1965	Using imported rhesus monkeys; using trypsin to digest and separate *in vivo* cells; monolayer cell culture	4570
1970	Using micro-SEPHADEX to replace cell adherent culture	1590
1975	Using captured bred monkeys to replace imported ones	463
1980	Using three-stage culture to replace primary culture; using *in vitro* trypsin digestion to replace *in vivo* digestion; improvement of infusion technology	67
Present	Using cell lines to replace cell culture	0

practice is to test whether compounds are likely to cause cancer mutagenicity using bacteria to do the Ames test. Some tiny organisms, such as yeast, are widely used as a vector for the expression of specific genes, which may encode specific antibody fragments, or even vaccine antigens. Transgenic plants can also sometimes be used for vaccine production. Another alternative use for lower organisms is to detect pyrogen using the limulus amoebocyte lysate (LAL) assay for endotoxin. Currently, the rabbit is still the standard animal for the detection of all non-intestinal medical pyrogens. After the rabbit receives an intravenous injection of a substance, a rise in body temperature is considered to be pyrogenic. In the LAL test, the amoebocyte lysate is obtained from the *Limulus polyphemus* (horseshoe crab) blood extraction, and can be converted into gel by one of the most important pyrogenic bacterial substances, endotoxin. Although most product testing at present cannot fully use the LAL test for technical reasons, the LAL application has significantly reduced the number of rabbits used for pyrogen testing.

In recent years, fish and lower vertebrates are used to replace warm-blooded vertebrates. Research in toxicology, oncology, and gene mutation screening, has made great progress for blood and other organ diseases. For example, an inbred strain of zebra fish and other fish are used in the area of gene mutation screening as well as for the study of developmental flaws on a large scale. The application of swordtail fish in toxicity studies reduces the number of higher vertebrates used in these areas.

Drosophila has a short life span, and it is possible to evolve many generations of fruit flies in a relatively short time period. Another reason researchers use *Drosophila* is that specific pathophysiological phenotypes are readily evaluated in the fruit fly. For instance, there are straightforward ways to test the strength of a fly's cardiovascular or neurological system, and to test the impact of cardio or neurological drugs on flies, thereby replacing rodents. Using fruit flies as a model organism allows us to more rapidly unravel the complexity of biological networks, many of which have similar structure and dynamics across flies, mice, and humans. Thus, fruit flies are an extremely powerful tool for the development of effective disease therapies.

2.4.2.3 Immunological Technologies

Immunological technologies are the basis of a number of *in vitro* methods, particularly useful in diagnostic tests, vaccine quality control, and basic immunology research. Well-known technologies include enzyme labeled immunosorbent assays (ELISAs), hemagglutination tests, and radioimmunoassay tests (RIAs). Although these *in vitro* detection methods are very sensitive, in some cases, there is lack of specificity such as in distinguishing related antigens and antibodies. Therefore, animal tests sometimes cannot be avoided.

In 1975, Georges Köhler and César Milstei fused the cells produced from antibodies with myeloma to form hybridoma cells, which may be subcultured. After selection and cloning, each cloned hybridoma cell can produce an antigen-specific antibody, known as a monoclonal antibody. They invented the technology used to create monoclonal antibodies. Many animals, especially mice, have been used for the production of monoclonal antibodies in the past. Ten to fourteen days after intraperitoneal injection of hybridoma cells into these animals, ascites containing the monoclonal antibody can be recovered. However, as ascites fluid increases the pressure on abdominal and intrathoracic organs, the animals are in pain.

Now, this *in vivo* production process can be replaced by several *in vitro* techniques such as the culture of hybridomas through the *in vitro* fermentation system or hollow fiber system. Through continuous improvement of culture systems, monoclonal antibody production has become very competitive in production yield and cost compared with the *in vivo* production method. Thus, in some European countries, such as the Netherlands, Switzerland, and the United Kingdom, regulatory guidelines have been enacted, except in a few rare exceptions, to limit the production of monoclonal antibodies from animals.

2.4.2.4 Chemical–Physical Methods

For alternative biomedical research methods, if there is no physical–chemical method, it is incomplete. These methods are generally used to resolve components from a complex multicomponent mixture. High-pressure liquid chromatography (HPLC) is the most typical example of this technology. Such methods to replace testing with animal experiments is applied in hormone quality control. Until recently, the titer determination of many products, such as insulin, calcitonin, and oxytocin, requires animal models, and a large number of animals were used for detection. Now, most of the natural hormones are produced with recombinant DNA technology, and all products have very high quality and purity. This makes HPLC technology very useful. Indeed, in some hormonal preparations, HPLC for titer determination has been officially designated as the standard technology.

2.4.2.5 Mathematical and Computer Models

We know that there is a certain relationship between molecular structure and physical or chemical properties, and the biological activity of a compound. It is possible to use this knowledge to predict the biological activities of many types of new compounds, including their toxicities, or to improve the efficacy of drugs by introducing small changes to the molecular structure. When analyzing a series of relevant compounds, this method can also be used to screen and sort. After using this method to screen from a large number of back-up compounds, animal testing is used to determine the remaining amount of the drug.

In the development of new drugs, profound and rich knowledge is very important. For example, three-dimensional structures of biological receptors, the biological process causing the efficacy of the medicine, the atoms of specific groups, the placing in the molecule, electric charge, etc., will all affect the biological activities of a drug. This knowledge must be understood and put into practice. It is known through the study of known substances with animal experiments and human experience that the biological activity in living systems can only be determined by the interaction reactions between compounds, and *in vitro* studies further support this view. The computer is then used to design new structures and properties of the desired compounds. This technology is known as computer-aided drug design, or rational drug design, and has been applied in the development of drugs to treat AIDS. In the drug discovery phase, the process of *in vitro* and animal testing always follows. However, through the application of computer-aided drug design technology, after the initial screening of biological activities of the expected drug, the compounds to be used in animal experiments are markedly reduced.

Many processes in the organism can be expressed through mathematical equations, that is, mathematical models of many physiological, biochemical, pathological, and toxicological processes can be built. In most cases, these models are set up and used on the computer. Thus, they are also referred to as computer models or computer simulations. Physiologically based pharmacokinetic (PBPK) modeling can predict absorption, distribution, metabolism, and excretion of these drugs in the body, based on the organism's physiological parameters, the physical and chemical properties of the drug, and the possible metabolism of the drug. This method can also predict the effects of drugs on *in vivo* tissue as well as its possible efficacy and toxicity.

2.4.2.6 Human Models

Most of the results of animal experiments will be inferred to humans. However, as there are differences in anatomy, physiology, metabolism, biological dynamics, and pharmacological and toxicological responses between animals and humans, dizziness and mood changes specific to humans cannot be or is difficult to be detected in animals. The extrapolation of data from animals to humans may cause some problems, meaning humans are the best model for experiments and tests, but due to ethical and legal reasons, such experiments with humans are opposed. In an increasing number of cases, however, humans or their tissues can rightfully serve as test objects.

More human materials are being used for *in vitro* experiments. For example, human skin and liver models have aroused great interest. The organ-type skin model was created from human skin tissue, and used for testing and basic research. As there is a large difference between the human and animal metabolisms, human liver studies are used for drug testing and development. A method of pyrogen screening using human blood is currently undergoing legalization, and may soon be able to replace rabbit and LAL testing. The principle is that when the pyrogenic substance is added to human blood, the white blood cells begin to produce a class of cytokines, which can be detected by ELISA, bioassays, or RNA-detection methods.

The disadvantage is that the human material is not always available or is not available in the necessary quantities. Building human tissue banks to keep tissues may be a solution to regulate supply and demand, and in this way, the human tissue supply problems in scientific research can be solved. Another problem is that, similar to laboratory animals, humans must meet strict legal requirements as the objects of study. During processing, risks must be excluded and the benefits obtained must be manifested. When healthy volunteers are used for clinical trials, the risks must be reduced to a tolerable level and the treatment must have no irreversible side effects. The safety of modern medical technologies, such as magnetic resonance imaging (MRI) and magnetic resonance spectroscopy (MRS), is high, therefore, humans can be employed to carry out such research. Patients and healthy human volunteers must be fully informed prior to the experiment, and the consent and authorization of the medical ethics committee must be obtained.

2.4.2.7 Other Alternatives

- **Telemetry.** Using telemetry, several parameters can be continuously measured from free-moving animals. The measuring apparatus, which can emit radio waves, is fixed in the body of animals, and data are collected through wireless receiving devices. These devices may be used to measure body temperature, blood pressure, heart rate, and electrocardiogram readings. Telemetry technology enables the continuous measurement of animals and human-induced stress on animals for a long time without interference. This technology also reduces the number of animals used and is painless, making it a good method.
- **Tiered analysis.** For each new method, alternatives to animal experiments are limited and they cannot completely replace animal experiments. An alternative method generally simplifies the reality, and has limits in comprehensively reflecting overall processes. Therefore, many alternative methods are used mostly in the beginning phases of detection schemes. Based on the findings of these methods, whether to conduct or how to conduct the following study is decided. In the following further studies, it is possible to use animal experiments as well as other alternative technologies. This idea formed the application of one or more appropriate programmed and stepped research programs, referred to as tiered analysis. Here is an example of evaluation of the skin corrosion of a chemical substance using tiered analysis. Originally, a computer model is used to evaluate the physical and chemical properties of this compound. If the result is positive, the compound is labeled as corrosive. If the result is negative, its pH is determined. If the pH value is above 11.5 or below 2.0, it is considered corrosive. If the pH value is between 2.0 and 11.5, the effective, nonanimal alternative model is used to test the compound; if the result is still negative, then animals are used for further testing. Then, skin irritability of this compound is tested in eye irritation tests. In this simple stratified analysis

method, before it is necessary to do animal testing, three pretest methods are used. The application of this tiered analysis in the pharmaceutical industry can reduce laboratory animal use by 50%.

- **Others.** In some cases, such as teaching demonstrations, organs from a slaughterhouse can replace laboratory animals. In addition, bovine eyes from the slaughterhouse can provide corneas for eye irritation tests, thereby replacing the Draize eye test carried out with rabbits. Some parameters of rabbit eye irritation tests can be obtained from slaughtered chicken eye *in vitro* trials. If models of other aspects that can reflect the tests, such as inflammation and damage recovery, can be found, the rabbit eye irritation test can be completely replaced. To achieve this goal, non-animal tests need to be extended for use in other methods such as chorioallantoic membrane and cell culture systems. Another example to replace animal experiments with slaughterhouse material is the detection tests for skin irritation and penetration with pig ears.

2.4.2.8 Storage, Exchange, and Sharing of Research Data

In many cases, the decision of whether animal experiments should be carried out is made on the basis of results of previous animal experiments. This means that, as in the case of other subjects, the ease of collecting research data is a prerequisite for progress. There is no need to repeat previous research without much scientific value. Furthermore, if there is no need for necessary research to be done on animals, it has even less value. Therefore, for scientists and animals, easily saving and sharing relevant data is very important. Scientific journals contain the latest research results, and are therefore the most important and primary source of information. Currently, a growing number of journals have turned to digital format, which can be easily and quickly obtained. Information from academic conferences is also a primary source. Secondary sources of information include books, seminars, review articles, reports, and networking. Reference manuals, reference data, and databases also fall into this category. In addition to these two categories, there is a class of so-called "gray literature," unofficially published reports such as informal talks, international reports, government documents, etc. Articles in preparation to be published also belong to this category.

Digital information is becoming increasingly important. The rapid development of computer technology has improved the storage, exchange, and access of this kind of information. The latest information can be obtained remotely via the Internet, and it is very easy to get many comparisons and references from databases. Databases on many aspects of animal experiments and alternative studies have been set up, and it is simple to get relevant information. For example, we can search online information of animal experiments and alternative studies from Google Scholar, PubMed, Scopus, and Web of Science databases.

2.4.2.9 Alternative Methods in Teaching

Animal experiments for teaching purposes can often be replaced by alternative methods. In teaching, the purpose of animal experiments is not to verify scientific hypotheses, but to acquire knowledge and develop skills. Therefore, in animal experiments for education, the tests on biological processes and properties have been done before and will be repeated again and again. For these tests, alternative methods are relatively easy to find. Animals here are learning tools, thus, simulated animal experiments can be carried out based on the goal of clarifying the theory and skill training. Some specific alternative methods are effective such as chemical, physical or three-dimensional models, stored samples involved in research, and teaching through audio–visual materials and computers. Noninvasive techniques are safe for students and are favorable for the development of alternative methods. Alternative methods for audiovisual materials are most widely used, which are very good for the development of ethical values. Through new computer technology, a number of different alternative methods have been introduced to teaching such as interactive learning processes, analog and digital video, and websites. These technologies can meet many learning purposes. Multimedia can be used to present the results of experiments, and can be very realistic.

Animals are also used to train technical skills, where alternative approaches are not necessarily valid. However, simulated blood flow devices with slaughterhouse material may be used to train laparoscopic technicians, and commercially available rabbits can be used to train microsurgical techniques.

2.4.3 Validation of Alternative Methods

Over the past three decades, research on alternative methods for laboratory animals has been given special attention and emphasis. From theory to practice, many methods have been developed to replace, reduce, or refine animal experiments. Whether an alternative approach can be recognized and applied depends on many studies and evaluations of the effectiveness of this method. The definition of validity refers to the reliability and credibility of the method to achieve a specific purpose. Reliability means that the model can give the correct measured parameters. Therefore, *in vitro* mutagenicity tests should be able to distinguish between mutagenic and nonmutagenic compounds. To verify the reliability, the results coming from the alternative methods are often compared with those of the classical methods, referred to as the "gold standard." Credibility refers to the sensitivity, specificity, reproducibility, accuracy, and usefulness, based on comparison with the gold standard. The credibility of the new method means that this method can be applied to make mutual comparison studies within the laboratory and between laboratories, under precise and standard conditions to perform the evaluation.

Several international organizations, such as the World Health Organization (WHO) and the Organization for Economic Cooperation and Development (OECD), have released guidelines on the efficiency of general research trials. The guidelines by the European Centre for the Validation of Alternative Methods (ECVAM) have been widely recognized. According to these guidelines, validity studies should be divided into several successive stages. The first is a validity pre-stage, where the main goal is to refine the replacing models and methods, to produce standard operation procedures, and to convert test results into the format of *in vivo* toxicity so that the experiment can be carried out easily in other laboratories and the test operation procedure process can be evaluated with the standard reference compounds. The second stage is the formal validity study. In this stage, the experiments with numbered compounds are done independently in several laboratories and the results are analyzed. The final stage is the evaluation of method acceptability, and test method guidelines are drafted. The guide is then passed to a recognized authority, and will be verified in research work. In short, the verification of method validation is a study based on international cooperation and requires the participation of many laboratories. In all data submitted, the error of results within the laboratory and between laboratories, and the reliability of experimental methods are stressed.

Cooperative research is time-consuming, expensive, and a logically complex task. These issues with efficiency studies are precisely why only a minority of the current alternative methods is accepted. Another reason is there is still doubt as to whether the alternative method for the relevant animal model (the gold standard) is effective. These animal models are built on different test parameters or on a large range of data, and thus, it is more complex to demonstrate a good correlation between them and the alternative methods.

2.4.4 Impact of Alternative Methods on the Use of Laboratory Animals

Since the 1980s, in most European countries, the number of animals used in experiments has declined. This may be the result of synergistic effects of several factors: the revision of regulations controlling and limiting the use of animals, increased efficiency, improvement of the quality of laboratory animals, economic factors, as well as further strengthening of animal ethics such as the activities of animal ethics committees. Application of alternative methods is another reason. In the Netherlands, the amount of animals has been reduced by approximately 50%, mainly due to the application of alternative methods.

The recognition of some alternative methods is limited. The reason might be that compared with a complete organism, they are incomplete. Physiological processes, such as absorption, biotransformation, and secretion, can only be simulated to a certain extent in tissue culture, and computer simulation simplifies reality. Therefore, the extrapolation possibility of the results from these methods and the application of the results to humans are limited. When the results of laboratory animals are extrapolated to humans, the differences between species must be considered. Computer simulation, lower species, and *in vitro* system models are different from humans. This means that in order to verify the results obtained from alternative methods, animal experiments are sometimes unavoidable. To consider applying alternative methods, the research nature must be decided, either basic research or applied research. In general, in basic research, the researchers are free to choose the research method according to their purpose, and in this case, the 3Rs principle is relatively easy to apply. Its recognition is typically based on the efficacy within the field and being officially published in journals. Applied research, especially the recognition of routine testing methods, is based on a series of conditions and guidelines that must be met. Alternative methods can only be applied under recognition by organizations with international obligations (e.g., The OECD). Recognition is based on extensive study of its efficacy in multiple laboratories. Therefore, the implementation of the 3Rs principle in routine testing is a tedious and time-consuming process. Even after successful efficacy research, due to lack of coordination on the guidelines among organizations, in routine testing, alternative methods are also hindered from wide application. However, once an alternative method is recognized in conventional testing, the number of animals used will be significantly reduced.

The impact of the increasing number of GM animals used in animal research on the entire number of animals is still under investigation. Overall, the establishment of a new strain of GM animals itself requires many animals, and more fields require the use of transgenic animals to do research, indicating an increase in the number of animals in biomedical research. On the other hand, the use of GM animals may refine scientific research, resulting in the reduction of the number of animals. Moreover, GM animals can breed humane animal models, thereby reducing the barriers between the extrapolated results in animal models and humans.

2.4.5 Future of Alternative Methods

Tissue culture, computer technology, immunological technologies, and physical and chemical methods were developed as alternative methods to animal experiments, and have made rapid developments. Through the application of gene chips or next-generation, high-throughput sequencing technology in human tissue, knowledge about disease-related genes or RNAs can be obtained. In this field, the use of animals can be partially avoided. Introduction of ethics is not the only reason for the development in this field; the desire to explore suitable alternative methods to achieve the replacement is another reason. Animal experiments are expected to remain indispensable in the future, but the target of animal experiments may change from initial research to validation of the *in vitro* test data. Moreover, in biomedical research, final validation studies should be conducted in humans or other animals.

It is difficult to predict the impact of alternative methods on the future use of laboratory animals. There may be fewer animals used to solve scientific problems, but on the other hand, the alternative methods may raise new problems that require animal testing for validation.

2.5 HUMANE END POINTS OF ANIMAL EXPERIMENTS

"Pain is a common symptom that all patients experience; no pain relief has harmful effects on the patients' body or physiology." The above sentence explains that pain occurs in all patients, and that the medical field and society have a very clear understanding of the importance of relieving and alleviating pain. Similarly, many years ago, the importance of pain management in laboratory animals was already recognized. Targets for pain management in humans and animals were sought to minimize the occurrence of pain. However, because some medicines to reduce pain cannot be used in some animal experiments, this has been a problem without a suitable solution for a long time.

Laboratory animal pain comes from disease induction, experimental operation, or poison administration. Although there is the provision in the relevant laws that sedatives, analgesics, and anesthetics should be properly administered if the operation causes transient or mild pain and distress to animal, pain caused in the animal experiments often cannot be relieved with drugs because these drugs interfere with the experimental objective or results. Therefore, some regulations require that no relief of pain and distress be allowed during the test period to achieve the scientific target, and these animals experience severe or prolonged distress.

Is there a way to reduce the number of animals experiencing nonrelieving pain and distress? Recently, it has been agreed that the experiment may be ended as soon as possible without affecting the implementation and effectiveness of the scientific objectives. The standard for the most ideal humane end point is to finish experiments before animals experience pain and distress. In the case that pain relief drugs cannot be used, the development and use of humane end points can reduce and shorten the extent and time of the animals' pain.

2.5.1 Determination of Pain and Distress in Laboratory Animals

As previously mentioned, pain and tension can cause distress, and its biological effects can interfere with the experimental results. For this reason, the effects induced by pain and tension in laboratory animals during the experiment are controlled and mitigated. Pain and tension are part of life, and cannot be completely eliminated. How do animals feel pain from nonthreatening tension? Here, we focus on the biological significance of pain, tension, and stress, and their determination and evaluation guidelines.

2.5.1.1 Pain

Pain-inducing behaviors include limb-withdrawal reflex, wowing, facing the direction of stimulation, escape and attack, trembling, friction, grasping, biting, licking stimulated site, as well as some reactions in the autonomic nervous system such as elevated blood pressure, increased heartbeat, erected hair, and dilated pupils.

2.5.1.2 Distress

After peripheral nerve or central nerve tissue is damaged, or with persistent disease, such as cancer, pain will persist as chronic pain. Pain occurs when an adaptable value is no longer present.

2.5.2 Moribund State as the End Point of the Experiment

Many experiments are associated with high mortality, or produce progressive or serious illness, thus resulting in death of the animals. Once moribund state is determined, it will serve as the end point of the effective experiment, avoiding death as the end of the experiment to reduce unnecessary pain and distress.

If moribund state is accurately confirmed, it is beneficial to animal welfare, and at the same time, because the test objectives are coherent, it also directly benefits the experiment itself. If one can accurately predict the time of death of the animals, euthanasia can be arranged according to schedule. In this way, time can be specified accurately for collecting unobtainable samples due to the accidental death of animals. For example, in the recent study of rats with leptomeningeal tumors, if it is known that hind limb paralysis is the end point of the experiment, tissues that can be used to study tumor scope and for histological assays can be collected. In addition, approaching death can change important physiological variables, and the data collected under this condition may become abnormal or even uninterpretable in the whole study. For example, when microorganism-infected mice are close to death, there will be significant hypothermia as well as abnormal electroencephalograph results, thereby making these mice physically unfit for some experiments.

Therefore, preemptive euthanasia has several benefits to research: After severe disruption of physiological functions, data collected may be somewhat useless or misleading for some experimental purposes, and tissues to be collected and analyzed after death may disappear. If the research team has full understanding of the significance of euthanasia at the end point of the study, then in accordance with the existing standard, the end point of the experiment can be established without difficulty. In the end point, a clear definition of mortal status can also improve the management of animal health and efficient collection of high-quality data by veterinarians and animal management personnel.

Mortal status can be defined through distinguishing/identifying the values of several different variables before the animals' impending death, and these values can be used as a preemptive euthanasia signal. To be applied well in practice, these values should come from certain specific experimental models, and the values evaluated should come from part of the normal data obtained. Data studied generally are carefully studied by research teams, and the increase of the premonitory changes of key variable values can be identified. Thus, this strategy makes the routine collection of information, including information related to death, part of the experiment, rather than redundant work.

Monitoring many clinical symptoms is labor-intensive, and the use of evaluation variables required by the generalized system for researchers may be more arbitrary or have no correlation with the severe physical condition of the animals. Following preset standards may make it difficult to make a reference point. However, if the measured variables are associated with the objectives of the study, then the researchers are more likely to achieve the objectives. Furthermore, many irrelevant variables confuse which factors affect what, as occurred in the evaluation of pain systems. The most useful approach is to determine specific variables and measure their predictive values. It is important to know which factors in a particular case are the most important.

Comparison of data collected from dead animals with data from live animals may reveal the changes in experimental variables before approaching death, which may become effective indicators of death or mortal status. Some key observations or measurements of frequency should be done, and when particular conditions change and death occurs, relevant frequency information is useful. Such evaluation can often be guided in small-scale pilot experiments, but before practical indicators of death are confirmed. It is necessary for people who use animal models for long periods to become familiar with these indicators.

2.5.3 How to Select and Use Humane End Points

Choose appropriate end points that are objective and relevant for the assessment of pain/distress in the species. These may include body weight changes, external physical appearance, behavioral changes, and physiological changes (e.g., body temperature, hormonal fluctuations, clinical pathology), etc.

Examples of humane end points are given as follows:

- **Deteriorating body condition score:** Objective and easy to use for assessing the condition of animals used in the study, especially studies where animals may experience some degree of debilitation as the study progresses. Scoring methods have been developed for many species including mice. Researchers should retrieve relevant literature for related content.
- **Weight loss:** Rapid loss of body weight of 15%–20% within a few days. This requires frequent monitoring of body weight. Several reasons can lead to gradual weight loss (emaciation): certain debilitating conditions such as tumor growth and ascites; the inability to rise or ambulate; correlates with inability to access food or water; visually obvious, objective, and easy to assess.
- **Tumor size:** Tumor size can be measured as diameter of the mass or percentage of body weight (i.e., greater than 1.2 cm in mice or 2.5 cm in rats, or 1.5 and 2.8 cm, respectively, for therapeutic studies; or greater than 10% of the body weight).
- **The presence of labored respiration:** The animal exhibits an increased respiratory rate and/or effort. Labored respiration is often accompanied by strong abdominal breathing.
- **Dehydration:** The skin loses its elasticity. Skin pinched over the back should return to its normal position after it is released. In a dehydrated animal, the skin will remain tented.
- **Ulcerated, necrotic, or infected tumors:** Animals have large open wounds.

BIBLIOGRAPHY

Agar N. *Life's Intrinsic Value: Science, Ethics, and Nature.* New York: Columbia University Press, 2001.

Alehouse P, Coghlan A, Copley J. Animal experiments—Where do you draw the line? Let the people speak. *New Scientist* 1999;162:26–31.

Flecknell PA, Avril WP. *Pain Management in Animals.* Philadelphia: WB Saunders, 2000.

Griffin DR. *Animal Thinking.* Massachusetts: Harvard University Press, 1985.

He Z, Li G, Li G et al. *Laboratory Animal Welfare and Animal Experimental Science.* Beijing, China: Science Press, 2011.

Keating SC, Thomas AA, Flecknell PA et al. Evaluation of EMLA cream for preventing pain during tattooing of rabbits: Changes in physiological, behavioural and facial expression responses. *PLoS One* 2012;7(9):e44437.

King JC, Anthony EL. LHRH neurons and their projections in humans and other mammals: Species comparisons. *Peptides* 1984;5:195–207.

Langford DJ, Bailey AL, Chanda ML et al. Coding of facial expressions of pain in the laboratory mouse. *Nature Methods* 2010;7:447–9.

Liu E, Yin H, Gu W. *Medical Laboratory Animals*. Beijing, China: Science Press, 2008.

National Research Council. *Recognition and Alleviation of Pain and Distress in Laboratory Animal*. Washington: National Academy Press, 1992.

National Research Council. *Science, Medicine and Animals*. Washington: National Academy Press, 1992.

National Research Council. *Guide for the Care and Use of Laboratory Animal*, Eighth Edition. Washington: National Academy Press, 2011.

Regan T. *The Case for Animal Rights*. New York: Basil Blackwel, 1985.

Russell WMS, Burch RL. *The Principles of Humane Experimental Technique*. London, UK: Methuen, 1959.

Singer P. *Animal Liberation: A New Ethics for Our Treatment of Animals*. New York: Random House, 1975.

Sotocinal SG, Sorge RE, Zaloum A et al. The rat grimace scale: A partially automated method for quantifying pain in the laboratory rat via facial expressions. *Molecular Pain* 2011;7:55.

Stokes WS. Humane endpoints for laboratory animals used in regulatory testing. *ILAR Journal* 2002;43(Suppl):S31–8.

Ullman-Culleré MH, Foltz CJ. Body condition scoring: A rapid and accurate method for assessing health status in mice. *Laboratory Animal Science* 1999;49:319–23.

Van Zutphen LFM, Baumans V, Beynen AC. *Principles of Laboratory Animal Science*, Second Edition. Amsterdam, Netherlands: Elsevier Science Publishers, 2001.

CHAPTER **3**

Quality Control on Laboratory Animals

Enqi Liu and Jianglin Fan

CONTENTS

3.1 GENETIC STANDARDIZATION OF LABORATORY ANIMALS

The genetic background of laboratory animals is one of the main factors that influences the variation of experimental results. Laboratory animals can be divided into two types: homogenic and heterogeneic animals. The genetic background of all individuals in homogenic strain is identical or similar. Heterogeneic animals have differing genetic backgrounds. Using high-quality and standardized animals in biomedical research can minimize the influence of the animals themselves so as to obtain accurate, reliable, and repeatable results.

3.1.1 Animal Breeding System

3.1.1.1 Inbreeding

Inbreeding means mating between blood-related members, usually within seven generations. The most common approaches in breeding laboratory animals include mating between the following members: cousins, brother and sister, father and daughter, and mother and son. A smaller number of reproductive animals in a group in reality, use of a strain in a small area, different male-to-female proportions in animal reservations or breeder selection can all cause a certain extent of inbreeding.

Inbreeding can affect animals in many ways. Genetically, inbreeding generally can reduce the heterozygous rate and increase the homozygous rate, create inbred strains with varying homozygous levels through continuous inbreeding, decrease the ability of buffering and automatic regulation, and finally cause inbreeding depression characterized by poor adaptability, decreased vitality and fertility, aggregation of detrimental genes, increased genetic disease, and hindered growth.

Inbreeding depression is the reduced biological fitness in a given population as a result of inbreeding or breeding of related individuals. Population biological fitness refers to the ability to survive and reproduce independently. Inbreeding depression is often the result of a population bottleneck. In general, the higher the genetic variation or gene pool within a breeding population, the less likely it is to suffer from inbreeding depression.

3.1.1.2 Coefficient of Inbreeding

The coefficient of relationship is a measure of the degree of consanguinity (or biological relationship) between two individuals. A coefficient of inbreeding can be calculated for an individual as a measure of the amount of pedigree collapse within that individual's genealogy. The measure is most commonly used in genetics and genealogy. In general, the higher the level of inbreeding, the closer the coefficient of relationship approaches a value of 1, expressed as a percentage, and approaches a value of 0 for individuals with arbitrarily remote common ancestors. Figure 3.1 shows the inbreeding coefficients under different mating or breeding systems.

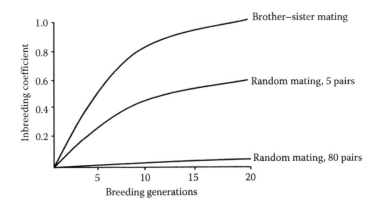

Figure 3.1   Inbreeding coefficients in different mating and breeding schemes.

3.1.1.3 Hybridization

When an animal is bred with an animal from a different strain, stock, species, or genera, the process is known as hybridization or crossbreeding.

Depending on the parents, there are a number of different types of hybrids:

- **Single cross hybrids** is from the cross between two true breeding organisms (such as an inbred strain) and produces an F1 generation called an F1 hybrid (F1 is short for Filial 1, meaning "first offspring"). The cross between two different homozygous strains produces an F1 hybrid that is heterozygous; having two alleles, one contributed by each parent and one is generally dominant and the other recessive. Typically, the F1 generation is also phenotypically homogeneous, that is, all are similar to each other.
- **Double cross hybrids** are from the cross between two different F1 hybrids.
- **Outcrossing** or **outbreeding** is the practice of introducing unrelated genetic material into a breeding strain. It increases genetic diversity, thus reducing the probability of an individual being subject to disease or reducing genetic abnormalities.
- **Backcrossing** is a crossing of a hybrid with one of its parents or an individual genetically similar to its parent, in order to achieve offspring with a genetic identity which is closer to that of the parent. It is used in animal breeding and in production of gene knockout organisms.
- **Intercrossing** is breeding two strains having a common ancestry with one another. Intercrossing and backcrossing are often used to create congenic inbred strains.
- **Population hybrid** is the result from the crossing of plants or animals in a population with another population. These include crosses between organisms such as interspecific hybrids or crosses between different breeds.

The offspring of the hybrid have hybrid vigor or outbreeding enhancement, which is the improved or increased function of any biological quality in a hybrid offspring. An animal exhibits hybrid vigor if its traits are enhanced as a result of mixing the genetic contributions of its parents. Hybrid vigor is mainly due to excellent dominant gene complementation and the increase of heterozygosity in the population, leading to the inhibition or reduction of "bad" gene expression. In this way, hybrid vigor improves the average dominance and upper effects of the whole population, which is marked by increased vitality and fertility, and decreased deformity and death.

When a population is small or inbred, it tends to lose genetic diversity. Inbreeding depression is the loss of fitness due to loss of genetic diversity. Inbred strains tend to be homozygous for recessive alleles that are mildly harmful (or produce a trait that is undesirable from the standpoint of the breeder). Hybrid vigor, on the other hand, is the tendency of outbred strains to exceed both inbred parents in fitness.

3.1.1.4 Random Mating

Random mating is where all individuals are potential partners. This assumes that there are no mating restrictions, neither genetic nor behavioral, upon the population, and that all recombination is therefore possible.

In genetics, random mating involves the mating of individuals regardless of any physical, genetic, or social preference. In other words, the mating between two organisms is not influenced by any environmental, hereditary, or social interaction. Hence, potential mates have an equal chance of being selected. Random mating is a factor assumed in the Hardy–Weinberg principle and is distinct from lack of natural selection: in viability selection for instance, selection occurs before mating.

3.1.2 Inbred Strains

Individuals in inbred strains or F1 hybrids have the same genotype, which is isogenic. The inbred strains include common inbred strains, coisogenic inbred strains, congenic inbred strains, recombinant inbred (RI) strains, and others.

3.1.2.1 Definition

For most laboratory animals, an inbred strain (also called inbred line or pure line) refers to a strain of animals that has been inbred by consecutive brother–sister mating (or the equivalent of brother–sister mating) for more than 20 generations. An inbred strain is essentially homozygous at all loci and each member of the strain is genetically identical to other members (except sex differences), all individuals in an inbred strain can be traced back to a common pair of ancestors. The inbreeding coefficient of inbred strains is greater than 98.6%. Inbreeding can not only increase gene homozygosity, and but also fix some traits due to extended inbreeding. Inbred strains of animals are frequently used in laboratories for experiments where reproducibility of conclusions for all the test animals should be as similar as possible.

3.1.2.2 Substrains

The genetic changes caused by residual heterozygosity and mutations are the formation of a substrain or subline. A substrain refers to the inbred strain that has fixed genetic differences from the other branches. Substrain differentiation usually occurs under the following three conditions: a substrain formed between 20 and 40 generations of brother–sister mating; a branch is separated from other branches for breeding for more than 100 generations; a heritable change in a substrain has been identified, which may result from residual heterozygosity, mutation, or genetic contamination. Substrain nomenclature should be updated if it is caused by genetic contamination because its hereditary difference from the original strain is relatively huge.

The following ways can be used to produce substrains: artificial technology (e.g., embryo transplantation, artificial feeding, foster nursing, ovarian transplantation, or embryo cryopreservation), or transferring the strain to another location for breeder reservation.

3.1.2.3 Stocks and Strains

The definition of a stock or strain is its breeding rather than animal taxonomy. Formation of a species is the result of natural selection, but that of a stock or strain is the result of artificial selection. Animals from the same species can be divided into various stocks or strains by their hereditary characteristics in breeding. A line of inbred animals with specific genetic characteristics is called a strain, such as the mouse strain C57BL/6J. Outbred animals are called a stock such as CD1 (ICR) or Kunming (KM) mouse stock. Animals from the same stock or strain should have a similar appearance, unique biological characteristics, stable heritability, common ancestors, certain genetic structure, and sufficient individuals.

3.1.2.4 Maintenance and Breeding of Inbred Animals

Inbred strains usually originate from inbreeding among a basic group of breeders (not inbred or wild type). The principle of determining a breeding method is to choose inbred animals that are congenic and homozygous. The original inbred animal should come with a clear genetic background and source, and relatively complete background information (e.g., strain name, generation of inbreeding, hereditary and biological characteristics).

The breeding system of inbred animals includes ancestors, foundation stock, pedigree expansion stock, and production stock (see Figure 3.2). Animals from the production stock are used for biomedical study.

3.1.2.5 Phenotypic Characteristics of Inbred Animals

All individuals within an inbred strain have an identical genotype, which is different between strains. Every inbred strain has a unique phenotype. For example,

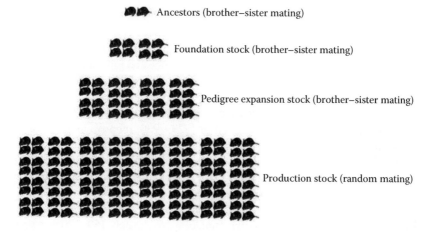

Figure 3.2 Breeding pattern of inbred animals.

senescence accelerated mice (SAM) experience rapid aging without any experimental manipulation; C57BL/6 mice are not sensitive to alcohol or narcotics; some inbred strains have advantages in research on genetically modified (GM) and embryonic stem (ES) technology; the bigger male pronucleus in fertilized eggs of FVB mice makes them suitable for DNA microinjection; gamete transmission is most likely to succeed in ES cells of 129 mice. BALB/c and C3H mice are prone to mutations while being treated with N-ethyl-N-nitrosourea. Information about inbred mice can be found on the Internet (Mouse mutagenesis consortium, http://www.mgu.har. mrc.ac.uk/mutabase/; German the Human Genome Project, http://www.gsf.de/isg/ groups/enu-mouse.html; The Jackson Laboratory, https://www.jax.org/).

Different inbred strain animals of different phenotypes are important when designing animal experimentations. Changes in animal characteristics and experimental study of inbred animal phenotypes may not be relevant, thus affecting the results. For example, C3H/HeJ mice and all other Jackson substrains are homozygous for the retinal degeneration 1 mutation (*Pde6b^rd1*), which causes blindness by weaning age, but lack the *nob5* allele of *Gpr179*, so they may not be suitable for any research related to eyesight. Therefore, the background information of an inbred strain should be well acknowledged before experiment.

Due to inbreeding depression, inbred strains tend to have less offspring and trouble with reproduction. The hybrid vigor of the descendants gained from a cross of two inbred animals is in proportion with how genetically distant the parents are. *Mus spretus* mice are an ideal choice, as they are separated from currently used laboratory mice (*Mus musculus*) by at least a million years in evolution. A cross between them creates infertile male F1 generation and fertile females.

Crosses between inbred animals of differing phenotypes can help locate the loci responsible for quantity or quality. As genetic mapping relies on parental polymorphisms, an analysis for complex traits of the F1 requires two inbred strains with distinct phenotypes and genotypes. More than 95% of the inbred mice are rarely used in biomedical study because their potential as parents is not very promising.

3.1.2.6 *Origin of Inbred Strains*

Work on inbred laboratory animal (especially mice and rats) development began in the early last century .

Dr. Clarence Cook Little led the initiation of inbreeding mice in 1909, and developed the DBA strain of mice, now widely distributed as the two major substrains DBA/1 and DBA/2, which were separated in 1929–1930. Inbreeding mice was then started on a much larger scale by Dr. Leonell C. Strong, leading to the development of strains C3H and CBA, and by Dr. Little, leading to the C57 family of strains (C57BL, C57BR, and C57L). Many of the most popular strains of mice were developed during the following decade and some are closely related. Evidence from the uniformity of mitochondrial DNA suggests that most of the common inbred mouse strains were probably derived from a single breeding female, approximately 150–200 years ago.

Many of the most widely used inbred strains of rats were also developed during this period, several of them by Curtis and Dunning at the Columbia University

Institute for Cancer Research. Strains dating back to this time include F344, M520, and Z61, and later ACI, ACH, A7322, and COP. Tryon's classic work on selection for maze-bright and dull rats led to the development of the TMB and TMD inbred strains, and later to the common use of inbred rats by experimental psychologists.

The similarity in genotype and phenotype between any two related inbred strains depends on when they were separated from the parents and has nothing to do with the contribution from other inbred strains to its parents. Access to http://www.informatics. jax.org/external/festing/ and http://genetics.nature.com/mouse/ can provide information about the origin of each inbred strain of mice.

Knowledge of the inbred mouse pedigree is crucial for biomedical research. A comprehensive review helps the researchers to understand the relationship between different inbred strains and phenotypic differences, and makes it more feasible to analyze results.

3.1.2.7 *Heredity of Inbred Animals*

A combination of pedigree analysis and molecular genetic analysis reveals the nature of allele separation. If some rare and identical genes exist in different inbred strains, they probably originated from a single mutation or multiple independent mutations occurring on an active locus. The inbred strain DBA/2 has a rare allele *Trp53*, and an identical mutation also took place in the inbred strain SM/j in a survey of 25 inbred strains. Do they share the same ancestors or did the same mutation happen twice? Pedigree analysis shows that the ancestors of SM/j are related to those of DBA/2. Therefore, the mutation of DBA/2 accounts for the gene *Trp53* carried by SM. In fact, SM/J, created by MacArthur in 1939 from crosses involving seven inbred strains—among them DBA—followed by inbreeding with selection for small body size, segregates white-bellied agouti (*Aw*) versus nonagouti (*a*) at the agouti locus. The hereditary relevance or irrelevance among inbred mice strains can be analyzed by research and investigation based on molecular genetic technique.

To obtain the pedigree information, massive amounts of molecular genetic research and mutation analyses between two inbred strains is required. The research on genetic similarity of inbred mice is based on biochemical markers and microsatellite DNA mutations (http://www.resgen.com/) or single nucleotide polymorphisms (SNPs). Increasing the accuracy of gene location markers can distinguish loci between similar strains or substrains such as C57BL/6 and C57BL/10.

3.1.2.8 *Digital Resources of Inbred Mice Strains*

Microsatellite DNA mutation is studied in at least 10% of the inbred strains, and a small fraction of the information is about DNA polymorphisms such as SNPs and randomly amplified polymorphic DNA markers. For a comprehensive understanding of inbred mice, please visit the mouse genome database (MGD) (http://www.informatics. jax.org), and mouse resource database (http://www.jax.org/pub-cgi/imsrlist or http:// imsr.har.mrc.ac.uk/).

The importance of laboratory animals to biomedical research is embodied by inbred mice. Collected data of standardized inbred mice can be compared in different studies. An investigation on the inbred strains' genotype and phenotype, and acquisition of useful information is needed for a full understanding and maintenance of mouse diversity.

3.1.2.9 Nomenclature of Inbred Strains

Mice and rats used in the laboratory are derived from a variety of sources. Production of inbred strains means that these backgrounds can be defined and thus require nomenclature conventions. It should be borne in mind that genetic drift means that there may still be unknown genetic differences between individuals within strains.

- Mice
  Most laboratory mice have contributions from both *M. musculus* and *Mus musculus domesticus*. There is evidence that smaller contributions may have also come from *Mus musculus molossinus* and *Mus musculus castaneus*. Therefore, they should not be referred to by species name, but rather as laboratory mice or by use of a specific strain or stock name. In addition, some recently developed laboratory mouse strains are derived wholly from other *Mus* species or other subspecies such as *M. spretus*. Mouse strain names should be registered through the MGD at http://www. informatics.jax.org/mgihome/submissions/amsp submission.cgi.
- Rats
  Laboratory rat strains are derived from the *Rattus norvegicus* species. Another species, *Rattus rattus*, is also used as an experimental model, but has not contributed to the common laboratory rat strains.
  Rat strain names should be registered through the rat genome database (RGD) at http://rgd.mcw.edu/tools/strains/strainRegistrationIndex.cgi.

In this section, nomenclature of mouse or rat inbred strains will be briefly described, the nomenclature for other strains or stocks of animals will be introduced in the following section brief, respectively.

An inbred strain should be designated by a unique, brief symbol made up of uppercase, Roman letters or a combination of letters and numbers beginning with a letter. For example, mouse strains: NZB, NZC, C57BL/6; rat strains: SR, SS. Note that some preexisting strains do not follow this convention; for example, mouse strain 129P1/J.

Care should be taken to see that mouse and rat strains do not overlap in strain designations. Inbred strains that have a common origin, but are separated before F20, are related inbred strains, and the symbols should reflect this relationship.

Substrains are given the root symbol of the original strain, followed by a forward slash and a substrain designation. The designation is usually the laboratory code of the individual or laboratory originating the strain. Examples: A/He, substrain of A mouse strain originating from Walter Heston; IS/Kyo, substrain of IS rat strain originating at Kyoto University.

Substrains may give rise to further substrains by continued maintenance by a different investigator or through establishment of a new colony. In addition, substrains arise if demonstrable genetic differences from the original substrain are discovered. In either case, further substrain designations are added, without the addition of another slash. Examples: C3H/HeH, mouse substrain derived at Harwell (H) from the Heston (He) substrain of C3H; SR/JrIpcv, rat substrain derived at the Institute of Physiology, Czech Academy of Sciences (Ipcv) from the John Rapp (Jr) substrain of SR.

3.1.3 Strains Made from Multiple Inbred Strains

3.1.3.1 Coisogenic Inbred Strains

Coisogenic strains are inbred strains that differ at only a single locus through mutations occurring in that strain. Strains containing targeted mutations in ES cells or other new gene editing techniques that are then crossed to, and maintained on, the same inbred substrain can be regarded as coisogenic, but the possibility of mutations elsewhere should be considered. Similarly, chemically or radiation-induced mutants in an inbred background can be considered coisogenic, although other genomic alterations could be present. A coisogenic strain may accumulate genetic differences over time by genetic drift unless periodically backcrossed to the parental strain.

The maintenance of a mutant gene depends on whether it is recessive or dominant, and its influence on animal's fertility and survival. If the mutation has such effects, we must choose a homozygote and a heterozygote for brother–sister mating to keep the coisogenic strain. For example, female homozygotes in athymic nude mice with the recessive gene nude (nu) lack lactation capacity, which makes it very difficult to keep the strain. This mutant inbred strain can be preserved by using heterozygous nude female mice (+/nu) to mate with homozygous nude males (nu/nu) capable of reproduction. On the other hand, we can produce the nude mice for biomedical research.

Coisogenic strains are useful for studying the effects of a mutation or transgene free of interference from segregation in the genetic background, although the expression of the gene may depend on the background.

Nomenclature of coisogenic strains should be designated by the strain symbol (and where appropriate the substrain symbol) followed by a hyphen and the gene symbol of the differential allele, in italics. Example: C57BL/6JEi-*tth,* the tremor with tilted head (*tth*) mutation in the C57BL/6JEi strain.

3.1.3.2 Congenic Inbred Strains

A congenic inbred strain refers to a new strain that is constructed by introducing a gene (or a particular marker) from the donor strain into an inbred strain through repeated backcrosses or intercross. This new strain has only one gene on a small segment of chromosome different from the original animals, thus called a congenic inbred strain. Congenic strains that differ at a histocompatibility locus and therefore resist each other's grafts are called congenic resistant (CR) strains.

A strain developed by this method is regarded as congenic when a minimum of 10 backcross generations to the background strain have been made, counting the first hybrid or F1 generation as generation 1. At this point, the residual amount of unlinked donor genome in the strain is likely to be less than 0.01 (note that the amount of donor genome linked to the selected gene or marker is reduced at a much slower rate, approximately equivalent to 200/N, where N is the number of back-cross generations for N > 5). Figure 3.3 shows the method of "cross–intercross–backcross" to create a congenic inbred strain.

By this way, the m gene in strain B can be introduced into inbred strain A. Marker-assisted breeding or marker-assisted selection breeding, also known as "speed congenics," permits the production of congenic strains equivalent to 10 back-cross generations in as few as 5 generations. Provided that the appropriate marker selection has been used, these are termed congenic strains if the donor strain con-tribution unlinked to the selected locus or chromosomal region is less than 0.01. Ideally, descriptions of speed congenic strains in first publications thereof should include the number and genomic spacing of markers used to define the congenic-ity of the strain. As speed congenics depend upon thorough marker analysis and can vary by particular experimental protocol, the inbred status of speed congenics should be regarded with caution.

The first successful creation of a congenic strain happened in 1948 when George Snell was studying the genes of histocompatibility. The key of breeding is to transfer a gene along with a segment of chromosome from inbred (or non-inbred) strain B to strain A, then replace the allele and segment of the chromosome on strain A. The new strain generated in this way is called a congenic inbred strain compared with

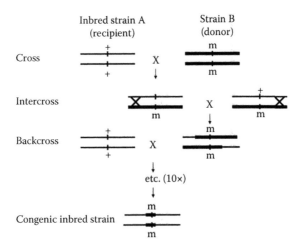

Figure 3.3 The sketch map of "cross–intercross–backcross" to create a congenic inbred strain.

strain A (Figure 3.3). Moreover, strain B which provides the target gene is called a donor strain, strain A receiving the gene is called a partner strain. A partner strain must be an inbred strain, while the donor strain can be any type as long as it owns the gene of interest.

The congenic strain itself is identical with the partner strain except for the gene of interest; during the gene-induced procedure, other closely linked genes may be introduced into the genome of the inbred strain with the target gene. It is different from the original strain not just in alleles, but also a short chromosome segment; congenic strains are widely used in research in order to study the effects of a gene without serious complications from a segregating genetic background.

Congenic strains are designated by a symbol consisting of three parts. The full or abbreviated symbol of the recipient strain is separated by a period from an abbreviated symbol of the donor strain, this being the strain in which the allele or mutation originated, which may or may not be its immediate source in constructing the congenic strain. A hyphen then separates the strain name from the symbol (in italics) of the differential allele(s) introgressed from the donor strain. Examples: B6.AKR-*H2k*, a mouse strain with the genetic background of C57BL/6 (B6) but which differs from that strain by the introduction of a differential allele (*H2k*) derived from strain AKR/J.

3.1.3.3 Segregating Inbred Strains

Segregating inbred strains are inbred strains in which a particular allele or mutation is maintained in a heterozygous state. They are developed by inbreeding (usually brother × sister mating), but with heterozygosity selected at each generation.

Segregating inbred strains can be used in comparison research for the performance of normal and mutant genes, which can minimize the error caused by other uncertain segregating genes. The alleles on two or more linked loci are segregated in segregating inbred strains, which provides a basis for genetic recombination. They can also be used to study lethal, infertile, and deleterious recessive mutant genes.

Segregating inbred strains are designated like other inbred strains as the segregating locus is part of the standard genotype of the strain. When segregating coat color alleles are part of the inbred strain's normal phenotype, they need not be included in the strain name. Examples: 129P3/J, this mouse strain segregates for the tyrosinase alleles albino (*Tyrc*) and chinchilla (*Tyrc-ch*).

The similarities and differences among coisogenic, congenic, and segregating inbred strains can be found below:

- **Similarities:** They are all inbred strains, involving controls over the gene on certain loci; they are almost the same in genetic characteristics except for several genes; these three are usually used in the study of individual genes.
- **Differences:** The method of generation is different; differences between them and their founder inbred strains and their respective linked genes are varied from the others. Coisogenic strains are identical to their founders besides one gene. However, in a congenic strain, some closed linked genes may be introduced into the genome

of the inbred strain with the target gene, and they are called passenger genes. Thus, it is different from the original strain for not just the target gene, but a short chromosomal segment. Differential genes in segregating inbred strains are in a heterozygous state and their linked genes are usually not homozygous.

3.1.3.4 Recombinant Inbred Strains

RI strains contain unique, approximately equal proportions of genetic contributions from two original progenitor inbred strains. Traditionally, RI strains are formed by crossing animals of two inbred strains, followed by 20 or more consecutive generations of brother × sister mating as shown in Figure 3.4. Alternate breeding schemes can be used, such as creating RI strain sets from advanced intercross lines, where F2 animals are nonsib mated for several generations, followed ultimately by 20 or more consecutive generations of brother × sister mating. Note that if backcrossing to one of the parental strains is involved, this will create recombinant congenic strains.

For example, RI strains (CXB1, CXB2, CXB3, etc.) generating from inbred strains C57BL/6 and BLAB/c. The parental strains of an RI strain is called a progenitor strain, and all the heritable traits of the recombinant strain are derived from that. In the RI strain, there is random separation and recombination of genes without linkage, as linked genes due to linkage distance (if in the ancestral linkage) are a fixed trend of inbreeding. RI strains come from the cross of two inbred strains, but they are very different from the founders: although the whole gene compositions are limited by the two founders, the presence of gene-free combination, exchange of chromosomes and the genetic composition of RI strains are not equal; RI strains are highly homozygous as any common inbred strain; gene recombination in RI strains is caused by gene-independent assortment of every chromosome and crossing-over on the same chromosome.

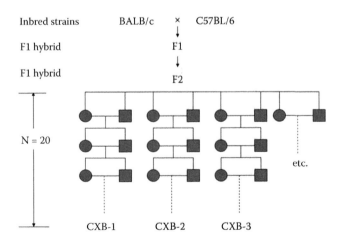

Figure 3.4 Breeding pattern of RI strains. RI, recombinant inbred.

The origins and history of RI strains are described by James Franklin Crow. While the potential utility of RI strains in mapping analysis of complex polygenic traits was obvious from the outset, the small number of strains only made it feasible to map quantitative traits with very large effects (quasi-Mendelian loci). RI strains are now widely used in systems genetics and to study gene–environment interactions. It is possible to accumulate extensive genetic and phenotypical data for each member of a family of RI strains under several different conditions. Each strain has a single fixed genome and it is also possible to resample a given genotype multiple times in multiple environments to obtain highly accurate estimates of genetic and environmental effects and their interactions.

As an example, Otterness and Weinshilboum (1987) found that the activity of the enzyme thiopurine methyltransferase differed between the strains AKR (low) and DBA/2 (high). A set of 23 RI strains had already been bred from a cross between these two strains followed by at least 20 generations of brother × sister mating. When these 23 strains were typed for activity of the enzyme they fell into two groups, with approximately half the strains having high activity and half having low activity. This is the pattern that would be observed if the activity is dependent on a single genetic locus. If the response was dependent on several loci and environmental influences, then all levels of activity would be observed.

RI strains were first developed using inbred strains of mice but are now used to study a wide range of organisms—*Saccharomyces cerevisiae* (yeast), *Zea mays* (maize), barley, *Drosophila melanogaster, Caenorhabditis elegans,* and rats. RI strains provide a powerful system in which to study complex phenotypes including disease-related traits. These strain panels provide a framework for integrating data across multiple phenotype domains spanning molecular, morphological, physiological, and behavioral traits.

The mouse is the most extensively used mammalian model for biomedical and aging research, and an extensive catalogue of laboratory resources is available to support research using mice: classical inbred lines, GM mice, selectively bred lines, consomics, congenics, RI panels, outbred and heterogeneous stocks, and an expanding set of wild-derived strains. However, these resources were not designed or intended to model the heterogeneous human population or for a systematic analysis of phenotypic effects due to random combinations of uniformly distributed natural variants. The collaborative cross (CC) is a large panel of recently established multiparental RI mouse strains specifically designed to overcome the limitations of existing mouse genetic resources for analysis of phenotypes caused by combinatorial allele effects. The CC is being constructed using a randomized breeding design that systematically outcrosses eight founder strains (A/J, C57BL/6J, 129S1/SvImJ, NOD/LtJ, NZO/HlLtJ, CAST/EiJ, PWK/PhJ, and WSB/EiJ), followed by inbreeding to obtain new RI strains. Five of the eight founders are common laboratory strains, and three are wild-derived. In the past years an increasing number of reports have established the value of the CC to provide insights into the genetic architecture of multiple traits, the identification of novel loci associated with them, and the characterization of functional novel alleles at known genes (http://csbio.unc.edu/CCstatus/index.py).

RI strains should be designated by uppercase one- or two-letter abbreviations of both parental strain names, with the female strain written first, and separated by an uppercase letter X with no intervening spaces. All members of RI sets involving the same two strains will be serially numbered regardless of whether they were created in one or more laboratories. Examples: BXD1, BXD2, and BXD3, members of the BXD set of mouse RI strains derived from a cross of C57BL/6×DBA/2.

3.1.3.5 Recombinant Congenic Strains

Recombinant congenic strains are formed by crossing two inbred strains, followed by a few (usually two) backcrosses of the hybrids to one of the parental strains (the "recipient" strain), with subsequent inbreeding without selection for any specific markers. Such inbred strains will consist of the background recipient strain genome interspersed with homozygous segments of the donor (the amount of donor strain genome depends on the number of original backcrosses, two backcrosses will give on average 12.5%).

Recombinant congenic strains should be regarded as fully inbred when the theoretical coefficient of inbreeding approximates that of a standard inbred strain. For this purpose, one generation of backcrossing will be regarded as being equivalent to two generations of brother × sister mating. Thus, a strain produced by two backcrosses (N3, equivalent to F6) followed by 14 generations of brother × sister mating (F14) would be fully inbred.

Recombinant congenic strains should be designated by an uppercase abbreviation of the two strains, recipient strain listed first, separated by a lowercase "c." Example: CcS1, CcS2, and CcS3, a recombinant congenic strain between BALB/c recipient and STS donor.

3.1.3.6 Chromosome Substitution or Consomic Strains

Chromosome substitution or consomic strains are produced by repeated backcrossing of a whole chromosome or its parts onto an inbred strain. As with congenic strains, a minimum of 10 backcross generations is required, counting the F1 generation as generation 1. For autosomes, it is necessary to genotype progeny to ensure that the selected donor chromosome has not recombined with the corresponding recipient chromosome.

The term chromosome substitution strain is a common designation for consomic, subconsomic, and conplastic strains. To create a chromosome substitution strain, transfer of a whole chromosome or a large chromosomal region is carried out, while in congenic strains, the transferred entity is a gene, marker, or genomic segment including a specific marker or interval.

Experience shows that on occasion it is impossible to transfer an entire chromosome from one strain to another due to lethal effects on a particular chromosome. For example, a consomic set on which PWD/Ph individual chromosomes were transferred to C57BL/6J revealed that Chr 11 and Chr X cannot be transferred intact. To designate "sections" of transferred chromosomes that contribute to a consomic set, regions can be indicated as a decimal 1, 2, 3, etc.

The generic designation for consomic strains is HOST STRAIN-Chr #^{DONOR} ^{STRAIN}. Example: C57BL/6J-Chr 19^{SPR}, in this consomic mouse strain, the *M. spretus* chromosome 19 has been backcrossed onto C57BL/6J.

3.1.3.7 Conplastic Strains

Conplastic strains are strains in which the nuclear genome from one strain has been introduced into the cytoplasm of another, either by backcrossing (in which the mitochondrial donor is always the female parent) or by direct nuclear transfer into an enucleated zygote.

The designation is NUCLEAR GENOME-mt$^{CYTOPLASMIC\ GENOME}$. Example: C57BL/6J-mt $^{BALB/c}$, a strain with the nuclear genome of C57BL/6J and the cytoplasmic (mitochondrial) genome of BALB/c. Such a strain is developed by crossing male C57BL/6J mice with BALB/c females, followed by repeated backcrossing of female offspring to male C57BL/6J. As with congenic strains, a minimum of 10 backcross generations is required, counting the F1 generation as generation 1.

3.1.4 F1 Hybrids

The F1 hybrids are the progeny of two inbred strains, crossed in the same direction and are genetically identical. These two inbred strains are called the paternal strain and maternal strain, which must be differing from each other in biological characteristics.

Features of F1 hybrids: (1) genetic isogenicity. All the F1 hybrids are genetically identical to each other like the members of an inbred strain, and their genotype can be determined by examining any individual; (2) long-term stable inheritance. F1 hybrids are as stable as inbred strains, and it will be more stable if the accumulated mutation is recessive. They will be influenced only if the same mutation occurs in two parents. In general, F1 hybrids change more slowly than parental strains; (3) F1 hybrids have phenotypic consistency like inbred strains.

Difference between F1 and inbred strains: (1) All F1 hybrids are heterozygous, which means the F1 hybrids will not undergo "true" breeding. Every F1 hybrid mouse or rat must be produced by cross of the parental strains whenever needed; (2) hybrid vigor owned by F1 hybrids makes them more adaptable to environmental changes.

Application of F1 hybrids in biomedicine: (1) for general biological research, F1 hybrids have vigor, genetically uniform and isogenic animals, and are more resistant to stress in experiments than inbred strains; (2) they are used to determine the mode of inheritance or studied only when a specific hybrid is to be observed. For example, NZB × NZW hybrids are widely used as a model of systemic autoimmune lupus erythematosus; (3) hybrid vigor is valuable in some research projects. They are very useful in high fecundity studies, and are used as surrogate mothers, for fertilized eggs or ovarian transplant recipients; (4) F1 hybrids can also provide the genetic background for some deleterious mutations. Hybrid vigor helps keep animals alive. Otherwise, those mutations would be lethal or cause early death in homozygous inbred strains; (5) F1 hybrid production is the easiest way to create multiple genotypes, thus the need of repeating experiments on a large number of different genotypes can be met by F1 hybrids. Crosses among inbred strains can produce n(n−1)/2 F1 hybrids (except backcrossing), and is more feasible than only a few inbred strains.

Nomenclature of F1 hybrids: for example, D2B6F1 is mouse that is the offspring of a DBA/2 mother and C57BL6/J father. A full F1 designation is (DBA/2N×C57BL/6J) F1; CB1BD22F1 is mouse that is the offspring of two RI strains, a CXB1 mother and BXD22 father; full F1 designation is (CXB1/ByJ×BXD22/TyJ)F1.

3.1.5 Outbred Strains and Closed Colonies

In biological evolution, the presence or absence of a single organism is meaningless. Every species' survival and development depends on group living. The hereditary differences among individuals originate from alleles. A colony consists of members who can intercross with each other. The different heredity among colonies is caused by their various allele frequencies on a locus.

There are two genetically undefined laboratory animal populations: outbred stocks and closed colonies.

3.1.5.1 Outbred Stocks

Outbred stocks are genetically undefined; that is, no two individuals from an outbred stock are the same. Outbreds are intentionally not bred with siblings or close relatives, as the purpose of an outbred stock is to maintain maximum heterozygosity.

An outbred strain is created by outbreeding for at least four generations, without introducing any foreign individuals. Outbred animals do not introduce any foreign members and are bred under confined conditions, thereby maintaining the general genetic characteristics and retaining heterozygosity. Outbred strains derived from inbred animals do not have any inbreeding, and those from a noninbred strain is a group of animals going through five years of continuous breeding. Outbred animals are isolated from the outside community. To avoid inbreeding and losing genes within the group, a closed environment and random mating are adapted to keep the gene frequencies within the populations stable so that genetic traits are kept within a certain range.

The principle of outbred animal reservation and breeding is to try to keep gene heterozygosity and polymorphism in order to avoid the inbreeding coefficient rising too fast with the generations. In this way, the gene and genotype frequencies between the two generations are relatively fixed. The founders should have a clear genetic background, source, and a generally complete profile. The number of induced animals should be sufficient to maintain gene heterozygosity and polymorphism. Before introducing any animals, the inbreeding "increment" should be determined to deduce the "Effective Population Size" of the group. Then, based on the conservation and mating schemes, the minimum number for outbred animals for conservation and breeding can be calculated, namely minimum number for breeding. Generally, each generation of inbreeding "increment" should be controlled less than 1% during outbreeding. To meet that requirement, unrelated male and female individuals should be selected for mating. With more pairs for random mating and the ratio between females and males coming closer to 1:1, the inbreeding coefficient increases more slowly. The inbreeding coefficient will rise faster in polygynous mating. Theoretically, the number of breeding pairs in outbred stock is as many as possible, but taking into account the scale and cost of breeding, the number of male and female animals for breeding to be based on the actual situation, as they are small groups, the "increment" for each generation of inbreeding is inevitable. In general, for small groups of outbred random mating, the ratio of male and female animals is 1:1, and the introduced animal number cannot be less than 25 pairs.

For outbred strains with a large population, the random mating method can be employed. However, they are not allowed to cross with other strains, only within the same group in completely closed conditions. In strains with rules of randomization, animal mating should follow a random number table or other objective ways, rather than any manner chosen by personal wishes. During random mating, there should be more than 25 pairs of female and male individuals. As the genes are heterozygous, outbred animals are more vigorous, have more offspring, shorter birth intervals, high weaning rate, more disease resistant, and are easier to breed.

Outbred animals exhibit some special characteristics:

- Heterozygosity, outbred stocks do not introduce any new genes from outside of the stock. The implementation of random mating leads to no change in the genes within population, thus a certain level of heterozygosity is maintained.
- Relative stability, relative closed state, and random mating result in gene and genotype frequencies of the outbred group, therefore the allele and genotype frequencies in a outbred stock remain constant from generation to generation, also known as the Hardy–Weinberg equilibrium. The stock will remain relatively stable for genetic characteristics within a certain range.
- Increased fertility and disease resistance, outbred animals acquire hybrid vigor through random mating, and fecundity and disease resistance exceed inbred strains.

The genetic composition of outbred animals is highly heterozygous, and bears resemblance to the genetic characteristics of heterozygosity in humans. Therefore, they are valuable in research for human genetics, drug screening, and toxicology. Outbred animals retain stable genetic traits and average reactions, which makes them suitable for screening therapeutic effects of drugs. Outbred stock have a strong fecundity, they are easy to produce with less husbandry costs, have a relatively long life span, are resistant to diseases, and are in large supply. They are widely used in pretests, teaching, and general experiments. They are useful for experimentation where genotype is not important and where a random genetic population is desired.

For nomenclature of outbred stocks, the common strain root is preceded by the laboratory code of the institution holding the stock. Examples: Tac:ICR, the ICR outbred stock maintained by Taconic Farms, Inc.; Hsd:NIH Swiss, the NIH Swiss outbred stock maintained by Harlan Sprague Dawley, Inc.

3.1.5.2 Closed Colonies

A closed colony contains limited genetic diversity and is maintained neither by brother × sister mating (inbred), nor by selective mating to maximize heterozygosity (outbred). All mating occurs within the colony members, but breeders need not be selected from specific parentage. No animals are introduced into the colony from outside the stock from generation to generation.

Closed colonies may be established as a way to more readily maintain a difficult mutation, where the desire is to maintain a reasonably uniform background, but poor mating performance prohibits use of brother × sister mating schemes. Note that closed colonies describe a permanent mating system and this does not apply, for example, if an inbred strain is out-crossed to a near relative in a single generation because of a temporary breeding crisis.

Closed colony designations consist of the strain of origin and appropriately designated mutations (if applicable), followed by [cc] to indicate closed colony. Example: C57BL/6Tac-*Bmp4tm1Blh*[cc], a closed colony of mice originating from the C57BL/6Tac inbred strain and carrying the *Bmp4tm1Blh* targeted mutation.

3.1.6 Comparison of Inbred Strains, F1 Hybrids, and Outbred Stocks

3.1.6.1 Homozygosity

Inbred strains are defined by the homozygosity of all the members. If a gene locus is based on a neutral theory of molecular evolution, the possibility of a locus in any individual being homozygous is at least 98.6% in full brother and sister mating for over 20 generations. However, "residual heterozygosity" still exists in inbred strains; it occurs when a locus is super dominant for an animal's vitality.

F1 hybrids are heterozygous on all different loci between the parental strains. Many loci of F2 hybrids are heterozygous as well, and not isogenic.

Homozygosity of outbred animals depends on the origin of the stocks. It is generally believed that genetically variable outbred populations experienced some inbreeding in their history. For example, today's commonly available outbred Swiss stocks were derived from two males and seven females obtained by Clara Lynch from Switzerland in 1926. The heterogeneity of many stocks is limited by the narrow genetic base of the founders, early selective breeding and inbreeding programs, and subsequent rederivation efforts that "reconstituted" an outbred stock from only a few or even from only one breeding pair.

3.1.6.2 Isogenicity

All the loci of inbred animals are homozygous, thus their offspring are genetically identical, that is, isogenicity. There are three important effects of isogenicity: animals can accept skin grafts from the same strain, monitoring of an individual is able to determine the genotype of the entire inbred strain, and genetically identical substrains can be established. F1 hybrid animals are genetically identical and isogenic.

3.1.6.3 Long-Term Stability

One of the most important features of inbred animals is their long-term genetic stability. For example, the C57BL strain first came out in 1921, and it is still very much the same as the original strain. Selection will not cause genetic mutation within an inbred strain, which will occur much more rapidly in an outbred strain, as this has been confirmed in many studies. An animal strain will not stay absolutely unchanged, due to mutations, residual heterozygosity, and genetic pollution. Theoretically, F1 hybrids are more consistent than an inbred strain. If the mutation does not occur on the same locus of two parental strains, it will not show in the F1 generation, even if it occurred on multiple loci, and the influence on the F1 generation is decreased.

3.1.6.4 Identifiability

Many inbred strain animals have special or unique genotypes, which can be identified by coat color, skin grafts, biochemical sites, immunological sites, or quantitative trait testing. These methods accurately identify to which strain they belong. Until now, a complete genetic monitoring method has not been formed for outbred strains. For example, no set of traits can be used to reliably distinguish the commonly used Sprague-Dawley and Wistar outbred rats. The tumor incidence in Sprague-Dawley rats was investigated from six different sources, and it was concluded that the differences among them resulted from different suppliers, like those from different strains.

3.1.6.5 Uniformity

Inbred animals are homogenic animals with phenotypic uniformity, which is expressed by acceptance of skin grafts within the strain. For quantitative traits, inbred strains are more consistent with the outbred stocks. Comparison of mandible morphological structure variations among inbred, outbred, F1 and F2 hybrid mice shows no significant differences between inbred and F1 hybrid animals, and an obvious difference between outbred and F2 hybrid mice, although the variation is not large.

Evidence shows that inbreeding can lead to lower animal development stability, causing an increased sensitivity to environmental factors. Therefore, inbreeding can cause more phenotypic variations under certain circumstances. Sometimes there may be more phenotypic variations if an inbred strain is sensitive to environmental changes, which may be offset by their sensitivity to experimental treatments. Generally, F1 hybrids are more consistent than inbred strains, but no conclusion can be drawn when comparing inbred with outbred strains.

3.1.6.6 Individuality

Each inbred strain is a unique combination of genetic materials, resulting in a unique phenotype, referred to as individuality. A number of phenotypic characteristics are great assets to biomedical research, and serve as human disease models.

For example, DBA/2J is a widely used inbred strain that is valuable in a large number of research areas, including cardiovascular biology, neurobiology, and sensorineural research. Its characteristics are often contrasted with those of the C57BL/6J inbred strain. DBA/2J mice exhibit high-frequency hearing loss beginning roughly at the time of weaning/adolescence (between 3–4 weeks of age), which becomes severe by 2–3 months of age. This strain possesses three recessive alleles that cause progressive cochlear pathology initially affecting the organ of Corti. Decreasing anteroventral cochlear nucleus volume decreases and neuron loss parallel the progression of peripheral hearing loss. Young DBA/2J inbred mice are also susceptible to audiogenic seizures due to the *asp2* mutation; however, this susceptibility decreases as animals reach adulthood. Aging DBA/2J mice develop progressive eye abnormalities that closely mimic human hereditary glaucoma. DBA/2J mice also exhibit an extreme intolerance to alcohol and morphine.

C57BL/6 is the most widely used inbred strain. They are used in a wide scope of research areas including (1) a high susceptibility to diet-induced obesity, type 2 diabetes and atherosclerosis; (2) a high incidence of microphthalmia and other associated eye abnormalities; (3) resistance to audiogenic seizures; (4) low bone density; (5) hereditary hydrocephalus; (6) hair loss associated with over-grooming; (7) a preference for alcohol and morphine; (8) late-onset hearing loss; and (9) increased incidence of hydrocephalus and malocclusion.

BALB/c mice are particularly well known for the production of plasmacytomas following injection with mineral oil forming the basis for the production of monoclonal antibodies. Mammary tumor incidence is normally low, but infection with mammary tumor virus by fostering to MMTV+ C3H mice markedly increases tumor numbers and reduces age of onset. BALB/c mice develop other cancers later in life including reticular neoplasms, primary lung tumors, and renal tumors. BALB/c mice immunized with $PLP_{180-199}$ develop an atypical form of experimental autoimmune encephalomyelitis in which susceptibility is determined by location. In the spinal cord 60%–70% mice develop pathological lesions, and in the brain and cerebellum, 100% of mice develop severe lesions.

Although every inbred strain has its own unique set of characteristics, inbred strains are only used in some general, standard, or reproducible studies. Inbred C57BL, CBA, BALB/c, C3H mice, and F344, LEW, and PVG rats are ideal animals. An outbred strain also represents a unique genotype, but there is no evidence to show that an outbred strain is more representative of the entire species than an inbred strain, therefore, outbred strains cannot be used to make the results more accurate. Furthermore, its variability may seriously reduce the statistical value.

3.1.6.7 Vigor

Compared with most F1 hybrid or outbred animals, inbred animals have lower vigor and reproductive ability. Inbreeding depression will reduce the general performance of animals. For example, if inbreeding coefficient of a mouse increases by 10%, it will result in decreased average yield of 0.6 offspring per litter, and 0.58 g lighter female body weight at the age of 6 weeks. Groups within an outbred strain are

Table 3.1 Comparisons of Characteristics in Various Laboratory Animals

Characteristics	Inbred Strains	Outbred Strains	F1 Hybrids
Homozygosity	Very high	Low	Low
Isogenicity	High	Low	High
Long-term stability	High	Low	High
Identifiability	High	Low	High
Uniformity	Medium→high	Medium	Very high
Individuality	High	Low	?
Vigor	Low	Varied	Very high
International distribution	High	Medium	High
Background information	High	Medium	High

very different from each other, which is mainly determined by the impact of inbreeding and vigor influenced by homozygous genes. Inbreeding depression is the opposite of heterosis or hybrid vigor, and the hybrids generally show strong vitality.

3.1.6.8 International Distribution

Many inbred strains are widely distributed around the world, thus a comparative study can be conducted among nations. This suggests that researchers from different countries or regions may breed and use genetically identical standardized inbred animals, thus replication and verification of acquired results are possible. An individual from an inbred strain is totipotent, as all the animals within the strain are carrying the whole package of genes.

Main features of various laboratory animals are summarized in Table 3.1.

3.1.7 Genetic Quality Control

Genetic quality control of laboratory animals includes two aspects. One is animal scientific introduction, breeding, and production, that is, the control over the production process. The second is to establish a regular genetic monitoring system for the quality control of laboratory animals. There are many genetic monitoring methods for inbred animals, directly or indirectly, to detect changes in certain genes of animals.

3.1.7.1 Genetic Variations

Inbred animals' genetic "material" is constantly changing during the production and breeding process, due to genetic contamination, genetic drift, and mutations. Genetic contamination is the most common incident in animal management, usually caused by a cross with other strains. Genetic drift refers to a random genotypic change that may occur in the breeding process. This often results from separation of the residual heterozygous genes, thereby forming a new substrain. Mutation is caused by the replacement, deletion, or insertion of a certain nucleotide in the genome, leading to a change of the allele. Therefore, regular monitoring is of great significance to understand the genetic background of the laboratory animals.

3.1.7.2 Common Monitoring Strategies

Many strategies can be used to detect inbred animals' genome, such as conventional biochemical, immunological, morphological, and cytogenetic markers. One single monitoring method cannot reflect the whole profile of the genetic composition. Thus, multiple ways should be adopted to complement each other.

- **Biochemical markers:** Protein isoforms exist in animals (e.g., isozymes), which are controlled by different alleles on the same locus, shown in different phenotypes. Isozymes can be separated by electrophoresis, based on their moving speeds. This method can monitor the genes on a number of loci, involving more than a dozen chromosomes. Provisions of the National Standard (GB14923-2010, China) states that 16 biochemical loci of inbred mice and 11 of inbred rats should be monitored.
- **Immunological markers:** There are many glycoproteins on the membrane of T and B cells, which play a major role in the immune system. The complement and immunoglobulins in the serum also have genetic polymorphisms. Therefore, these serological components can be used as genetic markers. In addition, skin grafts can also be applied into identifying the similarities of genetic histocompatibility antigens.
- **Cytogenetic markers:** Each animal species has a fixed number and form of chromosomes in the nucleus, known as the karyotype. The mouse chromosome number is $2N = 40$, and includes 19 pairs of autosomes and 1 pair of sex chromosomes. After Giemsa and Quinacrine staining, these chromosomes are ready for morphological examinations and observation of heterochromatin (C band). The C band of inbred mice has a genetic polymorphism.

3.1.7.3 Molecular Genetic Markers

Molecular genetic markers generally refer to the markers at a DNA level, which are also called DNA markers. There is a large number of markers, and the number can be up to 10^8–10^{10} for an individual. Most of them are neutral mutations, which are genetically very stable and not affected by the animal's physiology or environmental factors. Also, they can be equally dominant or totally dominant. Several major DNA markers include the following:

- **Restricted fragment length polymorphisms (RFLP):** RFLP refers to the length differences of cleaved fragments containing homologous sequences after the individual's genome is treated with a restriction endonuclease. Each enzyme can recognize and cleave genomic DNA in a specific site. Once the nucleotide sequences of the site are changed, the site will no longer be recognized by the corresponding enzyme. Meanwhile, the nucleotide change on a former cleavage site may generate a new recognition sequence. Thus, the different distributions of the sites lead to length differences of cleaved fragments among genotypes in different DNA molecules. These DNA fragments can be detected by Southern hybridization.
- **Random amplified polymorphic DNA (RAPD):** This is polymerase chain reaction (PCR) amplification of genomic DNA by DNA polymerase with synthetic, randomly arranged short single-stranded DNA serving as primers. A set of discrete DNA fragments are obtained, wherein each of the products represents a genetic locus.

- **DNA fingerprints:** DNA fingerprints are the mapping of RFLP, generated by DNA fingerprinting probes composed of multiple RFLP bands, with a high degree of variability, specificity, and stable inheritance. DNA fingerprint mapping has multiple sites, high variability, and simple but stable heredity. The loci monitored are the highly variable sites in the genome, as the map consisting of the alleles on those loci is bound to have a higher variability. DNA fingerprints can also be stably inherited, that the distinguished parental band traits can be independently and equally assigned to the offspring. DNA fingerprints remain unchanged in body cells. Consistent results will be acquired with DNA from different tissues (blood, semen, muscle, organs, etc.) in an individual.
- **Minisatellite DNA markers:** Minisatellite DNA refers to a short repeating end-to-end sequence (10–60 bp) in the genome, which contains an identical or similar 10–15 bp core sequence. Due to the presence of thousands of these loci in the genome, core sequences of repeating parts on certain loci are identical or similar. Therefore, a minisatellite probe may simultaneously hybridize with the alleles on multiple minisatellites' loci, forming polymorphisms with a high degree of variability.
- **Microsatellite DNA markers:** This refers to a very short repeating end-to-end sequence (2–6 bp) in the genome. Their differences in repetition among different strains and species can be used to study polymorphisms among populations and individuals.
- **Single nucleotide polymorphism (SNP):** An SNP is a variation in a single nucleotide that occurs at a specific position in the genome, where each variation is present to some appreciable degree within an inbred strain. For example, a set of 1638 informative SNP markers easily assayed by the Amplifluor genotyping system were tested in 102 mouse strains, including the majority of the common and wild-derived inbred strains available from The Jackson Laboratory. Amplifluor assays were developed for each marker and performed on two independent DNA samples from each strain. The mean number of polymorphisms between the strains was 608 ± 136 SD. Several tests indicate that the markers provide an effective system for performing genome scans and quantitative trait loci analyses in all but the most closely related strains.

3.2 MICROBIOLOGICAL STANDARDIZATION OF LABORATORY ANIMALS

Quality control over microorganisms is for producing animals that reach the expected standard, which also should be followed in animal experimentation.

3.2.1 Reasons for Microbiological Quality Control on Laboratory Animals

3.2.1.1 Causing Diseases and Death in Animals

Infectious diseases are the main cause of diseases and death of animals, thus affecting their quality. Pathogens responsible for these diseases include viruses, mycoplasma, bacteria, and parasites. The susceptibility to a certain disease varies in different animals because of their genetic backgrounds after contacting with some

Table 3.2 Estimated Quantity of Pathogens Related to Rodents
           and Domestic Rabbits

	Mice	Rats	Guinea Pigs	Rabbits
Viruses	25	20	15	10
Mycoplasma	3	3	2	2
Bacteria	25	20	15	15
Parasites	25	35	20	25

kind of pathogen. For example, mousepox is fatal for some strains of mice, while others are resistant to it. Compared with F344 rats, Lewis rats are more susceptible to *Mycoplasma pulmonis*. C57BL/6 mice are more easily contaminated with *Streptobacillus moniliformis* than other mice.

Table 3.2 provides the possible quantities of several main pathogens, which may lead to an infection. Clinical symptoms do not necessarily appear after contamination (potential or subclinical infection). Previous research has shown that bacteria or viruses often cause respiratory and gastrointestinal infections. A mixed infection of viruses and bacteria will inevitably lead to worse conditions in animals. For example, rodents can carry lung mycoplasma without any clinical symptoms for their whole life, but they will suffer from a fatal pneumonia after a secondary infection of the Sendai virus. Respiratory viruses can affect the functions of pulmonary macrophages, inhibit their ability to remove bacteria from the respiratory tract, which leads to pneumonia. A nonrespiratory virus can also induce secondary bacterial pneumonia. For example, reovirus type 3 infection can not only cause hepatitis in rodents, but also damage their immunity in the lungs. Factors other than pathogens can also play an important role in the development of animal diseases. For example, a high concentration of ammonia can restrain the movement of respiratory cilia, limiting the clearance of pathogens from respiratory tract.

For the reasons mentioned above, specific pathogen-free (SPF) animals are suggested to be used in biomedical research.

3.2.1.2 Interference with the Results of Animal Experimentation

In most cases, animals do not display clinical symptoms after an infection. However, potential infection can seriously affect the results of study. A Sendai virus contamination can abate the response of B and T lymphocytes to antigenic stimuli, increase the yield of interferon, and reduce the level of the third complement factor in the plasma. The mouse hepatitis virus can inhibit phagocyte activity in the mesh endodermis system and suppress lymphocyte cytotoxic activity, inducing a higher level of aspartate aminotransferase (ASAT), alanine aminotransferase (ALAT), and many other liver enzymes in the plasma. Moreover, during a transplantable tumor study, animals can easily catch an infection of lactic dehydrogenase elevating virus (LDHV), causing a sharp increase of lactic dehydrogenase and corticosteroids in the plasma and a delay in graft rejection. These demonstrate that potential infectious pathogens can seriously influence the studies. Therefore, a full knowledge of the microorganisms carried by the animals can help to speculate their impact on the results of animal experimentation.

3.2.1.3 Zoonosis

Zoonoses are infectious diseases of animals (usually vertebrates) that can naturally be transmitted to humans. Major modern diseases such as the Ebola virus, Salmonella, and influenza are zoonoses. Zoonoses can be caused by a range of disease pathogens such as viruses, bacteria, fungi, and parasites; of 1415 pathogens known to infect humans, 61% were zoonotic. Most human diseases originated in animals; however, only diseases that routinely involve animal-to-human transmission, like rabies, are considered as zoonoses. Zoonoses have different modes of transmission. In direct zoonosis, the disease is directly transmitted from animals to humans through media such as air (influenza) or through bites and saliva (rabies). In contrast, transmission can also occur via an intermediate species (referred to as a vector), which carry the disease pathogen without getting infected. When humans infect animals, it is called reverse zoonosis.

Zoonoses may be derived from many different pathogens and all phases of infections ranging from subclinical states to fatal. Animals bred or caught in the wild may possibly carry pathogens that are harmful to animals and humans. SPF animals are not likely to be infected. However, some pathogens can be detected in animals that have undergone a cesarean section or hysterectomy, such as *S. moniliformis*, a common kind of bacterium in a healthy rat's nasopharynx. Once bitten by rats with *S. moniliformis*, humans may catch rat-bite fever. Drinking contaminated water or milk may result in Haverhill fever. Both diseases could be life threatening if not treated in time.

Trichophytosis is also an easily spread zoonosis, mainly caused by *Trichophyton* sp. and *Microsporum* sp. infection. These infections are often present with subclinical symptoms. People with trichophytosis will exhibit circular skin lesions.

Rats often have a latent infection of Hanta virus, which can spread through the respiratory tract, intestinal secretions, feces, and urine. People can be easily infected by direct or indirect contact with contaminated animals, biological products, or materials and equipment, and then exhibit severe acute interstitial nephritis, namely hemorrhagic fever with renal syndrome. In that case, kidney function is completely lost and it may be fatal if the situation worsens.

Due to their need in biomedical research, pathogens are sometimes inoculated to animals as they are certain sources of those diseases. Preventive protection must be used to limit the spread of those pathogens.

3.2.1.4 Affect the Quality of Biological Products

The serum, vaccines, and other biological products used on humans must be safe and without any contamination. Thus, the minimum requirement for animals used for their production is that they carry no zoonoses. If live or attenuated viral vaccines for human use are produced by contaminated animal cells, it is possible to infect people with a common disease. Vaccine products and other biological products require strict monitoring prevention systems. Good manufacture practice (GMP) is aiming to control all the production stages, and ensure safe and high-quality products. If the animals are used in the production of biological products, the potential of zoonoses or specific pathogens must be removed before human use.

SPF animals can be used to evaluate the efficacy and safety of a given product. For example, one of the many monitoring events of viral vaccines is to exclude the existence of a nonrelated virus. Animals are inoculated with vaccines and whether they are carrying some kind of virus can be determined by detecting its corresponding antibody in the serum.

3.2.2 Contamination Sources and Routes of Transmission

Laboratory animals can be contaminated by pathogens from several sources via different routes of transmission. The main sources of contamination during animal husbandry and studies are other animals, products from animals, pets, personnel, materials, and equipment.

3.2.2.1 Laboratory Animals

Infected laboratory animals are an important source of microbiological contamination. Although great efforts are made for microbiological quality control, the animals brought from outside can possibly bring pathogenic microorganisms into a new facility. Generally, preventive health care in animal study is much less rigorous than animal breeding, and those animals are not as heavily monitored either. During an experiment, animals from different sources are often kept in the same room. There are also overlaps of events that should be done at different times. This means that the animals in the study are likely to be contaminated with several kinds of pathogens. Therefore, the risk of contamination while introducing animals from outside ought to be considered.

3.2.2.2 Biomaterials

In current biomedical research, biomaterials usually are derived from the serum, ascites, cells, tissues, or organs of laboratory animals. Animals are used as the "medium" for those microorganisms, which cannot be cultivated *in vitro*. These animals can also be contaminated by those products from infected animals.

In general, viruses (possibly mycoplasma or bacteria) mainly existing in living cells cause contamination. According to an investigation conducted in the United States, one or more viruses are detected in more than 90% of transplanted tumors and in at least 70% of virus-related reagents. Some infectious microbe is caught during continuous passages of the biomaterial. For example, a tumor cell strain in animal ascites can be infected with the pathogens carried by the animal. Domestic rabbits used for the passage of *Treponema pallidum* may also catch coronavirus, which can cause the death of more than 40% of the rabbits. Usage of contaminated biomaterials is the main cause of mice ectromelia virus outbreaks.

3.2.2.3 Pets

Pets can also pose a serious threat to the health of laboratory animals. Personnel working on animal husbandry and experiments should not keep any pets, as well as their family members.

3.2.2.4 Personnel

Humans are the main factors responsible for transmitting diseases between two groups of animals. They serve as carriers and temporary hosts of pathogens after contact with infected animals.

The correct way to handle this is to isolate the personnel working in facilities for different classes of animals. For example, animal breeders cannot enter the laboratory room. Although some employees are not exposed to contaminated laboratory animals, it may be a source of pollution. In the case of tuberculosis, it can spread among almost all mammalian species. If anyone in the personnel (animal care or study) has symptoms, such as diarrhea, skin rash, or chronic respiratory diseases, the chance of transmitting pathogens (potential) will increase, which would facilitate the contamination of animals.

One hypothesis is that a variety of pathogenic microorganisms in humans cannot infect animals because they are different species. However, contamination can induce temporary production of antibodies in a small proportion of animals.

3.2.2.5 Materials and Equipment

Materials and equipment can be the media to transmit infections in experiments as well. Food and bedding are likely to be contaminated by wild rodents during production, harvest, and storage. Water is also a source. Animals with lower immunity are susceptible to *Pseudomonas aeruginosa* in the water. Cages and the use of surgical equipment are no exception, which require sterilization. The air entering the animal facility should be filtered to remove the pathogens from the air.

3.2.3 Classification of Laboratory Animals According to Carry Microorganisms

Laboratory animals can be divided into different levels based on the microorganisms they carry. For example, according to the National Standard (China): *Laboratory Animal—Microbiological Standards and Monitoring* (GB 14922.2 2010), laboratory animals can be classified into four classes: conventional animals, clean animals, SPF animals, and germ-free animals.

3.2.3.1 Conventional Animals

Animals kept in an open environment with minimum microbiological control are called conventional animals, and they do not carry any zoonoses or deadly infectious diseases. They are still widely used in biomedical study, especially for special experimental research, teaching, or preliminary experiments. If the microbiological status of the animals is unknown or suspicious, they should be seen as conventional animals. As the conventional animals are bred without any preventive health care, they ought to go through a period of quarantine before use. Length of the quarantine time depends on the shortest time needed for ruling out possible contamination.

3.2.3.2 Clean Animals

Clean animals do not carry any harmful or interfering pathogens in addition to the pathogens ruled out in the conventional animals. They are classified based on the National Standard (China) (GB 14922.2 2010). Clean animals must be derived from SPF or germ-free animals, and kept in a semibarrier facility.

Clean animals are only suitable for short term or part of a scientific experiment. They are healthier than conventional animals, can easily reach the quality standard of SPF animals, and are less likely to be influenced by animal diseases. Clean animals are a transitional class of animals under current circumstances.

3.2.3.3 SPF Animals

SPF animals do not carry specific (potential) pathogenic microorganisms or parasites. Besides the pathogens that should be ruled out in the conventional or clean animals, they do not have any pathogens that lead to infection or interfere with scientific experiments. SPF animals must come from germ-free or gnotobiotic animals.

SPF animals are considered as the standard laboratory animals. Monitoring the (potential) pathogenic microorganisms can help to speculate what types of microorganisms do not exist in SPF animals.

SPF animals have been widely used in biomedical research. The main reasons are as follows: drug safety evaluation and animal experimentations are free of interference from any infection; possibility of contamination is essential for whether the experiment can be continued; the risks of long-term experiments are higher than short-term ones; gerontological research requires SPF animals as they generally live longer than other animals; the average survival time of rats and animal tumor incidence are influenced by interventional infection, which SPF animals can avoid. SPF animals are also used in immunological studies. Animals like T-cell deficiency nude mice and severe combined immunodeficiency (SCID) mice, with both T and B cell deficiency, exhibit decreased immunity due to genetic factors or immunosuppressants, and they can only be conserved or reproduced in SPF level or above.

3.2.3.4 Germ-Free and Gnotobiotic Animals

Germ-free and gnotobiotic animals belong to the higher-level animals. There are no living microorganisms or parasites in any part of a germ-free animal detected by existing monitoring technologies. The microorganisms are viruses, rickettsia, bacteria (including helix, mycoplasma), fungi, and protozoa. Gnotobiotic animals are animals with known bacteria or microbial flora, which are implanted in the germ-free animals. Based on the number of induced bacteria, they are divided into monoxenic, dioxenic, trexenic, and polyxenic animals.

Germ-free and gnotobiotic animals can be obtained by hysterectomy, cesarean section, or embryo transfer from conventional animals. The core of this technology is to exclude any potential pathogenic microbial flora. The time for a hysterectomy is earlier than the normal delivery time. With a sterile surgery, the uterus is removed from the donor animal and put into an isolator via a sterilization tank full of disinfectant solution. The offspring are then taken out of the opened womb, and fed by people or lactating animals.

Most animals coming from hysterectomy and cesarean section are germ free. Occasionally, one could carry one or more microorganisms, which are transmitted to them via placenta or other vertical ways during gestation. Vertical transmission means that the microorganisms are directly passed on to the next generation by the mother. This happens when the mother has been contaminated with an active infectious disease during pregnancy. Pathogens in the blood can penetrate the placental barrier to the fetus.

Autochthonous flora are normal microbes residing on the skin, oral mucosa, respiratory tract, urinary tract, and gastrointestinal tract of conventional animals. This flora can help animals resist potential infection. Species and quantity of the autochthonous flora are not certain yet. According to preliminary statistics, there are more than 500 kinds of mouse intestinal flora. Every gram of intestinal contents contains 10^{10}–10^{11} bacteria. They are closely related to the host and are able to benefit from symbiosis.

Compared with conventional animals, germ-free animals generated from hysterectomy lose all autochthonous flora and show many abnormalities in morphology and physiology. For example, germ-free animals have an enlarged cecum and thinner intestinal wall compared with conventional animals. If mice and rats are provided with the autochthonous flora, the germ-free animals will be normalized with a considerably narrowed cecum and thickened bowel wall.

Gnotobiotic animals can be used to produce vaccines. Human vaccines can be produced by cells harvested from animals, which come from stocks without any exotic infection. Gnotobiotic animals comply with this requirement. Another application is in the research of the effect of bacteria on compound transformation via mouth, related to simple or complex gut bacteria. Moreover, gnotobiotic animals are suitable candidates for studying the effects of a nonlethal dose of radiation or other immunosuppressors used in cancer research. Due to frequent infections, this kind of research is almost impossible to conduct in conventional animals, or even SPF

Table 3.3　Relation between the Animal Microbial Quality and Breeding Facility

Animal Class	Breeding Facility
Conventional animals	Open system
Clean animals	Semi-barrier system
SPF animals	Barrier system
Germ-free and gnotobiotic animals	Isolators

Table 3.4　Characteristics of Different Classes of Laboratory Animals

Type	Germ Free	SPF	Clean	Conventional
Infectious diseases	No	No	No	Yes or possibly yes
Parasites	No	No	No	Yes or possibly yes
Experimental results	Definite	Definite	Almost definite	Questionable
Required animals	A few	A few	Several	Many
Statistical value	High	High	Relatively high	Not accurate
Long experiment time	Possible	Possibly good	Possibly good	Difficult
Mortality	Very low	Low	Relatively low	High
Long-term survival rate	Approximately 100%	Approximately 90%	Approximately 80%	Approximately 40%
Standardized experiments	Possible	Possible	Possible	Impossible
Value of experiment results	Very high	High	Relatively high	Questionable

animals. Gut ecological studies, including the pathogenesis of infection, the intestinal immune system, and the function of normal flora are nearly impossible without gnotobiotic animals.

Tables 3.3 and 3.4 show the comparisons of animal breeding environments and their characteristics.

3.2.4　Microbiological Quality Monitoring of Laboratory Animals

Microbiological quality control is divided into control over the barrier system and laboratory animals. Microbiological control should adhere to the principle of prevention being the most important. Feasible measures are set based on the three basic parts of animal epidemiology: source of infecting microorganisms, means of transmission for the microorganism and susceptible animals. In addition, quality monitoring of the animals in microbiology and parasitology on a regular basis are required.

The monitoring of pathogenic microorganisms in animals is quite important. Most of the attention has been placed on the periodical screening of healthy animals. Autopsy of sick or dead animals defines the cause of illness or death. It is assumed that the disease is caused by pathogenic microbiology and can be detected by direct methods such as microscopic examination or culture.

3.2.4.1 Monitoring Methods

According to the different objects, monitoring can be divided into four types: virological, bacteriological, mycological, and parasitological. All work should be done by professionals.

- Virus serological and etiological examination. Serological tests are suitable for regular inspection and disease census of all kinds of animals. Commonly used methods are blood hemagglutination (HA) test, hemagglutination inhibition (HAI) test, immunofluorescence assay (IFA), immunoenzymatic assay (IEA), and enzyme-linked immunosorbent assay (ELISA). Etiological examination is suitable for the detection of a virus or its status as responsible for an epidemic. This includes the isolation, cultivation, and identification of a virus, detection of the virus particles, antigens, or nucleic acid, the activation of latent virus and antibody production test.
- Common bacteriological monitoring methods are to isolate and culture pathogenic bacteria. Some pathogens like *Salmonella typhimurium, Corynebacterium kutscheri, Bacillus piliformis,* and *Mycoplasma gallisepticum* can be confirmed by serological tests, but a certain diagnosis still requires the results of isolation and culture. Meanwhile, pathogens like *B. piliformis,* which are unable to grow in artificial medium, should be examined with attached preparation and microscopy, and diagnosed with pathological examination.
- The mycological monitoring method at present is to isolate and culture in Sabouraud's agar. Skin fungi are cultured in 25°C, fungi in deeper parts are cultured in 37°C. Different fungi have different colony characteristics, combined with a specific straining. Sometimes a final diagnosis still needs biochemical results and immunological methods.
- Parasitological monitoring. External parasites can be observed with the naked eye. These parasites and their eggs can be examined by hair sampling with transparent sticky tape. Intestinal parasites can be seen with the naked eye in feces. The monitoring of parasites in blood requires microscopic examination after staining the thick and thin smears of peripheral blood. Monitoring of parasites in tissues requires attached preparation and sectioning of the suspected parts of parasitic infection.

The quality of laboratory animals and experimentations are mainly based on a full understanding of the different microorganical status of the animal stocks. Animal monitoring should be conducted under the guidance of a veterinary microbiologist because animals and humans are very likely to be infected by infectious microorganisms or contaminated biomaterials.

3.2.4.2 Gnotobiotic and SPF Animals

Gnotobiotic animals carry known flora, which depends entirely on the monitoring methods. In other words, animals may be recognized as germ free or gnotobiotic animals if no pathogens are detected or implanted by the adopted method. The possibility of vertical transmission is very low if the gnotobiotic animals originate from SPF stock. However, if they are from conventional animals, the monitoring range should be widened as the carried microorganisms can be transmitted directly. If there is no evidence to suggest a vertical infection, further examination can be limited to those invading microbes due to barrier damage (e.g., spore-forming bacteria). In the serum of animals from hysterectomy, there are antibodies from their mother for a variety of microorganisms. Those antibodies can be detected 2–3 months after birth.

SPF animals must be checked regularly for pathogens. The monitoring frequency ranges from 2 to 12 times a year. The number of animals used in each monitoring are not equal, generally between 5 and 25. On an average, 10 animals or 10 samples of serum are enough. The number for random sampling largely determines the possibility of detecting a contamination. If 50% of SPF animals are actually contaminated, the accuracy of monitoring is about 50% if five animals (sample size) are examined. This means that infection prevalence of less than 50% is 95% likely to be ignored. Most of the pathogens do not exist in the stocks breeding under classical SPF conditions. Microbial monitoring is difficult on a healthy SPF animal. Many pathogenic microorganisms can be only found by autopsy, rather than in healthy animals. The interpretation of the monitoring data collected in SPF animals is complex. The barrier and husbandry records should be taken into consideration while analyzing those data. As microbial infections may be episodic, infection may also occur during the animal transfer process.

3.2.4.3 SPF Animals in Studies

In general, management of animal experimentation is less strict than animal production. Thus animals are more likely to be infected during experimentation than the production and breeding groups. In longer-term animal experimentation, the use of SPF animals is often conventionalized with the managing method of conventional animals. The "conventionalization" will influence the animals in many aspects, such as physiology, thus affecting the results. Moreover, in toxicological studies, the total counts of white blood cells and classification, immunoglobulin levels (IgM and IgG), and relative weight of lymphoid organs (thymus and spleen) are often examined. If the values of these parameters vary widely in the control group, it may indicate that the SPF animal is infected.

3.2.4.4 Vaccination and Treatment

Usually, vaccination in laboratory animals is limited to large-sized animals such as cats and dogs. The inoculation of a small amount of virus (such as mousepox, Sendai) into rodents and rabbits is possible, but the effects are questionable. They could still fall ill after vaccination. Inoculation can induce antibody production, which seriously interferes with the monitoring of serological indexes. It is also difficult to distinguish whether the antibody was caused by vaccination or infection.

Antibiotics should be administered to protect valuable animal experimentation once there is a disease outbreak. However, this is risky, as the normal intestinal flora of rodents like guinea pigs and rabbits can be easily disrupted by antibiotics, which results in serious intestinal lesions. Preventive drugs can be given via food or drinking water, but any medication will affect the study results. Treating infected animals is not recommended. However, some situations, such as infection of Sendai virus, can cause devastating consequences in the stock. Such animals should be euthanized and disinfected because protection of the whole facility is more important than that of any individual's research results.

3.2.5 Biohazard and Prevention in Animal Experimentation

Biohazard is caused by misoperation by personnel in biomedical research, whereby the outside environment is contaminated by the spread of harmful pathogens, leading to the infection of surrounding people and animals.

Conventional and wild animals used in biomedical study may be a source of infection. A report suggested that 15 researchers caught influenza-like lymphatic choroid plexus meningitis due to exposure to rats. They had the symptoms of parotitis, meningitis, and unilateral orchitis. From 1970 to 1984, 126 scientific workers coming from 22 Japanese biomedical institutes were infected with hemorrhagic fever with renal syndrome, and one of them died. Moreover, in a research on analytical behavioral psychology conducted in Japan, many monkeys exhibited diarrhea, swollen lymph nodes, abscesses and weight loss, and subsequently died. *Tuberculosis* bacterium was detected in the lesions, and proven by microscopic examination and culture. The staff in the monkey facility were then checked for tuberculosis, and people with similar symptoms were positive in the tuberculin reaction.

Scientific researchers are the main part of the animal experimentations. During the whole study process, errors made by people who lacked comprehensive knowledge of pathogens, animals, or equipment, or who were not skilled enough, lead to infection. The routes of infection include wounds, oral, aerosols causing respiratory infection, and insects. Defective husbandry, isolation, and disinfection facilities can also increase the risk of biohazards.

Strengthening the management of animals, using qualified animals, and standardized operation can prevent biohazards from happening in an experiment.

3.3 ENVIRONMENT AND FACILITIES OF LABORATORY ANIMALS

According to the dramatype theory of Russell and Burch, in 1959, an animal's genetic structure is determined by the reproductive cells from its parents. During the period of embryonic development, the fertilized egg tends to express its phenotype under the influence of the environment (the mother's body and lactation after birth). A young animal presents as dramatype with continuous changes caused by its growing environment. The result of an animal experiment is the animal's overall reaction to the treatment and surroundings, under the premise of a certain phenotype and growing environment. The environment has a huge impact on the animal reactions when the treatments are determined. Thus, the standardization of animal environmental conditions and animal facilities are necessary.

3.3.1 Influence of Environment on the Animals

3.3.1.1 Temperature and Humidity

The body temperature of mammals is constant, but the changes in environmental temperature still cause fluctuation of the animal's temperature, affecting their physiological functions.

A low temperature can lead to delay in animal sexual cycles. When the temperature is just over 30°C, male animals will experience testicular atrophy or reduced ability in forming sperm. The females may have irregular menstrual periods, lactate less, or refuse to breastfeed. For example, when the temperature is below 18°C, golden hamsters would eat their babies and temperatures lower than 13°C can cause the death of the babies. When the temperature is close to 4°C, golden hamsters go into hibernation after a short period. In addition, a high or low temperature will not only affect the animal's farrowing rate, ablactational rate, survival rate of newborn animals, but the animals will also be less resistant to illness.

Humidity refers to the moisture content in the atmosphere. The amount of water vapor present in a unit of air (usually expressed in grams per cubic meter) is called absolute humidity. Relative humidity is the ratio of the partial pressure of water vapor to the equilibrium vapor pressure of water at a given temperature. Relative humidity depends on temperature and the pressure of the system of interest. Less water vapor is required to attain high relative humidity at low temperatures; more water vapor is required to attain high relative humidity in warm or hot air.

High humidity facilitates the growth of microbes, mildew of food and bedding, and significantly increases the number of bacteria in the air and ammonia concentration, leading to respiratory diseases. When the humidity is too low, it results in a dusty environment, which is harmful to the animal's health. For example, when the temperature is 27°C and the relative humidity is under 40%, rats experience ringtail disease. It is generally believed that the low humidity accelerates water loss in the tail and results in narrowed tail veins, causing blood circulation disorder.

Animals have a high rate of metabolism in low temperatures, which exerts a great influence on the animal's viscera weight. The heart, liver, and kidneys of mice are larger at low temperatures and smaller at high temperatures, which implies a significant negative correlation between them and the environmental temperature. The same phenomenon is observed in rats.

It is well known that in acute toxicity tests on mice and rats, the 50% lethal dose (LD_{50}) varies because of different environmental temperatures. The relations between temperature and toxicity can be divided into three types (A, B, and C) in terms of the drugs. Class A agents (e.g., acetylcholine, digoxin, strychnine, methanol, convection phosphorus, salicylic acid salt, and ephedrine) are more toxic at high temperatures. Class B (e.g., methacholine, pentobarbital, chlorpromazine, and armour piperidine oxazine) is the strongest at low temperatures, and less toxic with increasing temperature. Class C (e.g., procaine) is equally toxic at high temperature and room temperature, but their effects are markedly enhanced at low temperatures.

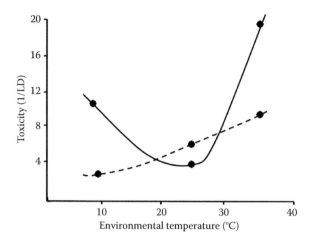

Figure 3.5 Relations between the toxicity of salicylate, ephedra, and environmental temperature. LD—lethal dose.

Figure 3.5 shows the relations between the toxicity of salicylate, ephedra, and environmental temperature, in which LD refers to the lethal dose and the unit of body weight is g/kg. The toxicity of salicylate changes with environmental temperature in a "U" shape curve, which is quite noteworthy. The toxicity of ephedrine is increased with the increasing study temperature. Therefore, laboratory temperatures should be standardized in order to reduce the variation of drug toxicity.

Humidity also has an important influence on the study results. For example, food intake of rats in an environment of 21°C, with 35% relative humidity, increases by approximately 5%, compared with that at 21°C, with 75% relative humidity. Mice are also more active in low relative humidity. Even at the same temperature, animals tend to eat more and be more active to generate extra heat because of increased dissipation in low relative humidity.

Table 3.5 shows the heart rate, breathing rate, and body temperature of mice when exposed to low, normal, or high relative humidity environments with temperatures at 15°C, 25°C, 30°C, and 35°C for 60 min. When the temperature was set at 15°C and 25°C, there were no significant changes caused by relative humidity. However, their heart jumped faster at 30°C and 35°C, and body temperature varied significantly under different humidity conditions at 35°C. The fluctuation caused by relative humidity was more obvious at high temperatures.

Sendai virus in mice is more easily transmitted under highly humid conditions, and the polio virus, and adenovirus type 4, 7 are also multiplied in abundance in such environment. However, influenza, type 3 parainfluenza, smallpox virus, and Venezuelan equine encephalitis viruses proliferate better with low humidity. Moreover, the amount of allergens in the room air also falls with the humidity. Thus, animal physiological changes according to the environment, and the characteristics of microbes interact with each other to exert comprehensive effects on animals.

Table 3.5   Heart Rate, Breathing Rate, and Body Temperature of Mice Exposed
             to Different Temperatures, and Humidity for 60 min

Environment Temperature (°C)	Relative Humidity (%)	Heart Rate (heartbeat/min)	Breathing Rate (times/min)	Body Temperature (°C)
15	40–50	563.3 ± 78.0	207.1 ± 31.5	30.0 ± 1.4
	60–70	602.2 ± 77.4	206.4 ± 22.0	30.9 ± 2.0
	85–95	950.6 ± 62.0	222.0 ± 16.1	30.2 ± 1.5
25	40–50	731.4 ± 53.4	180.8 ± 31.3	35.7 ± 1.0
	60–70	695.3 ± 47.6	188.3 ± 29.0	35.1 ± 1.1
	85–95	697.5 ± 42.6	195.6 ± 18.7	35.4 ± 0.9
30	40–50	701.4 ± 41.3	171.0 ± 21.3	36.9 ± 0.7
	60–70	668.7 ± 40.2	185.2 ± 14.8	37.1 ± 0.7
	85–95	646.3 ± 48.2	187.3 ± 22.5	37.5 ± 0.7
35	40–50	650.6 ± 49.0	205.3 ± 30.4	38.6 ± 0.7
	60–70	687.8 ± 49.0	189.8 ± 32.0	38.5 ± 0.5
	85–95	777.6 ± 7.50	227.9 ± 43.9	40.4 ± 0.5

Note: 9–10-week-old ICR male mice were used, the value is mean ± SD.

Temperature has a great influence on animal reproduction and animal studies, therefore, control on the laboratory temperature is necessary for obtaining reliable and repeatable results. However, experience and practice varies among scholars from different nations, and their geographical environment is different as well, thus the standard temperature is not always the same.

It has been suggested that the best, targeted, recommended value and permissible range can be used as standards for animal facility. The best value is a temperature range in which an animal is in the best state, but in terms of an individual, it is impossible to determine the most suitable temperature for a room full of animals from different species, strains, genders, ages, and stocking densities. The target value is the temperature presented by the air conditioners. For example, it can be set at 23 ± 2°C for mice and rats. A recommended value is the temperature recognized by facility management, which is 23 ± 3°C for mice and rats. The permissible range is a range of temperature in which animals can be safe and healthy, normally between 18° and 28°C. The National standard in China (GB14925-2010) states that the ideal temperature for laboratory animals is 19–26°C, and relative humidity is 40%–70%.

3.3.1.2 Air Flow, Speed, and Air Changes

The sensible air speed is 0.2–0.25 meter/second (m/s) for humans and 0.13–0.18 m/s for animals. Air distribution in laboratory animal facilities generally adopts turbulent flow to ensure uniform distribution of fresh air and cost reduction in construction and operating costs. Constant air flow and speed can make the temperature, humidity, and chemical composition consistent, and facilitate the emission of foul gas.

Air flow and speed are closely related to the dissipation of animal body heat. Air flow, speed, temperature, and humidity are not the sole factors impacting animals, as their interactions are what affect animals. When the room temperature is high, air flow facilitates thermal convection, which is good for the animals. When the room temperature is low, the air flow accelerates animal heat dissipating and strengthening the influence of cold. As most laboratory animals are small with a large ratio of body weight to surface area, they are more sensitive to air flow and speed. Slower air flow causes poor air circulation, animal anoxia, and a room full of foul smells, difficult heat dissipation and discomfort, sometimes leading to diseases and suffocation of the animals. Faster air flow facilitates animal heat dissipation and leads to more food intake.

Appropriate air changes can provide sufficient fresh air to the animals. However, too many air exchanges make animals consume more energy to make up for the heat loss caused by fast air flow. Air changes 10–20 times per hour is recommended for most animals.

Pathogenic microorganisms spread with the air flow. Static pressure within the facilities determine the direction of air flow. In a barrier system, air flows from the clean area (cleaning corridor or animal rooms) to the pollution area because of differences in static pressure, which is higher in the cleaning corridor or animal rooms. However, in an infectious or radioactive animal facility, static pressure in the animal rooms is lower to prevent the spreading of microbes or radioactive materials.

3.3.1.3 Physical and Chemical Factors

3.3.1.3.1 Dust

Particles floating in the air are called aerosols, and they are divided into dust, fumes, mist, fog, and smoke based on their state and physical and chemical forming processes, as well as size. The dust brought from the outside and animal hair, dander and bedding have certain effects on the animal room. Airborne bacteria, viruses, and rickettsia are treated as dust in physical terms, but they are very important in biology, and are attached on particles bigger than 5 μm floating in the air.

The effects of dust on humans and its relation with working environments are studied in depth, as dust can induce allergies. In an allergic reaction, the serum, fur, dander and urine of mice, rats, guinea pigs, and rabbits can all be antigens. Therefore, the dust in an animal room should be monitored like zoonoses.

3.3.1.3.2 Odors

It is reported that more than 400,000 types of substances have scents, about 4,000 of which can be sensed by humans. Organic chemicals containing nitrogen and sulfur are the decomposition of amines, thiols, butanes, and proteins. In addition, phenol, cresol, butyl phenol, amyl phenol, and higher fatty acids also fall into this category.

Observation of different odorous substances found that ammonia is of the highest concentration in all the rooms. Ammonia concentration is closely linked with temperature, humidity, and animal breeding density. Indoor ammonia is generated by the bacterial decomposition of urea in manure, which can be verified by a very low concentration of ammonia in germ-free animal facilities.

Most researchers believe that ammonia can cause respiratory mucosal abnormalities, resulting in respiratory diseases. In addition, it can also induce severe rhinitis, otitis, bronchitis, and mycoplasma pneumonia.

3.3.1.3.3 Noises

Noises refer to all unwanted sound with a high frequency, high sound pressure, huge impact, and complex waveform.

Auditory organs of mice are generally considered to be formed 14 days after birth. People react to a sound frequency ranging from 20 to 20,000 Hz, and the optimum value is 20,000 Hz. Mice, rats, hamsters, dogs, and cats can hear ultrasonic frequencies, which are inaudible to humans.

The effects of noises on animals include decreased birth rate, increased bite kills, reduced lactation, and even audiogenic spasms in several mouse strains. In addition, noises can also influence the repeatability of animal experiments by affecting the animal's heart rate, respiration, and blood pressure.

3.3.1.3.4 Illumination

Many physiological activities are cyclical such as heart rate, respiration, body temperature, neural activity, DNA replication, as well as estrus, ovulation, and farrowing. They manifest in lunar, circadian, seasonal, and yearly rhythms. Rhythms in heartbeat, breath, and body temperature usually fluctuate with environmental changes. Animals can still keep their original rhythms after a period of adjustments for time differences caused by migration. However, the peak and low values will be influenced by the new surroundings, as is also the case for yearly rhythms.

Among all the environmental factors affecting the circadian rhythms of animals, the most important is illumination.

Under natural conditions, golden hamsters experience reduced plasma gonadotropin and genital degradation in winter. Thus, there must be 12.5 h of illumination for golden hamsters every day to prevent testicular atrophy and maintain normal sperm production. According to our observations, golden hamsters reproduce and grow the best when the light to dark ratio is 1.4:1. The estrus of SD rats lasts 4 days with 12 h of light and 12 h of dark, and 5 days with 16 h of light and 8 h of dark. The sexual cycle will become disordered with 22 h of light and 2 h of dark.

3.3.1.4 Living and Biological Factors

3.3.1.4.1 Animal Breeding Methods

The basic husbandry method for small laboratory animals in the barrier system is caging, which are then placed on a shelf. It can be further divided into the following ways: cages for mouse, rat, and hamster production and experimentation (Figure 3.6); washing method that does not require replacement of bedding as animal manure can

Figure 3.6 Rabbit breeding system.

be rinsed off with water directly, but the room humidity still requires maintenance. From the perspective of controlling microorganisms, there are open shelves with static flow (with positive or negative pressure), barriers of static flow and isolators.

3.3.1.4.1.1 Cages — The material and construction of the cages can influence the experimental results in mice and rats. For example, in the study on carbon tetrachloride toxicity, animals living in cages made of metal meshes are heavier than those living in plastic cages, and exhibit less changes in liver function and histological examinations.

3.3.1.4.1.2 Bedding — Bedding keeps the animals warm, comfortable, and clean in cages. The most common beddings used are wood shavings, corn cob, and paper. When wood shavings are used as bedding, the chemical properties of the wood should be noted. An adequate safety assessment of wooden bedding containing aromatic hydrocarbons (like pine or cedar) is necessary.

3.3.1.4.2 Biological Factors

There are biological factors within a species or between species, the latter refers to the relationship between airborne bacteria and animals that has been discussed previously. Here is a brief introduction of the issues on biological factors within a species.

An animal society is formed when two animals are together, resulting in superior and inferior relationships among individuals. Social status of animals can be roughly divided into strain type and despotic type.

A strain can represent superior or inferior status of an animal in a strain type. The first one is the leader, who rules over the others. The second can rule the ones below it, and so on. Monkeys, rabbits, dogs, chickens, and pigs belong to this type.

In a despotic type, the leader is more superior to the rest. Fighting is not commonly seen among the animals other than the leader. Rats, mice, and cats fall into this type. Fierce fights are common in the process of forming such a society. This strive for hegemony can be seen among male mice kept in the same cage.

Animal fights and social status also affect their endocrine system. In a group of male mice coming from the wild, the accessory adrenal gland of inferior animals is heavier than that of the superior one. These features should be taken into consideration in determining stock density.

3.3.1.4.2.1 Stock Density — Animal stock density is the number of animals or live animal weight in a specific area for a specific period of time. Stock density is essentially animal concentration, and is a tool used to accomplish many experimentation goals in biomedical study.

Important considerations for stock density include the age and sex of the animals, the number of animals to be cohoused, and the duration of the accommodation, the use for which the animals are intended (e.g., production vs. experimentation) and any special needs they may have. For example, adolescent animals, which usually weigh less than adults, but are more active, may require more space relative to body weight. Group-housed, social animals can share space such that the amount of space required per animal may decrease with increasing group size, thus larger groups may be housed at slightly higher stocking densities than smaller groups or individual animals. Socially housed animals should have sufficient space and structural complexity to allow them to escape aggression or hide from other animals in the pair or group. Breeding animals will require more space, particularly if neonatal animals will be raised together with their mother or as a breeding group until weaning age. Space quality also affects usability.

Consideration of floor area alone may not be sufficient in determining adequate cage size; with some species, cage volume and spatial arrangement may be of greater importance. Cage height should take into account the animal's typical posture and provide adequate clearance for the animal from cage structures such as feeders and water devices. Some species—for example, nonhuman primates and cats—use the vertical dimensions of the cage to a greater extent than the floor. For these animals, the ability to stand or to perch with adequate vertical space to keep their body, including their tail, above the cage floor can improve their well-being.

Space allocations should be assessed, reviewed, and modified as necessary by the Institutional Animal Care and Use Committee (IACUC) considering the special needs determined by the characteristics of the animal strain or species (e.g., obese, hyperactive, or arboreal animals) and experimental use (e.g., animals in long-term studies may require greater and more complex space). At a minimum, animals must have enough space to express their natural postures and postural adjustments without touching the enclosure walls or ceiling, be able to turn around and have ready access to food and water. In addition, there must be sufficient space to comfortably rest away from areas soiled by urine and feces.

The stock density recommendations should be considered the minimum for animals housed under conditions commonly found in laboratory animal housing facilities. Adjustments to the amount and arrangement of space recommended in the following tables should be reviewed and approved by the IACUC and be based on performance indices related to animal well-being and research quality as described in the preceding paragraphs, with due consideration of the animal welfare regulations (AWRs), Public Health Services (PHS) policy, and other applicable regulations and standards.

For example, the *Guide for the Care and Use of Laboratory Animals* (Eighth edition, 2011) recommended the minimum space for commonly used laboratory rodents housed in groups. If they are housed singly or in small groups, or exceed the weights in the table, more space per animal may be required, while larger groups may be housed at slightly higher densities.

Minimum space for commonly used laboratory animals are discussed in Chapter 4.

Studies have recently evaluated space needs and the effects of social housing, group size, and density, and housing conditions for many different species and strains of rodents, and have reported varying effects on behavior (such as aggression) and experimental outcomes. However, it is difficult to compare these studies due to the study design and experimental variables that have been measured. For example, variables that may affect the animals' response to different cage sizes and housing densities include, but are not limited to, species, strain (and social behavior of the strain), phenotype, age, gender, quality of the space (e.g., vertical access), and structures placed in the cage. These issues remain complex and should be carefully considered when housing rodents.

3.3.1.5 Environmental Enrichment

The primary aim of environmental enrichment is to enhance animal well-being by providing animals with sensory and motor stimulation through structures and resources that facilitate the expression of species-typical behaviors, and promote psychological well-being through physical exercise, manipulative activities, and cognitive challenges according to species-specific characteristics. Examples of enrichment include structural additions such as perches and visual barriers for nonhuman primates, elevated shelves for cats and rabbits and shelters for guinea pigs, as well as manipulable resources such as novel objects and foraging devices for nonhuman primates, manipulable toys for nonhuman primates, dogs, cats, and swine, wooden chew sticks for some rodent species, and nesting material for mice. Novelty of enrichment through rotation or replacement of items should be a consideration; however, changing animals' environment too frequently may be stressful.

Well-conceived enrichment provides animals with choices and a degree of control over their environment, which allows them to better cope with environmental stressors. For example, visual barriers allow nonhuman primates to avoid social conflict, elevated shelves for rabbits and shelters for rodents allow them to retreat in case of disturbances, and nesting material and deep bedding allow mice to control their temperature and avoid cold stress during resting and sleeping.

Enrichment programs should be reviewed by the IACUC, researchers, and veterinarian on a regular basis to ensure that they are beneficial to the animal's well-being and consistent with the goals of animal use. They should be updated as needed to ensure that they reflect current knowledge. Personnel responsible for animal care and husbandry should receive training in the behavioral biology of the species they work with to appropriately monitor the effects of enrichment as well as identify the development of adverse or abnormal behaviors.

Like other environmental factors (such as space, light, noise, temperature, and animal care procedures), enrichment affects animal phenotype and may affect the experimental outcome. It should, therefore, be considered an independent variable and appropriately controlled.

3.3.2 Animal Facilities

A well-planned, well-designed, well-constructed, properly maintained and managed facility is an important element of humane animal care and use as it facilitates efficient, economical, and safe operation. The size of an animal facility depends on the scope of institutional research activities, the animals to be housed, the physical relationship to the rest of the institution, and the geographic location.

The size, nature, and intensity of an institutional program will determine the specific animal facility and support functions needed. In facilities that are small, maintain few animals, or maintain animals under special conditions—such as facilities used exclusively for housing gnotobiotic or SPF colonies or animals in runs, pens, or outdoor housing—some functional areas listed below may be unnecessary or may be included in a multipurpose area. Space is required for the following: animal housing, care and sanitation; receipt, quarantine, separation, and/or rederivation of animals; separation of species or isolation of individual projects when necessary; and storage.

According to the degree of control over microorganisms, animal facilities can be divided into the following: (1) open system for conventional animals (Figure 3.7a). The facility is usually connected to the outside by only one entrance/exit. Some important factors, such as temperature, humidity, and atmospheric dust, are under the influence of the outside world; (2) barrier system (Figure 3.7b), which is for housing SPF animals or performing experimentations; (3) isolators (Figure 3.7c), which provide a germ-free environment for the animals, in which all the care and experimental personnel cannot contact the animals and all the animals are kept in isolators.

3.3.2.1 Barrier Facilities

Barrier facilities are designed and constructed to exclude the introduction of adventitious infectious agents from areas where animals of a defined health status are housed and used. They may be a portion of a larger facility or a free-standing unit. While once used primarily for rodent production facilities and to maintain immuno-deficient rodents, many newer facilities incorporate barrier features for housing SPF mice and rats, especially valuable GM animals and SPF animals of other species.

Barrier facilities typically incorporate airlock or special entries (e.g., air or wet showers) for staff and supplies. Staff generally wear dedicated clothing and foot-wear, or freshly laundered, sterile, or disposable outer garments such as gowns, head

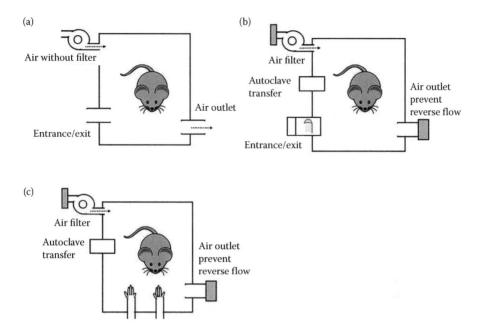

Figure 3.7 Structure schema of (a) open facility, (b) barrier facility, and (c) isolator.

and shoe covers, gloves, and sometimes face masks prior to entry. Consumables, such as feed or bedding, that may harbor infectious agents are autoclaved or are gamma irradiated by the supplier and surface decontaminated on entry. Drinking water may be autoclaved or subject to specialized treatment (e.g., reverse osmosis filtration) to remove infectious agents. Caging and other materials with which the animals have direct contact may be sterilized after washing before reuse. Strict operational procedures are frequently established to preclude intermingling of clean and soiled supplies and personnel groups depending on work function. Only animals of defined health status are received into the barrier, and once they leave, they are prohibited from reentering without retesting. Personnel entry is restricted and those with access are appropriately trained in procedures that minimize the introduction of contaminants.

Engineering features may include high-level filtration of supply air (e.g., high efficiency particulate air [HEPA] or 95% efficient filters), pressurization of the barrier with respect to surrounding areas, and directional airflow from clean to potentially contaminated areas. Specialized equipment augmenting the barrier may include isolator cages, individually ventilated cages (IVC), and animal changing stations.

In some countries, people can enter the facility without showering, which is also questioned by some scholars. The preventive measures taken for SPF animals are not as strict as gnotobiotic animals, but the only purpose is to prevent the intrusion of potentially pathogenic microorganisms. Thus, in theory, disinfection of the materials and equipment is sufficient. But in reality, the materials and equipment are also sterilized to be more secure. Staff need special management, and the number of people entering the breeding room should be limited to a minimum. Showering before entering the clean zone is a basic requirement, and although the shower increases the shedding of skin normal flora, it removes potentially pathogenic microorganisms. Thus, water and air showers exert positive effects on the maintenance and management of the facilities, and should be retained.

The layout of the barrier facility changes with factors, such as the location, construction area, purpose, strains and number of the animals, investment and management, and a uniform layout pattern is impossible. Based on the requirements of the breeding process and the barrier system, an animal facility can be divided into two parts: breeding and management area.

The animal breeding area can be divided into three units in terms of function: a breeding unit that consists of a general preparation room, washing room, separation processing room (shower, dressing), clean preparation room, clean corridors, antechamber of animal room, animal room, back chamber (buffer), a nonclean corridor (dirt corridor), and air conditioning room; an isolator unit includes a clean preparation room, isolator room, and isolator washing room; an experimental unit includes a general preparation room, clean room, and operating room.

A management area includes a duty room, waiting room, control room, power room, data processing room, document library, toilet, maintenance room, and storeroom.

The two areas and their functional units above are arranged to constitute a general layout. There are many plans such as single, double, or multiple layers, and some units may be used in combination to reduce construction expenses. Designs should be made in line with the specific requirements and situations.

3.3.2.2 Modified Barrier Facility

Preventive measures for SPF animals during experimentation come from the classic barrier system. The actual measures depend on the awareness of contamination risk and the results of a damaged barrier. For long-term toxicology studies, the barrier system may copy the classic SPF barrier system, but for short-term pharmacological studies, strict preventive measures are not necessary. For this situation, an independent ventilation cage (IVC) was developed in the 1990s (Figure 3.8). Animals can be kept in a cage with a protective cover (filter covers). Those covers are only removed while managing the animals and the cages are only opened on the static flow shelves.

IVC is a special kind of improved barrier system with a low cost and high efficiency. Every cage is equipped with continuous HEPA filters, and 99.999% of the air is filtered.

The invention of the IVC system is considered as a revolution in barrier facility construction. IVC is an ideal sustained prevention barrier, which provides a high-level germ-free environment and absolutely reliable separated prevention barrier for the rodents to prevent cross infection. Air in the cage can be discharged in lateral flow form through devices equipped on the cage rack. A variety of laboratory animals can be bred separately, but managed in the same working area. Also the frequency of

Figure 3.8 A system of IVC for mouse breeding.

changing cages and sterilization is reduced. The transmission and spread of harmful substances and contaminated air is prevented. Even in a poorly equipped workplace, IVC cages are completely qualified for general scientific research. They are particularly suitable for breeding immunodeficient animals, GM animals, or for other short-term experimentation using SPF animals.

It should be noted that the IVC system is not ideal for large-scale production of laboratory animals, but quite suitable for study using rodents. The IVC system is preferably placed in a barrier facility, and the bedding should be replaced in a special enclosure.

3.3.2.3 Barrier Facility for Infectious Animals

Biomedical studies sometimes require infectious animals, thus the environment should be protected from the pathogens carried by the animals. The barrier system used for breeding these animals is actually the inverse conversion of a classic barrier system to avoid infecting humans. For example, before leaving the area, the waste needs to be purified, personnel should take a shower, and the air flowing out must undergo hazard-free treatment.

An infectious animal experimental facility is a special barrier facility, guiding principles and the starting point for its construction and management are completely different from general barrier facilities. When designing and constructing general barrier facilities and management, the main consideration is how to avoid pollution caused by the external environment, internal laboratory animals, and facilities. The infectious animal laboratory facilities primarily consider how to prevent infected animals carrying pathogens from leaking out of facilities to the outside, and protection for laboratory personnel, the environment, and the local community.

The main differences between infectious animal facilities and general barrier facilities are as follows: an infectious animal barrier facility is in negative pressure compared with the outside world, while the barrier facility usually requires a positive pressure; the air must be filtered or sterilized whether it is in or out in an infectious facility, but only the incoming air is filtered in a barrier system; all the items entering or leaving the facility must undergo strict sterilization, but only the items taken inside require sterilization in a normal barrier system; all the liquid or solid waste generated within the facility must be sterilized before being discharged into sewage; the facility is designed for pathogens that are harmful for humans and other animals; goods and animals that are biological hazards should be marked when entering the laboratory; the doors must be equipped with glass windows so the working process and animals can be seen from the outside; all the researchers need to follow the standard operation procedure (SOP) in the facility; calling devices shall be provided inside and outside, in case there is any incidence needed to be reported to supervisors; the installation of monitoring and alarm systems are also necessary.

The protection required for infectious animal experimentation is defined as biosafety levels. A biosafety level is a set of biocontainment precautions required to isolate dangerous biological agents in an enclosed laboratory facility. The levels of containment range from the lowest biosafety level 1 (BSL-1) to the highest at level 4 (BSL-4). In the United States, the Centers for Disease Control and Prevention (CDC) have specified these levels. Facilities with these designations are also sometimes labeled as P1 through P4 (for Pathogen or Protection level), as in the term "P3 laboratory."

At the lowest level of biosafety, precautions may consist of regular hand washing and minimal protective equipment. At higher biosafety levels, precautions may include airflow systems, multiple containment rooms, sealed containers, positive pressure personnel suits, established protocols for all procedures, extensive personnel training, and high levels of security to control access to the facility.

The four biosafety levels were developed to protect against a world of select agents. These agents include bacteria, fungi, parasites, prions, rickettsial agents, and viruses, the latter being probably the largest and most important group. In many instances, the work or research involves vertebrate animals, everything from mice to cattle.

BSL-1 is suitable for work with well-characterized infectious animals, which do not cause disease in healthy humans. This level of biosafety is appropriate for work with several kinds of microorganisms including nonpathogenic *Escherichia coli*, *Bacillus subtilis*, *S. cerevisiae*, and other organisms not suspected to contribute to human disease.

BSL-2 is suitable for work involving agents of moderate potential hazard to personnel and the environment. This includes various microbes that cause mild disease to humans or are difficult to contract via aerosol in a laboratory setting. Examples include Hepatitis A, B, and C viruses, human immunodeficiency virus (HIV), pathogenic *E. coli*, *Staphylococcus aureus*, *Salmonella*, *Plasmodium falciparum*, and *Toxoplasma gondii*.

BSL-3 is appropriate for work involving microbes, which can cause serious and potentially lethal disease via the inhalation route. BSL-3 is commonly used for research and diagnostic work involving microbes, which can be transmitted by aerosols and/or cause severe disease. These include *Francisella tularensis*, *Mycobacterium tuberculosis*, *Chlamydia psittaci*, Venezuelan equine encephalitis virus, Eastern equine encephalitis virus, SARS coronavirus, *Coxiella burnetii*, Rift Valley fever virus, *Rickettsia rickettsii*, several species of *Brucella*, chikungunya, yellow fever virus, and West Nile virus.

BSL-4 is the highest level of biosafety precautions, and is appropriate for work with agents that could easily be aerosol-transmitted within the laboratory and cause severe to fatal disease in humans for which there are no available vaccines or treatments. BSL-4 facilities are used for diagnostic work and research on easily transmitted pathogens, which can cause fatal disease. These include a number of viruses known to cause viral hemorrhagic fever such as Marburg virus, Ebola virus, Lassa virus, and Crimean–Congo hemorrhagic fever.

3.3.2.4 Absolute Barrier System

Gnotobiotic and germ-free animals can only survive in an isolated environment, and the equipment for those animals is called an isolator. Commonly used isolators are made of plastic or stainless steel (Figure 3.9). A complete isolator comprises a support, shell, glove, sterilization tank, internal and external blocks, garbage bag, filter, and rubber bands. All items entering the isolator, such as cages, materials, feed, and bedding, must be thoroughly soaked or sterilized by peroxyacetic acid steam.

Isolators are very important for breeding germ-free and gnotobiotic animals, and indispensable to scientific research in clinical and experimental medicine. Correct assembly, reasonable monitoring, and scientific use of isolators are the key to maintaining the sterile properties of the barrier.

Figure 3.9   Isolator for breeding mice or rats.

3.3.3 Facility Management

The core of managing a laboratory animal facility is to minimize contact with the personnel, materials, and equipment, therefore reducing the chance of infection, by specific measures including cleaning, disinfection, and sterilization.

3.3.3.1 Cleaning

Animal room, cages, and water bottles should be cleaned regularly. Open animal rooms should be rinsed with plenty of hot water, supplemented by high-pressure equipment if possible. After removing the dirty bedding, cages are placed in a special washing machine or washed by hands. Hot water and detergents are used to get rid of any remaining dirt from the cages and drinking bottles. Drinking bottles require thorough rinsing, scale deposits inside will be removed by mechanical force.

The most important aim of cleaning is to reduce the number of microorganisms so that disinfection and sterilization become more effective.

3.3.3.2 Disinfection and Sterilization

The removal of potentially pathogenic microorganisms is called disinfection, and the elimination of all living microorganisms is called sterilization. Disinfection and sterilization are equally effective methods. Compared with sterilization, disinfection treatment is less rigorous. If both methods are possible, sterilization is better. Disinfection does not guarantee that all potentially pathogenic microorganisms are killed.

Both physical and chemical methods can be used for disinfection and sterilization. Animal feed can be disinfected with 0.9 Mrad gamma radiation and sterilized with 2.5 Mrad gamma radiation. The possibility to achieve the expected effect depends on the quantity of microorganisms contained in the feed, as a specific processing method can only deal with microorganisms of a certain number. Whether it is disinfection or sterilization is determined by heating, intensity, time, and temperature. Disinfection asks for a short time of heating at 70°C, while sterilization requires 121°C or higher temperatures under positive pressure. The complexity of eliminating microbes varies with different treatment methods.

The effects of chemical disinfection on bacteria are determined by many factors, such as the concentration of detergents, temperature in processing, pH, and the presence or absence of organic matters. The sanitizers are chosen by the properties of the disinfected items and microorganisms, and several sanitizers (e.g., ethanol, halogens, phenols, aldehydes, and biguanides) can be applied for different kinds of disinfection. Halogen activity depends on the concentration of free chlorine, and many microorganisms can be effectively suppressed by it. However, biguanides, such as chlorhexidine, only have effects on vegetative bacteria. Peracetic acid is a highly effective antimicrobial agent and is considered as a sterilizing agent. Peracetic acid steam is widely used in the elimination of known bacteria. The use of chemical

Table 3.6 Sterilizing Methods for Different Items

Sterilizing Methods	Items and the Process of Sterilizing
Pre-vacuum autoclave	General items, 121°C, 20 min Feed 121°C, 15 min Liquids and animal carcasses, 121°C, 30 min
Peracetic acid	Space and materials, 2%–3%, atomization 0.05%–0.2%, scrubbing, pH = 2.5–3.0
Ethylene oxide	General items, 880–1500 mg/L, 26–32°C, 4 h, mainly used on electronics, precision instruments, paper, etc.
Formaldehyde	Space, 10–12 mL/m^3 General items, 15–20 mL/m^3
^{60}Co radiation	General items, 2.5–3.0 Mrad, >1 h Feeds, 3.0–4.0 Mrad, >1 h
Ultraviolet radiation	30 W/10–15 m^2, 40–120 min, absence of humans or animals

disinfectants and sterilizing agents must ensure that all their ingredients and the concentrations are harmless to mammals, including humans.

Table 3.6 is a brief summary of the sterilizing methods for different items.

The results of cleaning and disinfecting can be monitored with agar plates, by counting the bacterial colonies after proper cultivation. High-temperature sterilization can be monitored by measuring the temperature, pressure, and humidity used in the process, or by temperature-dependent color-changing test paper. Processed bacterial spore paper may also be used after incubation.

3.3.4 Control of Animal Facilities

Barrier facilities constitute a series of preventive measures that can be monitored with microbiological and physical methods. The effects of high-temperature disinfection can be verified by testing whether bacteria have been eliminated or not in the process, or by recording the duration, temperature, humidity, and pressure. Meanwhile, the quantity of the microorganisms in the animal feed, air, and surface are measured and evaluated. Pressure, air flow, cleanliness, and illumination in the animal room can be recorded, and deviation from the preset values can set off an alarm or be automatically adjusted.

3.4 NUTRITION AND ANIMAL EXPERIMENTATION

Animals obtain nutrition from feed. Nourishment is a biological process in which animals uptake, digest, absorb, and utilize the feed for nutrients to satisfy their physical needs. The necessary nutrients from feed for sustaining life are called nutrients, including protein, lipids, carbohydrates, minerals, vitamins, and water.

The food given to animals by humans is called feed. All six kinds of nutrients needed by animals come from feed. The feed contains all the necessary nutrients for animals, but they have different values as they are different substances.

3.4.1 Nutrients for Laboratory Animals

3.4.1.1 Protein

Protein is the basic raw material that constitutes animal tissues and cells, essential for maintaining the life, growth, and reproduction. It is acquired from the continuous feed supply.

Protein is composed of approximately 20 kinds of amino acids. Eight amino acids, which cannot be synthesized within animals or whose production is not sufficient are called essential amino acids. Those amino acids that can be synthesized abundantly within the animals are called nonessential amino acids.

Animals can utilize the feed better when the composition and proportion of amino acids contained in feed protein (especially the quantity and ratio of essential amino acids) are closer to those of the animals. Otherwise, the protein in feed will not be fully utilized.

3.4.1.2 Lipids

Lipids are a group of naturally occurring molecules that includes fats, waxes, sterols, fat-soluble vitamins (such as vitamins A, D, E, and K), monoglycerides, diglycerides, triglycerides, phospholipids, and others. Fat refers to triglycerides composed of fatty acids and glycerol. Phospholipids, sterols, and waxes are a combination of fatty acid, glycerin, and other nitrogen-containing substances. Fatty acids in the feed are divided into saturated and unsaturated fatty acids. Unsaturated fatty acids, such as linoleic acid and α-linolenic acid, are indispensable to animals as they cannot be synthesized. Fatty acids that must be supplied through diet are called essential fatty acids. Unsaturated fatty acids are abundant in vegetable fat, compared with in animal fat, as well as the essential fatty acids.

Fat can provide energy needed by animals. Excessive energy will be stored in the form of fat, as fats are the best energy reserve.

3.4.1.3 Carbohydrates

Carbohydrates, also known as sugar, are most abundant nutrient in vegetable feed and the most important part of animal feed.

3.4.1.4 Minerals

Animal tissue contains almost all the natural elements in the world. Other than carbon, hydrogen, oxygen, and nitrogen, which mainly exist in the form of organic compounds, the rest of the elements are collectively referred to as minerals. According to the content within animals, elements can be divided into macroelements and microelements. Although minerals cannot provide energy, they are indispensable for life.

3.4.1.5 Vitamins

Vitamins are divided into lipid-soluble and water-soluble. Vitamins in the animal's body neither provide energy, nor constitute any body parts, but assume the important role of regulating physiological functions. Other than several vitamins synthesized within the body, they are generally supplied in the feed.

3.4.1.6 Water

Water is the most important nutrient for animals. The water within the animals mainly comes from three sources: drinking, water in the feed, and metabolic water.

3.4.2 Energy

Energy is the basis of nutrition as it is indispensable for all the activities of animals. Life ends the moment energy metabolism stops. Carbohydrates, fat, and protein contained in feed are the three major nutrients for body heat, and the main energy sources for sustaining life and physiological activities.

The unit of energy used to be calorie or kilocalorie (kcal). Now joule (J), kilojoule (kJ), and megajoule (MJ) are the most commonly used units. The conversion is as follows: 1 cal = 4.184 J, 1 J = 0.239 cal. Heat generated from the nutrients in the feed can be calculated in the following way: 1 g carbohydrate→16.7 kJ (4.0 kcal), 1 g fat→36.7 kJ (9.0 kcal), 1 g protein→16.7 kJ (4.0 kcal). For example, while calculating the heat of a certain feed, assuming that the feed contains 23% protein, 5% fat, 5% ash, 4% crude fiber, 7% water, 55% soluble nitrogen-free extract (carbohydrate), the heat in 100 g feed is, $100 \times 23\% \times 4 + 100 \times 5\% \times 9 + 100 \times 55\% \times 4 = 357$ kcal = 1.493 MJ. More than 75% of the energy content of conventional diets used in laboratory rodent production comes from carbohydrates.

Animals require energy to maintain normal body processes, grow, and reproduce. Feed ingredients that supply energy are major components of all diets, and the quantity of a diet voluntarily consumed by an animal depends on its energy content.

Information about the energy content of feed ingredients is needed to formulate diets with specified energy concentrations for different animal species. It is used to calculate the amino acid (and other nutrient) levels relative to energy concentration. Being able to assess the energy content of diets for stock feed regulation purposes is also important.

The energy content of a feed may be broken down into the following components, some of which are more easily measured than others: gross energy is energy released on combustion of a feed, it is determined in a bomb calorimeter, and indicates the potential energy in a feed, but not necessarily the amount of useable energy; the energy in feed after subtracting the energy lost in feces is called digestible energy; metabolizable energy is the energy in feed after subtracting the energy lost in feces, urine, and gaseous emissions; net energy is the metabolizable energy value less the heat produced during digestion of the feed. Of all the definitions, net energy is most

Table 3.7   Estimated Amount of Metabolic Energy and Daily Consumption of Rats

Physiological Status	Body Weight (g)	Energy Requirements (MJ/day)	Feed Intake (g/day)
Basal metabolism	100	0.21	15
Growth	200	0.36	25
Life maintenance	400	0.23	16
Gestation	400	0.30	21
Lactation	400	0.65	46

Note: Energy density of the feed is 14.5 kJ/g.

closely related to the energy available for production. It is highly desired by nutritionists, but is the most elusive value to determine because many factors influence it.

In terms of the nutritional value of heat, animals' need for different feeds varies by species. Even animals from the same stock can have different demands for heat because of strain, gender, age, pregnancy, or other physical conditions. Table 3.7 shows the energy requirements of rats in different physiological conditions. They require more energy while growing or lactating. Energy needs are closely related to metabolic weight (weight $kg^{0.75}$), which can be calculated by the following ways:

Basal metabolic need $= 0.45 \times$ weight $kg^{0.75}$
Growth need $= 1.20 \times$ weight $kg^{0.75}$
Gestation need $= 0.60 \times$ weight $kg^{0.75}$
Lactation need $= 1.30 \times$ weight $kg^{0.75}$

All the energy is expressed by daily intake of metabolic energy (MJ), and the unit for weight is kg. The estimated values above include the minimum consumption needed for animal activities, but they are not suitable for any long time, heavy load exercises of animals.

Common natural animal feeds consist of the following energy (by the percentage of weight): 50% carbohydrate, 25% protein, and 5% fat. The energy density of this kind of feed is 14.5 kJ/g. Crude fiber contribution to energy can be neglected. Based on the energy needs of animals and the energy density of the feed, expected free feed intake can be calculated (Table 3.7).

Along with body weight and physiological conditions, the animal's state should be taken into account when considering food intake. For example, a thin animal generally consumes less feed; energy density is an important factor determining the food intake and the taste of the feed may also determine feed intake.

3.4.3 Nutritional Needs of Laboratory Animals

Nutritional needs of laboratory animals are the amount of nutrients needed for the growth and reproduction of animals. Specifically, they refer to every animal's daily needs of energy, protein, minerals, and vitamins.

The nutritional needs of an animal vary because of its species, strain, age, gender, growth, gestation, lactation, or other physiological conditions. Therefore, the purpose of research on laboratory animal's nutritional needs is to explore the animal nutritional requirements in different physiological activities and the changing rules, which can be used as the basis for making husbandry standards and reasonable diets. Namely, it is about a scientific breeding of laboratory animals in order to meet their physiological and reproductive needs. Feed can be utilized most economically and reasonably and animal production can be maximized at the same time.

The nutrients needed by the animals include maintenance requirements, which is for maintaining normal body temperature, respiration, heart rate, basal metabolism, other basic life activities, and free moving; production requirement, which is for animal growth, gestation, lactation, and meat and egg production. The nutritional needs in maintenance and production are closely related, and influence each other. The nutrient requirements for maintenance are lower than that for production.

Determination of animal nutritional needs is quite complicated. At present, there are mainly two ways. One is the general method, in which the nodes are approximately calculated through animal feeding trials, a metabolic rate test, nitrogen and carbon balance test, energy balance test, and slaughtering experiment. The other is the factorial method, which is to dissect the functional components of animal's nutritional needs. It provides an accurate understanding for animal nutritional needs of several activities and the sum is the total nutritional requirement.

The daily energy and the amount of all kinds of nutrients given to every animal are scientifically stipulated, based on species, gender, age, weight and physiological stages of the animal, combined with the metabolism experimentations of energy and other nutrients, and the result of feeding experiments. This rule is usually called the feeding standard, whose value is the supplied amount of nutrients. It is determined by the minimum requirement and the safe coefficient based on that requirement. There has been much systematic and in-depth research done on the nutritional needs of laboratory animals and breeding standards in countries like the United States, which is advanced in the area of laboratory animal science. They have respective standards to guide the production practice of laboratory animals. They have also standardized animal nutrition, improved animal quality, and promoted the development of related areas.

The requirement for nutrients varies among different kinds of animals. Every country's standard is not the same. The Chinese government stipulated in 1994, revised and enacted the national standards for breeding laboratory animals in 2010 (GB14924-2010). Table 3.8 shows the nutrient requirements of the common laboratory animals, summarized by the Nutrient Requirements of Laboratory Animals (1978, 1995) and Nutrient Requirements of Laboratory Rabbits (1977), which were drawn by the National Research Council (NRC), United States.

3.4.4 Animal Feeds

Feeds are the material basis for animal husbandry, and the cost accounts for over 60% of the total production cost. Therefore, understanding the type of feed and nutritional characteristics, and utilizing feed rationally is of great significance to animal breeding.

Table 3.8   Estimated Nutrient Requirements of Common Laboratory Animals

	Mouse	Rat	Hamster	Guinea Pig	Rabbit
Energy					
Digestible energy (kJ/g)	16.8	16.0	17.6	12.6	10.5
Fat (g/kg)	50	50	50	su	20
Fiber (g/kg)	ru	ru	su	150	110
Protein (g/kg)	180	150	150	180	160
Amino Acids					
Arginine (g/kg)	3	4.3	7.6	12	6
Asparagine (g/kg)	su	4	su	su	su
Glutamic acid (g/kg)	su	40	su	su	su
Histidine (g/kg)	2	2.8	4	3.6	3
Isoleucine (g/kg)	4	6.2	8.9	6	6
Leucine (g/kg)	7	10.7	13.9	10.8	11
Lysine (g/kg)	4	9.2	12	8.4	6.5
Methionine+cysteine (g/kg)	5	9.8	3.2	6	6
Phenylalanine+tyrosine (g/kg)	7.6	10.2	14	10.8	11
Proline (g/kg)	su	4	su	su	su
Threonine (g/kg)	4	6.2	7	6	6
Tryptophan (g/kg)	1	2.0	3.4	1.8	2
Valine (g/kg)	5	7.4	9.1	8.4	7
Glycine (g/kg)	su	su	su	su	Su
Minerals and Microelements					
Calcium (g/kg)	5	5	5.9	8	4
Chlorine (g/kg)	0.5	0.5	su	0.5	3
Magnesium (g/kg)	0.5	0.5	0.6	1	0.35
Phosphorus (g/kg)	3	3	3	4	2.2
Potassium (g/kg)	2	3.6	6.1	5	6
Sodium (g/kg)	0.5	0.5	1.5	0.5	2
Sulfur (g/kg)	su	ru	su	su	su
Chromium (mg/kg)	2	ru	su	0.6	su
Copper (mg/kg)	6	5	1.6	6	3
Fluorine (mg/kg)	su	ru	0.024	su	su
Iodine (mg/kg)	0.15	0.15	1.6	0.15	0.2
Iron (mg/kg)	35	35	140	50	ru.
Manganese (mg/kg)	10	10	3.65	40	8.5
Selenium (mg/kg)	0.15	0.15	0.1	0.15	su
Zinc (mg/kg)	10	12	9.2	20	ru
Vitamins					
Vitamin A (mg/kg)	0.15	1.2	1.1	7.0	0.17
Vitamin D (mg/kg)	4	25	62	25	ru

(Continued)

Table 3.8 (*Continued*) Estimated Nutrient Requirements of Common Laboratory Animals

	Mouse	Rat	Hamster	Guinea Pig	Rabbit
Tocopherol (mg/kg)	32	27	3	40	40
Vitamin K (mg/kg)	1	1	4	5	ru
Vitamin B1 (mg/kg)	5	4	20	2	ru.
Vitamin B2 (mg/kg)	7	3	15	3	su
Vitamin B6 (mg/kg)	8	6	6	3	39
Vitamin B12 (ug/kg)	10	50	10	10	nr
Nicotinicum acidum (mg/kg)	15	15	90	10	180
Folic acid (mg/kg)	0.5	1	2	4	su
Biotin (mg/kg)	0.2	0.2	0.6	0.2	ru
Pantothenic acid (mg/kg)	16	10	40	20	su
Choline (mg/kg)	2000	750	2000	1800	1200
Inositol (mg/kg)	ru	nr	100	nr	su
Vitamin C (mg/kg)	nr	nr	nr	200	nr

Note: Mouse nutrient requirements are expressed in an as-fed basis for diets containing 10% moisture and 16–17 kJ of metabolizable energy/g and should be adjusted for diets of differing moisture and energy concentrations. Unless otherwise specified, the listed nutrient concentrations represent minimal requirements and do not include a margin of safety. Higher concentrations for many nutrients may be warranted in natural-ingredient diets. ru—required but requirement unknown; su—status unknown; nr—not required.

3.4.4.1 Compound Feeds

Feed classifications have not been unified around the world. Based on nutritional characteristics of the feeds, the American scholar Harris (1956) put forward the classification of feeds, which was accepted by most countries. He divided the feeds into eight categories: forage roughage, pasture plants and greens feeds, silage, energy feeds, protein supplements, minerals, vitamins, and feed additives.

A single ingredient of feed often cannot meet the nutritional needs of animals. Therefore, a feed formula with full nutrients is developed to meet animal requirements, which is a collection of several kinds of ingredients in certain proportions. The mixed feeds are called compound feeds. The full nutrition compound feeds are usually used in animal production and experimentation. This balanced type of feed can provide all the necessary nutrients. A satisfactory result can be acquired using this kind of feed without any additional nutrients in breeding.

The main ingredient of the animal feeds should be natural animal products and plants and raw materials such as grain, forage grass, fruit, vegetables, fish meal, bone meal, etc. Feeds are formulated and manufactured according to the feeding standard. Under normal circumstances, this kind of feed can be used in animal reproduction and production. One of the diets is of open formula and clear ingredients (Table 3.9 gives a sample of natural-ingredient diets used for rats and mice in NIH), another kind is of secret formula and public ingredients. Most of the commercial feeds belong to the latter. While choosing a feed, one should note whether the ingredients

Table 3.9   Examples of Natural-Ingredient Diets Used for Rat and Mouse Breeding
            Colonies at the National Institutes of Health

Ingredient	Conventional (NIH-07)	Autoclavable (NIH-31)
Basic Diet, g/kg diet		
Dried skim milk	50.0	
Fish meal (60% protein)	100.0	90.0
Soybean meal (48% protein)	120.0	50.0
Alfalfa meal, dehydrated (17% protein)	40.0	20.0
Corn gluten meal (60% protein)	30.0	20.0
Ground #2 yellow shelled corn	245.0	210.0
Ground hard winter wheat	230.0	355.0
Ground whole oats		100.0
Wheat middlings	100.0	100.0
Brewer's dried yeast	20.0	10.0
Dry molasses	15.0	
Soybean oil	25.0	15.0
Salt	5.0	5.0
Dicalcium phosphate	12.5	15.0
Ground limestone	5.0	5.0
Mineral premix	1.2	2.5
Vitamin premix	1.3	2.5
Mineral Premix, mg/kg diet		
Cobalt (as cobalt carbonate)	0.44	0.44
Copper (as copper sulfate)	4.40	4.40
Iron (as iron sulfate)	132.30	66.20
Manganese (as manganous oxide)	66.20	110.00
Zinc (as zinc oxide)	17.60	11.00
Iodine (as calcium iodate)	1.54	1.65
Vitamin Premix, per kg diet		
Stabilized vitamin A palmitate or stearate	6,060.00 IU	24,300.00 IU
Vitamin D3 (D-activated animal sterol)	5,070.00 IU	4,190.00 IU
Vitamin K (menadione activity)	3.09 mg	22.10 mg
All-rac-α-tocopheryl acetate	22.10 mg	16.50 mg
Choline chloride	617.00 mg	772.00 mg
Folic acid	2.43 mg	1.10 mg
Niacin	33.10 mg	22.10 mg
Ca-d-pantothenate	19.80 mg	27.60 mg
Pyridoxine HCl	1.87 mg	2.21 mg
Riboflavin supplement	3.75 mg	5.51 mg
Thiamin mononitrate	11.0 mg	71.7 mg
d-Biotin	0.15 mg	0.13 mg
Vitamin B12 supplement	0.004 mg	0.015 mg

Note: Amounts listed for mineral and vitamin premixes represent the mass or IU of the specific
       mineral element or vitamin rather than the added compounds.

are consistent with the real nutrients. Minerals in the feed can be quite different because of the raw material being made in different places.

In addition to the natural ingredients, extracted and chemically synthesized ingredients are often added into the animal feed. For example, purified casein can be used as a source of protein, vegetable oils, or animal fat as a source of fat, chemically pure inorganic salt and vitamins, as well as chemically synthesized amino acids, sugar, fatty acids or glycerin, minerals, and vitamins. These components are reliable with a minimum possibility to cause mutation and pollution; however, they are very expensive. Synthesized materials may also contain other chemical substances, which are harmful to the animals. Special attention should be paid to the manufacturers, quality, and expiration date of the ingredients other than the natural ingredients.

At present, the domestic and foreign animal breeding standards are not completely the same. The NRC standard and Chinese Standard (GB14924-2010) have certain differences. Nutritional requirements are not the same among different stocks or strains within the same species. Therefore, choosing suitable animal feeding standards should be determined by a comprehensive analysis of the actual situation of the facility and region.

In developing countries, the trend of socialization, commercialization, and specialization in laboratory animal production has been preliminarily formed. Researchers working in general animal production and experimentation have been suggested to use qualified commercialized laboratory animal feeds in order to ensure the standardization of animals and experimentation. The study results will therefore not be interfered with by the nutritional factors and the reliability will be improved.

3.4.4.2 Feed Types

The shape of pellet feeds is generally the most suitable for rodents. Pellet feeds are easier to manage, store and feed, and are rarely wasted by the animals. No additional ingredients or drugs can be added into the processed pellet feeds unless those feeds are reground and reprocessed.

The results of feeding animal powder feed meal are not very satisfactory because this form of feed is easy to be wasted. Feeds stored in the form of powders are more likely to agglomerate and need special equipment for feeding. However, they are suitable for feeding while additional ingredients or drugs are necessary.

Feed type is decided by the animal experimentation. For example, when highly toxic ingredients are added into the feed, the feed form that produces the minimum dust should be chosen; when powder or toxic chemicals are added into the feed, half wet or stick-shaped feed is more appropriate because this type of feed is usually more delicious than dry feed. However, it facilitates bacteria proliferation and requires frequent feeding, asks for high working load, and is very difficult to manage.

3.4.4.3 Feed Disinfection

The feed sources of raw materials are very complicated, and they can be contaminated by the pathogenic microorganisms in any step such as harvest, storage, transportation, or feed processing. Therefore, feeds go through disinfection to meet the hygienic standard. It is very necessary to provide the animals with disinfected feeds of full nutrition. Most animal experimentation requires that the feeds used be disinfected to eliminate pathogens or be sterilized completely.

There are many ways to disinfect feeds. The specific method should be determined by the different requirements of animals, feed type, and actual conditions. However, some nutrients are lost in disinfected feeds, which must be accounted for while feeding.

- Dry-heat sterilization is to roast the feeds under temperatures of 80–100°C. Equipment used in this manner is rather simple, but the temperature is not very easy to handle. Feeds may be wasted because of the long duration, incomplete sterilization, nutritional loss, and even carbonization.
- Sterilization under high temperature and high pressure is to process the feeds under temperatures of 121°C and $1.0 kg/cm^2$ for more than 15 min. The usually maintained temperatures are 115°C, 30 min; 121°C, 20 min; and 125°C, 15 min. This sterilization method is cost effective, takes a shorter time, loses fewer nutrients than dry heat, but easily damages vitamin C, vitamin B1, vitamin B6, and vitamin A.
- Radiation sterilization uses ^{60}Co radioactive rays to disinfect grain feeds. This method destroys fewer nutrients and has the best effects, but the cost is very high.

3.4.5 Feeds and Animal Experimentation

3.4.5.1 Animal Feeding Regimes

The regime of animal feeding is decided by the goal of the study and the operability. Commonly used regimes are as follows:

- **Ad libitum feeding:** With this regime the animals have free access to food any time of the day or night. Rodents and domestic rabbits consume most of the feed at night in this system. For example, rats can eat 12 times in day, 8 of which is at night.
- **Meal feeding:** During fixed time periods, one or more periods per day, the animals are allowed to consume as much as they like. This regime is usually adopted by studies that ask for a strict control of the nutritional status. For example, animals are fed fixed hours after a meal.
- **Restricted feeding:** This regime involves limiting the food intake or underfeeding, but is not equivalent to malnutrition or the induction of nutrient deficiency. It involves the restriction of nutrients and energy at the same time. This method is often used to balance the feed intake of different animals such as the control group and the experimental group.
- **Pair feeding:** Pair feeding is a special form of restricted feeding. It forces the control and experimental groups to have the same feed intake through providing the control animals the same amount of food as laboratory animals whose intake

is calculated based on their feed consumption. When this regime is adopted, each animal in the experimental group should have a counterpart in the control group. During actual operation, the amount of feed consumed by the experimental group can be provided to the control group on the next day.

Under the conditions of free access to feed, animals are provided with feed every 2–4 days. Animals must be separated into different cages and fed by hand or automatic devices when they go through restricted feeding. If the animals are kept in groups, the superior individual may undergo *ad libitum* feeding, resulting in restricted feeding reducing their breeding.

When pair feeding or restricted feeding are used in study, there should be a group of animals with *ad libitum feeding* as an additional control group. Researchers should consider the fact that the control animals in *pair feeding* often suffer from insufficient food, which makes them consume faster than animals with unrestricted feeding. This can be partly compensated for by adopting different experimental technologies. For example, the limited amount of feed for control animals every day is provided at several times. All in all, every animal experimental research project requires a detailed record of the feed consumed by the animals, which will facilitate analyzing the results.

3.4.5.2 *Animal Experimentation and Energy Density of the Feeds*

Animals from the same strain, gender, or having close weight, age, and health status, consume almost equal energy when they are under *ad libitum* feeding. As the animals usually adjust their intake to meet their energy needs, they will stop eating when the energy is sufficient. Therefore, if the energy density in feeds is increased, the amount of feed consumed by animals will decrease and vice versa. That means when the feed used in a certain animal experimentation contains different ingredients, the proportion of the energy related parts must be the same in each group. Otherwise, under the condition of *ad libitum* feeding, animals of the control and experimental groups will absorb differently from the same diet. Thus, the energy density of the diet must be taken into account while making feeds containing different ingredients.

Sometimes researchers hope that when the amount of a certain ingredient in the feed changes, the consumption of other ingredients will also change. While making compound feeds, adding a huge amount of an ingredient like sugar or fat to satisfy the need of constant energy density is obviously wrong. Due to the fact that full nutrition compound feeds are used by the control group and the feed used by experimental groups containing ingredients waiting to be studied, nutrient concentration in the feeds of each group is not equal. It will affect the energy density of the feed, causing different consumption of feeds between the experimental group and control group. Observed different reactions or results may depend on the difference of a certain ingredient in the feeds of the experimental and control groups. Thus, data collected in this kind of study are not reliable. Adding only a small amount of an extra substance into the full nutrition compound feeds ensures that the impact of this on the experiment is limited to the minimum, which is exerted by the change of energy density caused by the new ingredient.

Feeds of the control and experimental groups can be made by adding certain ingredients into the commercial full nutrition compound feeds. However, one must bear the risk of reduction in the essential nutrient intake. Commercial full nutrition compound feeds usually include plenty of essential nutrients, allowing for appropriate dilution, which will not cause serious imbalances in animal nutrition. In general, 10%–20% additional matter can be added in the full nutrition compound feeds, but the researchers must evaluate the possible results brought by this kind of "dilution" seriously.

Table 3.10 describes the influence of low- and high-fat feeds on experimentation by using rats. The low-fat feed 1 contains 10% fat, 20% protein, 60% carbohydrates, vitamins and minerals mixture, and experimental components. In high-fat feed 2, the 20% carbohydrates are replaced by the same amount of fat and the rest is the same. The expected result of nutrition intake is listed beneath the feed composition.

Table 3.10 Expected Results of Feeding Rats with Low- and High-Fat Feeds

	Feed 1	Feed 2	Feed 3	Feed 4
	Low Fat	High Fat	Modified High Fat	Modified High Fat
Feed Components				
Protein(g)	20	20	20	20
Carbohydrates (g)	60	40	15	15
Fat (g)	10	30	30	30
Fiber (g)	4	4	4	4
Premixed minerals (g)	4	4	4	4
Premixed vitamins (g)	1	1	1	1
Experimental compound (g)	1	1	1	1
Inactive compound(g)	–	–	–	25
Total (g)	100	100	75	100
Energy (kcal/g)	4.10	5.10	5.47	4.10
Expected Feed Intake				
Energy (kcal/g)	82	82	82	82
Feed (g/day)	20	16	15	20
Protein (g/day)	4	3.2	4	4
Carbohydrates (g/day)	12	6.4	3	3
Fat (g/day)	2	4.8	6	6
Fiber (g)	0.8	0.64	0.8	0.8
Premixed minerals (g/day)	0.8	0.64	0.8	0.8
Premixed vitamins (g/day)	0.2	0.16	0.2	0.2
Experimental compound (g/day)	0.2	0.16	0.2	0.2
Inactive compound (g/day)	–	–	–	5

Note: "–" means no added ingredient.

The energy density of high-fat feed is 24% higher than that of low-fat feed. Rats have the same calorie intake with free feeding (assuming they consume constant energy). Therefore, we can deduce that rats fed with high-fat feed consume less than the animals fed by low-fat feed. In addition to the changes in the intake of fats and carbohydrates, the intake of protein, mineral premix, vitamin premix, fiber, and experimental compounds of rats fed with high-fat diet reduces as well. Therefore, when the experimental compound is added into the feed, the change in feed intake is also very important. In the given case, the amount of experimental component intake will be 20% different.

One way to rule out problems is explained by feed 3. It is an equal exchange of fats and carbohydrates: reducing some carbohydrates to compensate for the extra calories caused by the extra fat. This modified high-fat feed ensures that the intake of protein, vitamins, minerals, and fiber is equal to that of feed 1. Although ingredients of feed 3 are not increased to 100 g and the expected feed intake is less than that of feed 1, the intake of experimental compounds and all the nutrients other than fat and carbohydrate nutrients will be the same as that of feed 1.

In addition to fats and carbohydrates, rats fed with feed 1 and 3 will show different actual intakes such as the amount of intake. This will result in different statistics among the groups. In feed 4, all the ingredients in the feed reached 100 g by adding relatively inactive ingredients such as fiber. Feed 4 and 1 have equal calories and nutritional intake. This approach seems to be satisfactory, but there is no such thing as inactive ingredients. Therefore, an ingredient must be chosen to form a high-fat feed described in feed 3.

A similar case is the high fiber feed, which is low in carbohydrates. The energetic value of the feed of high fiber and low carbohydrates is expected to be lower than that of the feed of low fiber and high carbohydrates. As a result, the consumption of high-fiber feed is expected to increase. Combined with the changes in other nutrients intake, the comparison between animals fed with high- and low-fiber feeds will be very difficult.

In the above discussion of the effects of variable energy density of the feeds, we can assume that animals can maintain a constant energy intake. However, energy intake may change when animals have feeds of variable energy densities.

3.4.5.3 Changes of Feed Intake in Experimentation

In an animal experimentation of free feeding, the intake by experimental animals may be lower than that of the control group. Their reduced food intake may be attributed to the toxic effects caused by the questionable compound and has nothing to do with whether being given feed and lacking a certain nutrient. Food intake may decline if the ingredient in feed can reduce the appetite. The difference in food intake implies that the comparison between the monitored parameters is not direct. If both the treatment and feed intake have influence on the parameters, respectively, the effect of the treatment will not have a definite explanation. The observed results are partly caused by the different treatments between the control and experimental groups, partly caused by the feed intake difference. The difference in feed intake

Table 3.11   Influence of Feeding Regimes on Tumorigenesis in Rats (Percentage of Tumors)

Treatment	Feeding Regimes	Feed Intake (%)	A	B	C	D
Control	Free	100	30	30	30	30
Experimentation	Free	80	40	30	20	10
Control	Restricted	80	10	10	10	10

needs to be dealt with if one wants to study the special effects of the treatment. This can be solved by restricting feeding or ingredients.

The method of using restricted feeding to suppress tumor development and prolong the life of rats and mice has been verified. The explanation of the influence of the animal's restricted intake on experimentation has a special role in rat breeding and tumor biology identification. The effects of carcinogens will be underestimated if the rats are treated with carcinogens and their food intake is reduced at the same time. This often occurs in tumor biology identification when rats are under free feeding. Table 3.11 describes the hypothesized carcinogenesis of the control group under free feeding and the experimental group. The carcinogen causes a 20% reduction in feed intake, which often occurs in the actual situation. In the control group, 30% of rats will grow tumors, but a 20% reduction in feed intake will reduce the incidence of tumor to 10%. In experiment A, if the carcinogen increases the tumor incidence by 40%, when compared with the free feeding group, we can deduce that this carcinogen induces the occurrence of tumors. However, its effects are underestimated and can be clarified by the comparison between the experimental group and the control group of restricted feeding animals. In experiments B, C, and D with free feeding, the false conclusion that the carcinogen is not carcinogenetic or even anticarcinogenic will be reached. The latter conclusion often occurs in practice. An explanation from testing a carcinogen with free-feeding rats will be inconclusive. In theory, both the control and experimental rats should be restrictedly fed, which leads to the same amount of feed intake, or the pair feeding regime can be used.

BIBLIOGRAPHY

Haemisch A, Gärtner K. Effects of cage enrichment on territorial aggression and stress physiology in male laboratory mice. *Acta Physiologica Scandinavica, Supplementum* 1997;640:73–6.

Hedrich HJ. *The Laboratory Mouse.* Elsevier Academic Press, San Diego, 2012.

Liu E. *Laboratory Animal Genetics and Breeding.* Gansu Nationalities Publishing House, Gansu, China, 2002.

Liu E, Yin H, Gu W. *Medical Laboratory Animals.* Science Press, Beijing, China, 2008.

National Research Council. *Guide for the Care and Use of Laboratory Animal,* Eighth Edition. National Academy Press, USA, Washington, 2011.

National Research Council Laboratory Animal Management. *Rodents.* National Academy Press, Washington, 1996.

National Research Council (US) Subcommittee on Laboratory Animal Nutrition. *Nutrient Requirements of Laboratory Animals*, Fourth Edition. National Academies Press, Washington, DC, 1995.

Qin C, Wei H. *Laboratory Animal Science*, Second Edition. People's Medical Publishing House, Beijing, China, 2015.

Reardon S. A mouse's house may ruin experiments. *Nature* 2016;530:264.

Threadgill DW, Miller DR, Churchill GA et al. The collaborative cross: A recombinant inbred mouse population for the systems genetic era. *Institute for Laboratory Animal Research Journal* 2011;52:24–31.

Ullman-Culleré MH, Foltz CJ. Body condition scoring: A rapid and accurate method for assessing health status in mice. *Laboratory Animal Science* 1999;49:319–23.

CHAPTER 4

Anatomy, Physiology, and Husbandry of Laboratory Animals

Yi Tan and Dongmei Tan

CONTENTS

4.1 MICE

The laboratory mouse (Figure 4.1) is assigned to the genus *Mus*, subfamily Murinae, family Muridae, and order Rodentia. Laboratory strains are usually derived from mice bred by mouse fanciers and their genomes are a mixture of *Mus musculus musculus* (from Eastern Europe) and *Mus musculus domesticus* (from Western Europe). There are nearly 474 inbred strains, over 200 outbred groups, and more than 2000 mutant strains.

C57BL/6

B6C3F1 (C57BL/6 × C3H/He)

CD1-nude

BALB/c

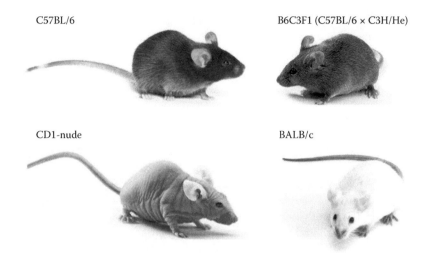

Figure 4.1   The laboratory mice.

Currently, the use of transgenic technology to produce genetically engineered (GM) mice is very important in biomedical research. For example, GM mice can be obtained by inserting foreign genes into the genome of mice; gene knockout mice can be obtained by specific site-directed mutagenesis of the mouse genome; gene knock-in mice can be obtained by the recombination of endogenous genes in the mouse genome. In the field of biomedical research, mice are the most studied and the most widely used of all laboratory mammals, and also have the largest number of different strains.

4.1.1 General Biological Characteristics

The mouse body is covered with hair and the face is cuspate with whiskers on both sides. Their ears are semicircular and they have big, bright red eyes. The general body length of the adult mouse is 10–15 cm; almost the same length as its tail. The tail is covered with short hairs and horny scales. There is a variety of hair colors such as white, gray, black, brown, yellow, chocolate, and cinnamon.

Mice are small and delicate with no sweat glands in the skin and have poor ability to adapt to the environment. Mice are docile and easy to handle. Mice like to inhabit a dimly lit environment and are nocturnal. Feeding, mating, and delivery of pups often occur at night. Mice are most active at 1–2 h after evening and during the predawn hours.

Mice are social animals, and grow faster when fed in groups than when single. However, overcrowding will inhibit their reproductive capacity. Sexual maturity is attained at an early age. Male mice will often bite other mice that they have not fathered. Whiskers are retained in dominant mice while the subordinate mice lack hair and whiskers. Mice are extremely sensitive to stimuli, such as light, noise, and odors, and overexposure can lead to neurological disorders, often resulting in cannibalism.

The genetic background of the mouse is one of the most studied of any animal used in research. The histocompatibility complex and coat color genes have been intensely studied in mice.

4.1.2 Anatomy

4.1.2.1 Skeletal System

The skeleton is composed of two parts: the axial skeleton, which consists of the skull, vertebrae, ribs, and sternum, and the appendicular skeleton, which consists of the pectoral and pelvic girdles, and the paired limbs. Normal mouse dentition consists of an incisor and three molars in each quadrant, whose dental formula is 2 (incisors 1/1, canines 0/0, premolars 0/0, molars 3/3) = 16. The third molar is the smallest tooth in both jaws; the upper and lower third molar may be missing in wild mice and in some inbred strains. The incisors grow continuously and are worn down during mastication.

4.1.2.2 Digestive System

The esophagus is slender and approximately 2 cm long, located on the back of the trachea. The esophagus of the mouse is lined by a thick, cornified squamous epithelium, making gavage a relatively simple procedure. The proximal portion of the stomach is also keratinized, whereas the distal part of the stomach is glandular. The capacity of the stomach is small (1.0–1.5 mL). Therefore, the dosage should not exceed 1.0 mL in gavage studies on mice. The intestine of mice is relatively shorter than herbivores, such as rabbits and guinea pigs, and their cecum is underdeveloped.

4.1.2.3 Respiratory System

The right lung is divided into four lobes: superior, middle, inferior, and postcaval. The left lung is a single lobe. However, there is a shallow ditch present on the left lung indicating that it is not a complete lobe. The trachea consists of 15 light, white rings of cartilage. The trachea and bronchi of mice are not robust, and they are not suitable models for experimental chronic bronchitis or efficacy experiments of asthma drugs.

4.1.2.4 Cardiovascular System

The heart consists of four chambers: left atrium, right atrium, left ventricle, and right ventricle. The apex cordis is located near the end of the sternum and the fourth intercostal rib. This is also the puncture site for exsanguination.

4.1.2.5 *Lymphoreticular System*

The lymphatic system consists of lymph vessels, thymus, lymph nodes, spleen, solitary peripheral nodes, and intestinal Peyer's patches. The lymphatic system in mice is highly developed, but there is no palate or throat tonsils. Hyperplasia of the lymphatic system may occur when stimulated by some factors and this may cause lymphatic system disorders. The spleen contains hematopoietic cells including megakaryocytes and primary hematopoietic cells.

4.1.2.6 *Reproductive System*

The female reproductive system consists of the paired ovaries and oviducts, uterus, cervix, vagina, clitoris, and paired clitoral glands (Figure 4.2). The ovary, surrounded by a membrane, is not connected to the abdominal cavity, thus preventing ectopic pregnancy. The female mouse normally has five pairs of mammary glands, three in the cervicothoracic region and two in the inguino abdominal region. The male reproductive organs consist of paired testes, urethra, penis, and associated ducts and glands. Testes are stored in the abdominal cavity before maturity and drop into the scrotum after sexual maturity. The prostate consists of dorsal and abdominal lobes.

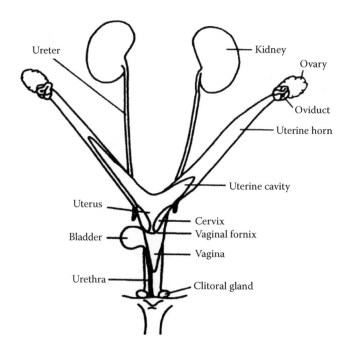

Figure 4.2 The mouse genital organs.

4.1.2.7 Bone Marrow

Bone marrow and splenic red pulp produce erythrocytic, granulocytic, and megakaryocytic precursors during the life of the mouse. Bone marrow is located in the protected matrix of cancellous bone and is sustained by reticular tissue rich in blood vessels and adipose cells.

4.1.3 Physiology

4.1.3.1 Growth

Mice are small animals that only weigh 1.5 g at birth. The weight increases to 12–19 g a month later, and up to 20–40 g when reaching 1.5–2 months of age. The ovum is fertilized by sperm in the ampulla of the fallopian tube. The morula that develops approximately 3 days following fertilization enters the uterus where the blastocyst is formed (about 5 days), initiating implantation. Gestation lasts 19–21 days. Newborn mice are hairless with red skin, closed eyes, and ears adhering to the skin. They are vocal and have tactile, olfactory, and taste sensations. When the mouse reaches 3 days of age, the skin turns white and the umbilical cord is shed. Acoustic perception is developed when aged 4–6 days and they begin to develop a furry coat. With open eyes, grown incisors, mice begin to eat and drink when they reach 12–14 days of age. They can live independently when 3 weeks old. Females are fertile at 4 weeks of age and the testicles of males descend to the scrotum to start generating sperm when 5 weeks old. The growth rate of mice is closely related to strain, nutritional status, health status, environmental conditions, maternal lactation capacity, and reproductive parity.

4.1.3.2 Reproduction

Mice have rapid rate of development. Sexual maturity is early and occurs at 6–7 weeks of age. Mature sperm can be found in the epididymis of males at 36 days of age, while ovulation occurs at 37 days of age in females. Males mature at 70–80 days of age, and females mature at 65–75 days of age. In general, mice can begin breeding between 65 and 90 days. In the female, the estrous cycle is 4–5 days, pregnancy is 19–21 days and lactation is 20–22 days. Female mice can produce 6–9 litters annually, with 6–15 pups each. Sexual activity can be maintained for approximately 1 year and the life span is approximately 2–3 years. Males begin to produce sperm and secrete androgens after sexual maturity. Accessory sexual glands (seminal vesicles, coagulation glands) secrete seminal fluid for movement of sperm. Semen will solidify after mating to form a vaginal plug for 10–12 h in the vagina and cervix of females to prevent leakage of semen, improving fertility. The vaginal plug is an important feature of mice and more obviously detected than in other rodents.

4.1.3.3 Temperature and Water Regulation

Mice have a relatively large surface per gram of body weight. This results in marked physiological changes in response to fluctuations in the ambient temperature. The mouse responds to cold exposure, for example, by nonshivering thermogenesis. A resting mouse acclimated to cold can generate heat equivalent to approximately triple the basal metabolic rate, a change that is greater than in any other animals. The mouse has no sweat glands and its ability to salivate is severely limited. It adapts to moderate but persistent increases in environmental temperature by a persistent increase in body temperature, a persistent decrease in metabolic rate, and increased blood flow to the ears to increase heat loss. This indicates that the mouse is not a true warm-blooded animal. Indeed, the neonatal mouse is ectothermic and does not have well-developed temperature control before 20 days of age. Thus, the environmental temperature has a great impact on mice. Fertility and resistance to disease may be decreased by hypothermia. Continuous high temperatures (above 32°C) often cause death or pathological effects in mice such as irreversible damage of some functions. There are repeated studies demonstrating that mice in a temperature range of 21–25°C grow faster, have larger litters, and have more viable pups than those maintained in the thermoneutral zone.

Due to the high ratio of evaporative surface to body mass, the mouse has a greater sensitivity than most mammals to water loss. Water conservation is enhanced by cooling of expired air in the nasal passages and by highly efficient concentration of urine. Its biological half-time for turnover of water (1.1 days) is more rapid than for larger mammals.

Major physiological parameters of mice are given in Table 4.1.

4.1.4 Laboratory Management and Husbandry

4.1.4.1 Sexual Differentiation

Sex is very easy to distinguish in adult mice because scrota of males are obvious, while vaginal opening and five pairs of nipples are easily visible in females. The sex of young mice can be determined by measuring the distance between external genitalia and anus; the smallest distance is a female while the larger is a male (see Chapter 6).

4.1.4.2 Health Features

Mice appearance is used to judge their health status, including appetite, activity and responsiveness, smooth hair, no scarring, straight tail, no external discharge, no deformity, and black and granular stool.

4.1.4.3 Cages and Housing

The popular shoebox cage used for housing and breeding mice is usually made of nontoxic plastic. Cage lids are stainless steel to facilitate cleaning and prevent rust.

Table 4.1   Physiological Parameters of Mice, Rats, Syrian Hamsters

	Mice	Rats	Syrian Hamsters
Adult Weight (g)			
Male	20–40	300–500	120–140
Female	25–40	250–300	140–160
Life span (year)	1–2	2–3	2–3
Heart rate (time/min)	300–800	300–500	250–500
Respiratory rate (time/min)	100–200	70–110	40–120
Body temperature(°C)	36.5–38.0	37.5–38.5	37–38
Chromosome number (2n)	40	42	44
Body surface area (cm^2)	20 g:36	50 g:130 130 g:250 200 g:325	125 g:260
Water intake (mL/100 g/day)	15	10–12	8–10
Adolescence (week)			
Female	5	6–8	4–6
Male	–	–	7–9
Breeding Season (week)			
Female	8–10	12–16	6–8
Male	8–10	12–16	10–12
Estrous cycle (day)	4 (2–9)	4–5	4
Estrus duration (h)	14	14	2–24
Gestation (day)	19 (18–21)	21–23	15–17
Litter size	6–12	6–12	6–8
Newborn weight (g)	0.5–1.5	5	2–3
Weight at weaning (g)	10	40–50	30–40
Weaning age (day)	21–28	21	20–22
Blood Parameters			
Blood volume (mL/kg)	76–80	60	80
Hemoglobin (g/100 mL)	10–17	14–20	10–18
Hematocrit (vol%)	39–49	36–48	36–60
White blood cells ($\times$1000/mm^3)	5–12	6–17	3–11
Blood glucose (mg/100 mL)	124–262	134–219	60–150

Metal cages are also used. Glass or plastic bottles can be used to make a watering trough with a cork equipped with metal or glass water pipes. Plastic cages are non-absorbent, corrosion-resistant, and easy to wash and dry. Its performance at high temperature is related to the plastic used to make the feeding boxes and can withstand more than 120°C temperatures. Solid-bottom cages should contain sanitary bedding, such as wood chips or ground corncob, for absorbing urine, warmth, and nesting. Bedding must be nontoxic, odorless, with no dust, inedible, and comfortable for the mice. Litter or waste must be sterilized to remove any potential pathogens or harmful substances.

4.1.4.4 Husbandry

The stomach capacity of mice is small and they forage continuously. Mice consume 4–7 g of feed per day after weaning and maintain this intake throughout life. Thus, food should be added in a small amount every time, 3–4 times a week. Limited access to food will reduce the waste when mice eat pellets for teeth grinding. Food particles and saliva may flow back into the bottle when mice drink water. To avoid microbial contamination of water bottles, water bottles and suction pipes should be carefully washed after being replaced. Drinking water should comply with the general level of health standards of urban drinking water, and sterilized water should be provided for specific pathogen-free (SPF) mice.

4.1.4.5 Recording

Scientific management must have many complete records. Work records should include genetic aspects, population, pedigree and strain, as well as individual records, such as breeding card, working diary, and environmental records, such as temperature and humidity, sterilization records, and experimental treatment and observation records.

Environmental requirements and recommended space for mice are provided in Table 4.2.

Table 4.2 Feeding Environmental Parameters of Mice, Rats, and Syrian Hamsters

	Mice	Rats	Syrian Hamsters
Environmental Requirements			
Temperature (°C)	18–26	18–26	18–26
Relative humidity (%)	30–70	30–70	30–70
Ventilation rate (time/h)	10–15	10–15	10–15
Light/dark cycle (h)	10/14 or 12/12	10/14 or 12/12	10/14 or 12/12
Minimal Feeding Space			
Floor area for single (cm²)	180	350	180
Floor area for breeding (cm²)	200	800	650
Floor area for group (cm²)	<10 g: 38.71	<100 g: 109.67	n/a
	10–15 g: 51.61	100–200 g: 148.37	
	15–25 g: 77.41	200–300 g: 187.08	
	>25 g: ≥96.77	300–400 g: 258.04	
		400–500 g: 387.06	
		>500 g: ≥451.57	
Height (cm)	12.70	17.78	12

4.2 RATS

The laboratory rat (Figure 4.3), *Rattus norvegicus,* is within the order Rodentia and family Muridae. The genus *Rattus* contains more than 130 species; however, the Norway rat, *R. norvegicus*, and the black rat, *Rattus rattus,* are the two species most commonly associated with the genus. Wild rats and wild albino rats have been used experimentally in Europe since the mid-eighteenth century.

The rat is second only to the mouse as the most frequently used mammal in biomedical and behavioral research.

4.2.1 General Biological Characteristics

The rat is larger in body size and weight than the mouse, but with a similar appearance. The body length of an adult rat is 18–20 cm. The tail is covered with short hair and circular horny scales. The skin of the rat lacks sweat glands, which are found only on the claws. The main mode of cooling is through the tail. Rats adapt well to new environments, and it is easy for them to accept the training of a variety of sensory instructions carried out by positive and negative reinforcement.

Foraging and mating often take place in the early morning and at night because rats are active during these periods. Rats are docile and easy to capture. However, they will become nervous and difficult to handle, and even become aggressive if treated roughly, are hungry, or hear the screams of other rats. Pregnant and lactating rats are more likely to exhibit aggressive behavior toward humans.

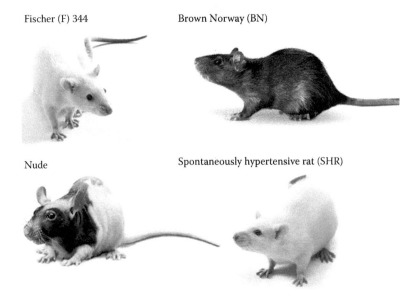

Fischer (F) 344   Brown Norway (BN)

Nude   Spontaneously hypertensive rat (SHR)

Figure 4.3   The laboratory rats.

4.2.2 Anatomy

4.2.2.1 Skeletal System

The skeleton is composed of the following bones: skull, vertebrae, sternum, ribs, forelimbs, and hind limbs. Rats have two incisors and six molars in the maxillae and mandible, whose dental formula is 2 (incisors 1/1, canines 0/0, premolars 0/0, molars 3/3) = 16. The incisors grow continuously during their lifetime, which often need to be abraded to keep their appropriate length. Their molar anatomy is similar to humans. Cariogenic bacteria and cariogenic foods can result in caries, making rats a model for studying human dental caries.

4.2.2.2 Digestive System

The stomach is divided into two parts: the forestomach or cardiac portion, which is nonglandular, and the corpus or pyloric portion, which is glandular. A ridge separates the two portions with the esophagus entering at the lesser curvature of the stomach through a fold of the ridge. This fold is responsible for the inability of the rat to vomit. The liver has four lobes: the median, which has a deep fissure for the hepatic ligament; the right lateral, which is partially divided; the left, which is large; and the caudate, which is small and surrounds the esophagus. The rat has high regeneration ability and is able to renew up to 60%–70% of its liver if surgically removed. Furthermore, 95% of liver Kupffer's cells have phagocytic ability that make the rat a suitable model for hepatic surgical experimental studies. The rat does not have a gallbladder, and bile ducts from each lobe form the common bile duct, which enters the duodenum approximately 25 mm from the pyloric sphincter.

4.2.2.3 Respiratory System

The lungs consist of the left lung, which is single lobed, and the right lung, which is divided into the cranial, middle, accessory, and caudal lobes. The rat does not have an adrenergic nerve supply to the bronchial musculature, and bronchoconstriction is controlled by vagal tone. The trachea is located in the ventral aspect of the esophagus, generally consisting of 24 "U"-shaped rings of cartilage. As the trachea and bronchus are both underdeveloped, the rat is not a suitable model for studying chronic bronchitis or asthma.

4.2.2.4 Cardiovascular System

The blood supply to the atria of the rat, unlike that of higher mammals, is largely extracoronary from branches of the internal mammary and subclavian arteries. The heart is located on a midline in the thorax, with its apex near the diaphragm and its lateral aspects bounded mainly by the lungs.

4.2.2.5 Urinary System

The right kidney is more craniad than the left, with its cranial and caudal edge at the level of L1 and L3 vertebra, respectively. As the kidney has only one calyx, the rat can be used to study cannulation of the kidney effectively.

4.2.2.6 Reproductive System

The male rat reproductive system has many highly developed accessory sex glands including large seminal vesicles, bulbourethral glands, coagulation glands, and prostate gland. The inguinal canal remains open throughout the life of the rat. The female rat has a bicornuate uterus depicted as a "Y" figure. The chest and abdomen have two rows of three pairs of nipples.

4.2.2.7 Endocrinology System

The rat pituitary loosely adheres to the lower part of the funnel. It can therefore be siphoned off with a pipette to provide a pituitary extraction model.

4.2.3 Physiology

4.2.3.1 Growth and Development

Newborn rats weight approximately 5.5–10 g, are hairless, with closed ears and short limbs. The ears open at 3–4 days and the incisors appear at 8–10 days. The eyes open at 14–17 days and the coat is fully developed at 16 days. Rat pups can be weaned at 20–21 days. The rate of growth and development in the rat is related to strain, nutritional status, health status, environmental conditions, lactation capacity, and reproductive parity of the female. Adult males weigh generally 300–600 g and females 250–500 g. The life span of rats is 2.5–3 years.

4.2.3.2 Reproduction

The testes descend into the scrotum 30–35 days after birth and produce sperm between 45 and 60 days. Male rats can mate 60 days after birth, but the optimal breeding time is 90 days after birth. The female vagina opens at 70–75 days, usually coinciding with the first cycle of ovulation. Optimum breeding occurs at 80 days. Ovulation is spontaneous, but in the nonestrus state, can be induced by forcing mating. The sexual cycle of the female is 4–5 days, which can be divided into early estrus, estrus, late estrus, and middle estrus. The stage of the estrous cycle can be determined by analysis of a vaginal smear. When females are kept in groups, estrus can be inhibited. Estrus will be induced by a male or its excreta. Postpartum estrus also exists in rats. Rat gestation is 19–23 days (an average of 21 days) with a litter size of 6–12 pups.

4.2.3.3 Nutrition

The rat is a good animal model for studying nutrition because it is sensitive to a variety of nutrient deficiencies. Vitamin A, vitamin E, vitamin K, riboflavin, and thiamine deficiency can cause infertility, skin disease, or bleeding. Rats can store fat-soluble vitamin B12 effectively, produce vitamin C, and supply most of its needs for vitamin B by eating fecal matter. In addition, the rat is commonly used for studying calcium and phosphorus metabolism.

Major physiological parameters of rats are given in Table 4.1.

4.2.4 Laboratory Management and Husbandry

Fluctuations or sudden changes in humidity in the environment may stress rats, which could promote outbreaks of infectious diseases caused by opportunistic pathogens. This affects the health of the rat and interferes with correct experimental results. Ring necrosis, described as a narrowing of skin of the rat tail (or toes) is caused by dry air (relative humidity less than 40%) and high temperature. This appears at the distal portion of the tail ring, which becomes narrow and appears gangrenous, and is likely to occur in rats that are preweaning or kept at the bottom of wire cages. Rat pathogens are carried and spread by aerosols. Dirty litter, overcrowding, or poor ventilation will lead to excessive ammonia produced in the breeding rooms, which will cause rat respiratory infections, especially mycoplasma disease. Rats are sensitive to noise. Loud noises may cause cannibalism or convulsion, and have effects on a variety of experimental parameters. Reproductive physiology and behavior are greatly influenced by light in rats. Light is a powerful stimulant and synchronizer of rhythms related to reproductive biology. Therefore, light timing should be used to provide adequate and appropriate diurnal variations in the light cycle in closed breeding cages.

Rats are omnivorous and feed at any time, but they are more active and eat more at night. Rats are fed with a complete feed in order to ensure nutrients supply for their growth and development. Rats are generally well adapted to all types of water supply equipment. The water bottle and suction tube should be cleaned and disinfected regularly.

Environmental requirements and recommended space for rats are provided in Table 4.2.

4.3 HAMSTERS

4.3.1 General Biological Characteristics

The hamster is a small rodent that is widely distributed in many places of the Eurasian continent. The laboratory hamster is domesticated from wild animals. Two types of hamsters serve as laboratory animals: the Syrian or golden hamster (*Mesocricetus auratus*) (Figure 4.4a) and the Chinese hamster or black lines

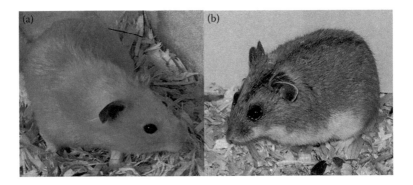

Figure 4.4   The Syrian hamster (a) and Chinese hamster (b).

hamsters (*Cricetulus griseus*) (Figure 4.4b). More than 80% of biomedical research uses the golden hamster.

The adult Syrian hamster usually grows to a length of 16–19 cm and weighs between 110 and 140 g, with light brown fur on the back, and white on the sides and belly. The hamster has a small blunt tail and smooth, short fur. The ears are pointed, with dark pigmentation, and the eyes are small, dark, and bright. The Chinese hamster is small, approximately 9.5 cm long, and its weight is approximately 40 g. The fur is taupe, and the eyes are big and dark. The hamster looks fat and has a short tail. There is a dark stripe from the back of the head to the base of the tail. Hamsters are nocturnal animals and are very active at night. Hamsters are sensitive to variations in temperature. They can hibernate when the temperature is 8–9°C, and the pups can freeze to death at below 13°C. Thus, the appropriate room temperature should be 22–25°C and the humidity between 40% and 60%. Hamsters are adapted for eating a plant-based diet and have an extensive diet. On both sides of the mouth there are well-developed cheek pouches, in which food and water can be stored and used when hibernating. Hamsters are ferocious, aggressive, and often attack each other. They have a high reproductive capacity, and the breeding season is in late spring and early autumn.

4.3.2 Anatomy

The dental formula is 2 (incisors 1/1, canines 0/0, premolars 0/0, molars 3/3) = 16. Incisors of golden hamsters grow throughout their life, and on each side of the mouth are cheek pouches with a depth of 3.5–4.5 cm and diameter of 2–3 cm. The main function of the pouch is to store food and transport nesting material (Figure 4.5). Moreover, the pouch has been termed immunologically privileged due to the lack of an intact lymphatic drainage system that contributes to extensive use for microvascular studies of tumor growth.

The spine of hamsters is composed of 43 or 44 vertebrae, including 7 cervical vertebrae, 13 thoracic vertebrae, 6 lumbar vertebrae, 4 sacral vertebrae, and 13–14 caudal vertebrae. The lung has five lobes: one left and four right lobes. The stomach consists of the forestomach and glandular stomach. The liver consists of six lobes:

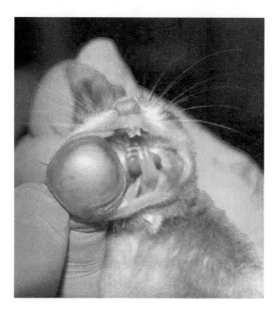

Figure 4.5   Eversion of the hamster cheek pouch.

two left lobes, three right lobes, and a small middle lobe. The length of the small intestine is 3–4 times as long as the body, 0.6 times the cecum, and 2.5 times the large intestine. The renal papillary is very long and extends into the ureter. Hamsters have 15 lymph organs and 35–44 lymph nodes. The testes of the golden hamster are large and mulberry-like in shape, and weigh 1.6–2 g. The testes lie in the abdominal cavity on the left side of the navel and under the stomach. The uterus is "Y" shaped with a pair of round ovaries on each side. Approximately 20 ova are produced from one ovulation. The female has 6–7 pairs of nipples. Male hamsters have a hip gland that produces secretions when in a state of sexual arousal.

4.3.3 Physiology

Sexual maturity of golden hamsters occurs between 30 and 32 days of age, and the sexual cycle lasts for 4–5 days. The sexual cycle can be divided into proestrus, estrus, post-estrus, and resting stage. Ovulation may happen on the evening of estrus day 1 and can last into the night. The hamster is a perennial estrus animal with characteristics of estrus after partus. Hamsters deliver 5–7 litters annually with 4–12 pups each. New born hamsters are hairless with closed eyes and ears. The ears open after 5 days, while the eyes open after 15 days, and the weaning time is 21 days after birth. The reproductive cycle of the golden hamster is approximately 15 (14–17) days, which is the shortest in rodents. Growth and development of hamsters is 1–5 years and the average life span is 2.5–3 years. Females are stronger than males and sexual maturation is early. Except in estrus, male and female hamsters must be housed separately, otherwise, the female will bite the males. Skin grafting is very special in golden hamsters. Skin can survive for a long time by transplanting between individuals in a closed population. Overall, 100% exclusion will occur, however, after transplanting skin between different populations.

Sexual maturity of the Chinese hamster occurs at 8 weeks of age, and the sexual cycle is 4.5 (3–7) days. Gestation lasts 20.5 (19–21) days, and lactation for approximately 20–25 days. Females have four pairs of nipples and live for 2–2.5 years. Chinese hamsters have 11 pairs of chromosomes, which are large and easy to be identified. The X chromosome is similar to that in humans, but the morphology of the Y chromosome is special. The islets of Langerhan of the Chinese hamster are easily degraded with cell atrophy and degeneration, which will produce true diabetes with glycemia that is 2–8 times higher than normal.

Major physiological parameters of Syrian hamsters are given in Table 4.1.

4.3.4 Laboratory Management and Husbandry

Hamsters are similar to mice in feeding and management, but golden hamsters have a much shorter pregnancy cycle and grow faster than mice. The body weight of mice at 21 days of age is 7.7 times that at birth, while for the golden hamster it is 18 times. The golden hamster is highly fertile, therefore special attention should be paid to content and quality of protein in the complete feed. The protein content should be 20%–25%, and an amount of animal protein (animal protein:vegetable protein as 1:2 or 2:3 ratio) must also be present in the food. Otherwise, they will deliver an increasing number of weak hamster pups. The daily diet for adult hamsters is 10–15 g of food, and 15–20 mL of water forage, such as carrots and cabbage, should be provided appropriately during the winter. Cucumber, cabbage, and rape should be given appropriately during summer. Chinese hamsters are similar to golden hamsters in their nutrition, but the daily diet for adult Chinese hamsters is 3–4 g of food and 3.5–5.5 mL of water. Hamsters have the habit of storing their food, therefore food should not be provided in excess during the rainy season and food residues in the cage should be cleaned up every day to prevent disease caused by consumption of moldy feed. Do not handle young hamsters directly during the management process otherwise the young hamsters may be killed and devoured by the adult female. Cages and animal room should be washed and disinfected regularly. Litter must be autoclaved as well.

Female Chinese hamsters are aggressive, and they should be separated after sexual maturity to prevent biting each other.

Environmental requirements and recommended space for Syrian hamsters are provided in Table 4.2.

4.4 GERBILS

4.4.1 Anatomy

The dental formula of the gerbil (Figure 4.6) is 2 (incisors 1/1, canines 0/0, premolars 0/0, molars 3/3) = 16. The spine consists of 7 cervical vertebrae, 12–14

Figure 4.6 Laboratory gerbils.

thoracic vertebrae, 5–6 lumbar vertebrae, 4 sacral vertebrae, and 27–30 caudal vertebrae. Gerbils have a large, ventral, sebaceous gland that is androgen-dependent, which secretes a special smell. The male gerbil abdominal marking gland is larger and appears earlier than the female, forming a hairless zone around the gland in adults (Figure 4.7). In groups, the dominator is the one who secretes most often. The female abdominal marking gland is small (difficult to find if not sheared), and its activity is enhanced during pregnancy and early lactation (Figure 4.6). In adult males, scrotal protrusions are obvious, and have dark pigmentation around the anus and scrotum. Comparing weight, the adrenal is 3 times larger than that in the rat. Steroids produced by the adrenal gland in the gerbil that metabolize sodium cannot be maintained if the adrenal gland is removed.

Gerbils have a very important anatomic character in that the posterior communicating artery of Willis' circle is deficient. There is no posterior communicating artery to connect the carotid artery system and vertebral artery system, which cannot

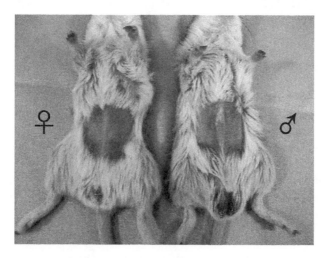

Figure 4.7 Ventral, sebaceous gland in the male gerbil is larger than that in the female.

constitute a complete Willis' circle. Gerbils are highly susceptible to cerebral infract following unilateral ligation of one common carotid artery and can be an ideal model to study the pathophysiology and treatment of cerebrovascular disease such as ischemic stroke in humans.

4.4.2 Physiology

Gerbils are small herbivores, whose average body weight is 77.9 (30–113) g in adults. Males are bigger than females with body length of 11.3 (9.7–13.2) cm and ear length of 1.5 (1.2–1.7) mm. The front fur of the ear is gray, and the top inside fur is short and small. The rest of the ear is hairless, the hair is gray and brown on the back, lighter on the sides and cheeks, and gray on the abdomen. The tail is soft, long, and thick, and the length is almost equal to the trunk, approximately 10.2 (9.7–10.5) cm. There are tufts of fur at the tip of the tail. Periods of activity alternate with resting periods both during the day as well as the night, but the highest activity takes place in the afternoon and night. Gerbils are very agile, with the ability to climb to an extent. Gerbils use jumping or hopping as a form of ambulation with the aid of muscular hind limbs. Their temperament is docile and fighting does not usually occur, but mixed groups of adult gerbils often lead to fierce fighting accompanied by injury and death.

In the wild, gerbils live as monogamous pairs in a self-built system of tunnels. If the lighting regime is kept constant with 12 h light, they can breed all year round. Reproduction rates fall slightly in winter. The female sexual maturity period lasts 9–12 weeks, while for males it last 10–12 weeks. Mean litter size ranges from 3 to 7 pups. Young gerbils suckle for approximately 21 days and begin to eat solid food at 16 days. The lifetime of adult gerbils is 2–3 years. Under laboratory conditions, females produce, at maximum, up to 14 births in a lifetime. Newborn pups have no fur and weigh between 1.5 and 2.0 g with closed ears and eyes. Ears prick up at day 3–4 and fur begins to grow at day 6. At day 8–9, the incisors appear, and eyes open at day 16–18. The suitable mating age is 3–6 months, with 15 months of breeding time for the female. The life expectancy is 2–4 years.

The sexual cycle is 4–6 days, with the vagina opening at 40–60 days. Sexual maturity occurs before 30 days. The gestation period for nonlactating females is 24–26 days, but lactating females always have a prolonged gestation of 27 days. Each litter size is 4–8 pups, while some litters can be up to 11. Lactation lasts 14–29 days and pups are weaned at approximately 21 days when the weight is 12 g. Vaginal swabs are usually not reliable for estrus detection. Behavior patterns of the two sexes are a better indicator and a copulation plug is generally not noticeable as it is small and retained deep inside the vagina.

Major physiological parameters of gerbils are provided in Table 4.3.

4.4.3 Management and Husbandry

Gerbils are generally housed in cages with a layer of sawdust. The cage should be shut tightly to prevent escape. Regular cleaning of the cage needs to be carried out.

Table 4.3   Physiological Parameters of Gerbils, Guinea Pigs, Rabbits

	Gerbils	Guinea Pigs	Rabbits
Adult Weight (g)			
Male	80–110	900–1000	2–5 kg
Female	70–100	700–900	2–6 kg
Life span (year)	3–4	5–6	5–6
Heart rate (time/min)	360	230–380	130–325
Respiratory rate (time/min)	90	42–104	30–60
Body temperature (°C)	38.1–38.4	38–40	38.5–39.5
Chromosome number (2n)	44	64	44
Body surface area (cm^2)	190 g:205	400 g:565	2–5 kg:1270
		800 g:720	4–8 kg:3040
Water intake (mL/100 g/day)	4–7	10	6
Adolescence (week)			
Female	9–12	4–5	16
Male	9–12	8–10	20
Breeding Season (week)			
Female	9–12	9–10	20–36
Male	9–12	9–10	24–40
Estrous cycle (day)	4–6	14–18	–
Estrus duration (h)	–	1–18	–
Gestation (day)	25–26	68(59–72)	30(28–35)
Litter size	4–6	1–6	4–10
Newborn weight (g)	2.5–3.0	70–100	30–100
Weight at weaning (g)	–	180–240	–
Weaning age (day)	20–30	15–28	35–56
Blood Parameters			
Blood volume (mL/kg)	66–78	69–75	60
Hemoglobin (g/100 mL)	13–16	12–15	10–16
Hematocrit (vol%)	44–47	38–48	36–48
White blood cells ($\times$1000/mm^3)	7–12	7–13	5–11
Blood glucose (mg/100 mL)	50–135	60–125	78–155

Separate feeding, replacing of bedding, hybridization, and handling of gerbils should be gentle. The use of solid-bottomed cages is preferable for the housing of gerbils. Gerbils often stand erect on their hind limbs, thus the height of the cage must be at least 15 cm. It is important that cages have a solid bottom and the floor-to-lid height be tall enough to allow for this behavior. At the same time, their general condition improves when they are housed on solid floors with a thick layer of bedding rather than when they are housed on wire mesh floors. Gerbils require a high protein diet under conditions of artificial feeding and are usually fed with complete pellet feed.

Table 4.4 Feeding Environmental Parameters of Gerbils, Guinea Pigs, Rabbits

	Gerbils	Guinea Pigs	Rabbits
Environmental Requirements			
Temperature (°C)	18–26	18–26	16–22
Relative humidity (%)	30–70	30–70	30–70
Ventilation rate (time/h)	10–15	10–15	10–15
Light/dark cycle (h)	10/14 or 12/12	10/14 or 12/12	10/14 or 12/12
Minimal Feeding Space			
Floor area for single (cm^2)	230	600	<2 kg:1390 2–4 kg:2790 4–5.4 kg:3720 >5.4 kg:≥4650
Floor area for breeding (cm^2)	1300 (a pair)	1200	1 kg:3000 3 kg:4000 5 kg:5000
Floor area for group (cm^2)		≤350 g:387.06 >350 g:≥651.55	
Height (cm)	15	17.78	35.56

Feed protein content should be adequate, usually at least 16%, and can be supplemented with fresh vegetables. The breeding environment should be quiet and ventilated because sudden noises may elicit epileptic seizures. Long-time cohabitation of males and females is preferred for breeding. Gerbils adapt to a wide range of ambient temperatures, but are generally maintained at 22–24°C. Due to their propensity for nasal dermatitis, the relative humidity should be above 50%, but low humidity is advisable.

Environmental requirements and recommended space for gerbils are provided in Table 4.4.

4.5 GUINEA PIG

4.5.1 Anatomy

4.5.1.1 Skeletal System

The skeletal system of guinea pigs (Figure 4.8) contains the skull, torso, limb, and bones. They are strictly herbivores, have transversely inclined molars, and the incisors are arcuated and extend to the jaw. The dental formula is 2 (incisors 1/1, canines 0/0, premolars 1/1, molars 3/3) = 20, with a diastema or gap between the incisors and premolars. All teeth are open rooted and grow continuously.

Figure 4.8 Laboratory guinea pig.

4.5.1.2 Gastrointestinal System

The oral cavity is small and narrow, making endotracheal intubation difficult and the stomach wall is very thin, with a capacity of 20–30 mL. Guinea pigs are monogastric, unlike other rodents. The stomach is undivided and is lined entirely with glandular endothelium. The intestine is longer, approximately 10 times the body length, which has cecal fermenters. The largest cecum can hold up to 65% of the total gastrointestinal contents.

4.5.1.3 Respiratory System

The trachea and bronchi are underdeveloped, and tracheal glands are only located in the throat. The guinea pig can be used as a model for lung-function impairment and bronchial reactions including airway hyper-responsiveness and reactions that resemble asthma in humans. Compared to other mammals, the guinea pig has the most prominent smooth muscle in the distal bronchi and muscle is spirally arranged. The lung is divided into seven lobes. The larger right lung has four lobes (upper, middle, lower, and side lobe), while the left lung has three lobes (upper, middle, and lower lobe).

4.5.1.4 Lymphoreticular System

The lymphatic system is developed and extremely sensitive to pathogenic microorganisms. There are many mononuclear leucocytes carrying a granular structure (Kurloff body) in the blood of guinea pigs. The Foa–Kurloff or Kurloff cell is an estradiol-dependent mononuclear leukocyte with natural killer activity that is found unique in the guinea pigs. Lymphoid tissue is particularly rich in the lungs. Lung lymph nodes are highly reactive and lymphadenitis occurs quickly with a small amount of mechanical or bacterial stimulation.

4.5.1.5 Nervous System

The brain matures in embryos between 42 and 45 days. The cerebral hemispheres have no obvious curve, with only a deep groove belonging to smooth brain tissue, and is more developed compared with other similar animals.

4.5.1.6 Reproductive System

Males have one pair of nipples in the inguinal region, same as females, but the nipples of the female are slender and located above the glands. Accessory sex glands in the male guinea pig include large transparent, smooth seminal vesicles, prostate, coagulating, bulbourethral, and rudimentary preputial glands. Testes remain in inguinal pouches with inguinal canals open for life. Adult males have large testes in obvious scrotal pouches.

Females have a U-shaped depression in the perineal tissue. The anus is located at the base of the U, and the branches are covered by a membrane. The uterus is bicornuate in females, which terminates into a single cervix. The vagina is closed during estrus.

4.5.2 Physiology

4.5.2.1 Growth

Newborn pups are born with hair, teeth, and open eyes and ears, and they can walk almost immediately within a few hours after birth. Average birth weight ranges from 45 to 115 g. The young guinea pig can eat solid food, grow rapidly, and can gain 2.5–3.5 g per day in 2 months after birth. Adult guinea pigs are generally 350–600 g. Lifespan is generally 4–5 years.

4.5.2.2 Reproduction

Sexual maturity is at an early age, and follicle development usually starts at 14 days and ovulation starts around 60 days in females. Guinea pigs are spontaneous ovulators, and under laboratory conditions, polyestrous breeders. The estrous cycle of the guinea pig lasts approximately 16 days (13–21 days). The estrous period is recognizable because at this time the vaginal membrane is open. Estrus lasts 8–11 h and is indicated by vaginal swelling and congestion. The average length of pregnancy is 68 days with 1–8 pups per litter, but usually 3 or 4. Young pups are usually weaned at 21 days of ages and the sow has a fertile estrus shortly after pups are weaned. Fertile postpartum estrus occurs from 2 to 10 h after parturition.

Male semen containing sperm and gonadal secretions, and vaginal secretions in the female form a coagulated vaginal plug, thus pregnancy can be recognizable after mating by the presence of a vaginal plug. On the other hand, pregnancy also can be detected by gentle palpation of the uterus. On the day of gestation, firm, oval swellings of approximately 5 mm in diameter can be felt in the uterine horns. During late pregnancy, abdominal distension becomes evident, and the pubic symphysis separates during the last week.

4.5.2.3 Blood Cells

Red blood cells, hemoglobin, and hematocrit of guinea pigs are lower than other rodents. Kurloff cells have large mucopolysaccharide, cytoplasmic bodies, which are metachromatic, and periodic acid-Schiff containing proteoglycans and hydrolytic enzymes. These cells are found primarily in the thymus and in the sinusoids of the spleen, liver, and lungs, with increased numbers in the peripheral circulation during pregnancy as well as in lung and spleen red pulp transferred to the thymus and placenta.

Major physiological parameters of guinea pigs are given in Table 4.3.

4.5.3 Management and Husbandry

Guinea pigs have sensitive vision. They are nervous and easily liable to panic, therefore, the housing should keep quite. The guinea pig is also very sensitive to temperature fluctuations and draughts. High environmental temperatures (above 28°C) in combination with high relative humidity (above 70%) are not well tolerated by guinea pigs. Sudden changes can lead to dramatic weight loss. Generally recommended environmental parameters for housing guinea pigs include an ambient temperature of 20–26°C, relative humidity of 30%–70%, and a 12 h light/12 h dark light cycle.

Commonly used caging systems for guinea pigs housed in research facilities include ground fence, microisolator cages, wire sides, and solid-bottom plastic caging in a ventilated rack. A solid bottomed cage with bedding material is better than a cage with a wire-mesh bottom as wire-mesh flooring may result in bone fracture and footpad inflammation in younger animals, as well as reduced production in breeding animals. Guinea pigs usually do not climb and the traditional feeding method is pool feeding. They can be housed without a lid if the cage sides are 40 cm high. Bedding materials need to be disinfected when in ponds or solid bottom cages. Bedding materials traditionally consist of corncob, hardwood chips, shavings, and paper products. Some bedding materials may interfere with animal test systems involving ascorbic acid depletion because of the presence of low levels of vitamin C, as well as cause an infection and obstruct mating, and in females it can cause vaginitis.

Guinea pig feed is usually crude fiber, with pelleted show containing approximately 18%–20% crude protein and 9%–18% fiber.

Guinea pigs do not have the ability to synthesize vitamin C and the feed must contain vitamin C. Any swelling of the joints, salivation, or lethargy could be indicative of a vitamin C deficiency. In addition, vitamin C can be derived from fresh, juicy fruits and vegetables as well as in the particle feed or drinking water containing vitamin C. The guinea pig's daily vitamin C requirement is 4–5 mg/100 g of body weight. During growth, pregnancy, lactation, or under stress, the actual daily needs are 30–40 mg/100 g of body weight.

Environmental requirements and recommended space for guinea pigs are provided in Table 4.4.

4.6 RABBITS

4.6.1 Anatomy

4.6.1.1 Skeletal System

Rabbits (Figure 4.9) have 275 bones, which contain the skull, vertebrae, ribs, sternum, and limb bones. A small pair of incisors are present directly caudal to the primary maxillary incisors and are referred to as the "peg" teeth. The dental formula is 2 (incisors 2/1, canines 0/0, premolars 3/2, molars 2–3/3*2) = 26 or 28. They have well-developed incisors, large molars; the molars do not have roots and are characterized by deep enamel folds. The teeth of rabbits erupt continuously throughout life, and will therefore continue to grow and lengthen unless normal occlusion and use are sufficient to wear teeth to a normal length.

4.6.1.2 Digestive System

The esophagus of the rabbit has three layers of striated muscle that extend the length of the esophagus down to, and including, the cardia of the stomach, as opposed to in humans. Rabbits are monogastric. The small intestine is shorter in comparison to other species. The large intestine includes the cecum, ascending colon, transverse colon, and descending colon. The cecum is coiled up like a spiral, and is very large with a capacity of approximately 10 times that of the stomach. The cecum ends in a blind sac, and the appendix contains a large number of bacteria and lymphoid tissue. At the joint of the ileum and cecum, there is a dilation known as the sacculus

Figure 4.9 Laboratory rabbits.

rotundus, which contains a large bulb of lymphatic tissue. The mucosa secretes an alkaline liquid to neutralize cecal microbial cellulose produced by the decomposition of organic acids, which is beneficial for digestion and absorption.

Rabbits have four pairs of salivary glands: the parotid, submaxillary, sublingual, and orbital glands. The parotid is the largest and lies laterally just below the base of the ear. The rabbit has a typical feature, the orbital glands, whereas other mammals do not. The pancreas is diffuse within the surrounding mesentery of the small intestine and enters the duodenum 30–40 cm distal to the common bile duct and appears to be the only laboratory animal with a single, separate pancreatic duct. The pancreas is pale pink with a texture like fat, and it is an irregular fat gland. Pancreatic ducts open away from the bile duct openings, which is another major feature of rabbits.

4.6.1.3 Respiratory System

The mouth of the rabbit is relatively small, and the oral cavity and pharynx are long and narrow, but the tongue is relatively large. These features make visual endotracheal intubation difficult to perform on the rabbit without specialized equipment. Nostrils of rabbits are well equipped with touch cells, and they have a well-developed sense of smell. Nasal breathing in rabbits is characterized by twitching of the nostrils at rates varying from 20 to 120 times per minute.

The lungs are spongy. In general, the right lung is larger than the left due to the presence of the heart in the left pleural cavity. Rabbit lungs consist of six lobes; both right and left sides have cranial, middle, and caudal lobes, with the right caudal being further subdivided into lateral and medial portions, while the left lung has two lobes (sharp lobes, heart septal leaflet), like a four-leaf right lung (pointed lobes, central leaf, the septal leaflet, middle leaf). There is a mediastinum in the middle of the left and right lung.

4.6.1.4 Circulatory System

The rabbit heart is divided into four chambers: the left atrium, right atrium, left ventricle, and right ventricle. The ability to transport oxygen is very strong making the body produce a lot of heat to maintain a constant body temperature. The chest structure is different from other animals with a mediastinal pleural cavity, which can be divided centrally without mutually unintelligible feeling about two and a half. The heart is separated by pericardial pleura. When the pericardium is opened for experimental operations after thoracotomy to expose the heart, the animals do not need artificial respiration as long as the mediastinum is not broken.

A unique feature of the cardiovascular system of the rabbit is that the tricuspid valve of the heart has only two cusps, rather than three as in many other mammals. A group of pacemaker cells can generate the impulse of the sinoatrial (SA) node in the rabbit, a feature that facilitates precise determination of the location of the pacemaker. The SA and atrioventricular (AV) nodes are slender and elongated, and the AV node is separated from the annulus fibrosus by a layer of fat.

4.6.1.5 Lymphatic System

The popliteal fossa of the hind legs has large oval popliteal lymph nodes, which are approximately 5 mm long. Chinchilla rabbits have larger lymph nodes. *In vitro*, it is easy to identify, and injection of drugs into the lymph nodes can be performed.

4.6.1.6 Nervous System

Rabbits have a depressor nerve as an independent branch in the neck. In humans, dogs, and cats, this nerve is not separate, but is in the sympathetic trunk or present in the vagus nerve. The rabbit neck neurovascular bundle has three different thicknesses. The largest, white one is the vagus nerve; the thinner, gray one is the sympathetic nerve; the thinnest one is the finest decompressed nerve, located between the vagus nerve and sympathetic nerve. They belong to the afferent nerve and their endings are distributed in the vessel wall of the aortic arch.

Rabbits have big ears making injection of blood vessels and blood collection easy. The iris of white rabbits completely lacks pigment. Due to the color of blood vessels within the eye, the eye appears to be red.

4.6.1.7 Reproductive System

The inguinal canal of male rabbits is wide and short, and is not closed during their life span. The testes can freely drop into the scrotum or retract into the abdominal cavity. The testes of the adult male usually lie within the scrotum. However, the inguinal canals that connect the abdominal cavity to the inguinal pouches do not close in the rabbit. Inguinal pouches are located lateral to the genitalia in both sexes. The reproductive tract of the female is characterized by a bicornuate uterus with two uterine horns and cervices that open independently into the vagina.

4.6.2 Physiology

Rabbits grow rapidly. The newborns are naked and remain in the nest for approximately 3 weeks. The eyes are closed, ears are occluded, and the toes are linked together at birth. After birth at 3–4 days, they start to grow fur. At 4–8 days, the toes are separated, the ear hole connects with the outside world at 6–8 days and the eyes open between 10 and 12 days. At 21 days, newborn pups begin normal eating. Around 30 days, fur is formed. Pups may begin consuming solid food at 3 weeks of age with weaning generally occurring at 5–8 weeks of age. When the pups are born they weigh approximately 50 g. After 1 month, the weight is equivalent to 10 times weight at birth. After 3 months, the speed of gaining of weight is relatively slow. Most species of male rabbits grow faster than females.

Sexual maturity of rabbits is early and occurs at around 3–4 months of age in small varieties, 4–5 months in medium-sized varieties, and 5–6 months in large varieties. Ovulation in rabbits is induced and occurs approximately 10–12 h after copulation. Generally, the sexual cycle is 8–15 days. Females do not have a distinct estrus cycle, but rather demonstrate a rhythm with an active phase with swelling, moistness, and reddening of the vulva that lasts 3–4 days. If mating fails, pseudo-pregnancy can occur because the formation of the corpus luteum after ovulation results in the formation of breasts and hysterauxesis. In such circumstances, ovulation is followed by a persistent corpus luteum that lasts 15–17 days. Gestation in rabbits usually lasts for 29–36 days, with an average of 32 days. Pregnancy can often be confirmed as early as day 14 of gestation by palpation of the fetuses within the uterus. Weaning is at 40–50 days with pups weighing 500–600 g in small rabbits and 1000–1200 g in larger ones.

Rabbits are strictly herbivorous with a preferred diet of herbage that is high in fiber, low in protein, and soluble carbohydrate. They should be fed pellets of complete feed twice a day. Rabbits have developed a cecum for the strong digestion of crude fiber. Reduction of crude fiber in feed can cause digestive diarrhea. Thus, crude fiber should be controlled to 10%–15%. The rabbit produces two types of feces: soft feces and the hard dry feces. Rabbits exhibit coprophagy, but they only eat soft feces directly from the anus. During the second half of the night and early morning, the contents of the cecum is transported to the colon and rectum virtually unaltered in small spherical particles surrounded by a mucous layer, then by coprophagy, protein, vitamin K, and other nutrients are reabsorbed.

For major physiological parameters of rabbits see Table 4.3.

4.6.3 Management and Husbandry

Stainless steel cages with a wire floor are commonly used for housing laboratory rabbits. Usually, experimental cage specifications are 50 cm × 40 cm × 30 cm and 80 cm × 50 cm × 38 cm for breeding cages with automatic flushing devices. Cages should be constructed of durable materials that resist corrosion, and the harsh detergents and disinfectants used in cleaning. They should be made in such a way that young animals do not get their legs trapped in the wire floor and the adult animals do not acquire damage to their foot pads. The toe nails must be clipped at regular intervals. The cage must be of a sufficient size to enable an adult rabbit to stretch its body to full length and to sit upright on its hind legs. Consequently, rabbit cages are most often constructed of stainless steel or plastics. Rabbits are usually housed in cages with mesh or slatted floors to permit urine and feces to drop through into a catch pan. A farrowing crate is made of wood, metal plates, or other materials. For young rabbits, the bottom should have a seam, and a crescent-shaped notch should be designed in front of the nest box. The crate size must accommodate doe and litter pups.

Due to the high concentration of crystals present in the urine of rabbits, which stick to the floor, the cages should be treated with acid solution periodically. Acid wash cycles or presoaks may be necessary to remove this scale prior to sanitation because normal alkaline wash cycles inadequately remove the scale.

Temperature is one of the most sensitive rearing conditions. The optimal temperature for rabbits is between 15°C and 29°C, and relative humidity is between 40% and 70%. Temperatures above 30°C in combination with high relative humidity leads to a risk of heat stress, which can cause panting, hair loss, infertility, and mortality. Rabbits are easily startled by sudden and loud noises, and noise should be controlled to 60 dB or less.

Environmental requirements and recommended space for rabbits are provided in Table 4.4.

4.7 DOGS

4.7.1 Introduction

Dogs (Figure 4.10) are mammals in the order Carnivora, suborder Caniformia, and family Canidae. The domesticated dog has been designated as *Canis familiaris*. Other members of the genus *Canis* include jackals, the coyote, the red wolf, and four gray wolf species. Nowadays, there are approximately 400 breeds, ranging in size and shape from the teacup Chihuahua to the large Irish wolfhound.

The dog (*C. familiaris*), which has a long history of living and is interdependent with humans, has been domesticated as a pet and has been widely used as a laboratory animal. Beagles (Figure 4.10) and mongrel dogs are the most frequently used as laboratory dogs. Research on beagles is more focused on pharmacokinetics, alternative drug delivery systems, and cardiovascular pharmacology. The next common areas of research are dental, and periodontal disease and surgery, orthopedic surgery and skeletal physiology, and radiation oncology. Other research areas that use beagles include canine infectious disease, imaging, prostatic urology, and ophthalmology.

4.7.2 Physiology

The habitus of *C. familiaris* is larger, the brain is developed, and the dog has instincts to obey commands of people. The nose of the normal dog is oily and moist,

Figure 4.10 The laboratory beagle dog.

which will feel cold when touched. Sweat glands are not developed and they dissipate heat mainly by accelerating the breathing rate, and by sticking out their tongue. They have no clear vision point because there are no macula lutea.

Dogs are not visually sensitive and are red–green colorblind. Dogs should therefore not be used to do conditioned reflex experiments with red and green as the color stimuli. The dog has a good nose that is full of highly sensitive olfactory nerve cells on the nasal mucosa, and the olfactory nerve is extremely rich. The sensitivity of smell exceeds 1000 times that of humans. Their hearing is very acute and is over 16 times that of humans. However, their sense of taste is insensitive.

The bitch has a monoestrus sexual cycle of 180 days, with clinical estrus occurring predominantly in January or February and again in July or August, followed by ovulation within 2–3 days. The estrus cycle consists of four stages: proestrus, estrus, diestrus, and anestrus. The average duration of proestrus is 9 days. Endocrinologically, proestrus is the follicular stage of the cycle and estrogen levels peak at this time. Estrus generally lasts 9 days, and the vulva is softer and smaller than in proestrus. Diestrus begins approximately 9 days after the onset of standing heat while the end of this stage is 60 days later. The duration of anestrus is approximately 4 months. Dog usually have an average of 6 pups per litter and lactation lasts for 60 days.

Pregnancy detection can be performed by abdominal palpation of the uterus 28 days after mating. The embryos and chorioallantoic vesicles form a series of ovoid swellings in the early gravid uterus. They are approximately 2 in. in length at 28–30 days, the time at which pregnancy is most easily and accurately diagnosed.

Major physiological parameters of dogs are given in Table 4.5.

4.7.3 Anatomy

The whole skeleton of the dog includes the skull, vertebrae, sternum, frame, front and rear upper limbs, and baculum. The penile bone is specific to canines. Dogs have typical teeth of carnivorous animals whose canine and molar teeth are well developed with strong tearing ability but poor chewing power. The deciduous teeth appear 10 days after birth and are gradually replaced by permanent teeth at 2 months later, and are fully developed within 8–10 months. The dog's stomach is large and the intestines are short, only 3–4 times the body length. The circulatory system is developed similar to the human. The heart is large, accounting for approximately 0.1%–0.5% of the body weight.

Female dogs have a bicornuate uterus. The ovary on both sides is completely surrounded by a serous pouch, which is connected to the short oviduct, thus dogs have no ectopic pregnancies in general. The ovaries are attached to the abdominal cavity by the caudal broad ligaments of the kidneys and are not palpable. The uterus consists of the cervix, uterine body, and uterine horns. The cervix is an abdominal organ, located approximately halfway between the ovaries and the vulva. The vagina is a long musculomembranous canal that extends from the uterus to the vulva.

Components of the canine spermatic cord include the ductus deferens, the testicular artery and vein, the lymphatics and nerves, and the cremaster muscle. The

Table 4.5   Physiological Parameters of Dogs, Cats, Miniature Pigs, Sheep, and Goats

	Dogs	Cats	Miniature Pigs	Sheep/Goats
Adult Weight (kg)				
Male	10–80	3–7	20–60	50–70
Female	10–60	3–4	15–60	50–60
Life span (year)	10–15	10–17	14–18	10–15
Heart rate (time/min)	80–150	100–120	60–90	70–80
Respiratory rate (time/min)	20–30	20–40	8–18	12–25
Body temperature(°C)	38–39	38–39.5	38–40	38.5–40
Chromosome number (2n)	78	38	38	54/60
Water intake (mL/100 g/day)	–	0.03–0.25	2–6	–
Adolescence (month)				
Female	8–14	6–8	5–7	6–10
Male	7–8	6.5–7	5–7	6–10
Breeding Season (month)				
Female	>12	10–12	>7	>10
Male	9–14	>12	>7	>10
Estrous cycle (day)	4–8 months	15–18	18–24	14–20/15–24
Gestation (day)	63–67	60–65	110–118	144–115
Litter size	3–6	3–5	11–16	1–2
Newborn weight (g)	200–500	90–130	900–1600	–
Weight at weaning (g)	1.5–4	0.6–0.8	6–8	–
Weaning age (month)	6–7	7	4–7	4–8
Blood Parameters				
Blood volume (mL/kg)	72–77	65–75	74	80/60–70
Hemoglobin (g/100 mL)	12–17	11–14	11–13	11–13/8–12
Hematocrit (vol%)	37–55	24–55	41	32/34
White blood cells($\times$1000/mm^3)	7–17	9–20	8–16	15–20/8–12
Blood glucose (mg/100 mL)	60–80	75–110	60–90	30–60

penis is a continuation of the muscular pelvic urethra and is attached to the scrotal arch by two fibrous crura. The accessory sexual glands of the dog consist of a well-encapsulated developed prostate gland surrounding the pelvic urethra and ampullary glands, which are at the termination of the vas deferens in the urethra. Males do not have seminal vesicles or bulbourethral glands, but have a unique baculum.

The central nervous system of dogs includes the brain and spinal cord. Brain weight is generally 1/40–1/30 of the body weight.

4.7.4 Management and Husbandry

4.7.4.1 Sex Identification

Sexual determination is easy, male dogs have testicles and a penis, whereas female dogs have nipples and a vagina.

4.7.4.2 Housing

Dogs can be housed indoors as well as outdoors and should preferably be housed in small groups with an indoor sleeping bed and an outdoor running area. The breeding place should be in a separate zone. The indoor area should have sound insulation and kennels need double door facilities. The indoor area should have a solid floor, a tender-foot floor or metal grid floor, and must be kept warm with no airflow. There should also be a dry insulated sleeping area, possibly with under-floor heating. The floor of the outside should have a suitable slope to ensure adequate drainage. If a dog should be kept alone, then cages can be used. When free breeding, the kennels should be built on higher ground and away from the location of the human living area.

Laboratory dogs are usually housed in cages, and the minimum dimensions of the cages are relative to the size of the dog. Usually the cage has a floor area of 1 m^2 and a height of 100 cm. New incoming dogs should be kept away from local groups, and extra care must be taken to ensure that they are not attacked. The environmental temperature for a group of dogs may be lower than that for housed individually. Newborn pups depend upon external heat for regulating their body temperature.

4.7.4.3 Feeding and Nutrition

Dogs naturally eat meat and chew bones. For long-term livestock, they can be fed with omnivorous or vegetation fodder, but the fodder should guarantee the basic needs of protein and fat.

Good nutrition and a sound, balanced diet are essential to the health, performance, and well-being of the animal. The basic nutrient requirements for dogs should be complied by the guidelines and represent the average amounts of nutrients that a group of animals should consume over time to maintain growth and prevent deficiencies.

Most commercially available balanced dog diets are "closed-formula" diets that meet the labeled specific minimum requirements for protein and fat, and the maximum values for ash and fiber. Ingredient composition varies depending on the cost relationships of the ingredients as the manufacturer attempts to achieve the label requirements at the lowest ingredient cost. An "open-formula" diet provides a more specified dietary control. In these diets, the ingredients are specified and the percentage of each ingredient is kept constant from batch to batch. "Semi-purified" diets provide the strictest control of ingredients and are formulated from the purified components: amino acids, lipids, carbohydrates, vitamins, and minerals.

Adult dogs should normally be fed once a day with commercial dog chow, either in a dry or wet form, and the breeding female and pups can be fed 2–3 times a day. Currently, dogs and laboratory dogs are fed with puffed granular feed containing full nutrition according to 4% of their weight, ensuring their needs for growth and development.

4.7.4.4 Sanitation

The feeding tableware, dog cage, and kennel grounds should be cleaned every day. The staff should pay attention to the mental state, feeding habits, mouth, eyes, nose, skin, genitals, behavioral pattern, etc., to ensure the health of the dog.

4.7.4.5 Exercise and Guide

Exercise can increase the metabolism of dogs to improve appetite and develop strong immunity. Dogs used for chronic experimental or conditioned reflex studies need to be properly trained in order to make the experiments go smoothly.

4.7.4.6 Quarantine

New incoming dogs should be quarantined at least 21–28 days before they are allowed to enter the facility. In this period, further experimental examination should be taken such as blood tests for potential infections.

Environmental requirements and recommended space for dogs are provided in Table 4.6.

Table 4.6 Feeding Environmental Parameters of Dogs, Cats, Pigs, Sheep, and Goats

	Dogs	Cats	Miniature Pigs	Sheep/Goats
Environmental Requirements				
Temperature (°C)	18–29	18–29	16–27	16–27
Relative humidity (%)	30–70	30–70	30–70	30–70
Ventilation rate (time/h)	15–20	15–20	–	–
Light/dark cycle (h)	10/14 or 12/12	10/14 or 12/12	10/14 or 12/12	10/14 or 12/12
Minimal Feeding Space				
Floor area for single (m²)	<15 kg:0.743 15–30 kg:1.115 >30 kg:≥2.230	≤4 kg:0.273 >4 kg:≥0.372	<15 kg:0.743 15–25 kg:1.115 25–50 kg:1.394 50–100 kg:2.230	<25 kg:0.929 25–50 kg:1.394 >50 kg:1.858
Floor area for breeding (m²)	–	2	–	–
Floor area for group (m²)	1–4	0.2–0.6	<25 kg:0.557 25–50 kg:0.929 50–100 kg:1.858 100–200 kg:3.716	<25 kg:0.790 25–50 kg:1.161 >50 kg:1.579
Height (cm)	60–180	50	50–80	1200/2000

4.8 CATS

4.8.1 Introduction

Cats (Figure 4.11) have been used in animal experiments since the end of the nineteenth century. Although the number of cats used in experiments has markedly declined during the last decade, many countries are breeding SPF and germ-free cats specifically for the purpose of experimental study such as neurology, ophthalmology, retrovirus research, inherited diseases, and immunodeficiency diseases.

4.8.2 Physiology

Cats prefer to live alone, have good eyesight and acute hearing, and are extremely sensitive to the photoperiod. They communicate with each other or with other animals and humans through different sounds and behavior patterns.

Cats usually defecate and urinate in the same place, and cover their feces with soil or sand immediately. Their teeth and claws are very sharp, and good for catching and climbing. Molting appears at the transition between spring and summer, and autumn and winter. Cats prefer to eat fish and meat, and are sensitive to environmental changes.

On an average, females reach puberty or experience their first estrus cycle between 5 and 9 months of age. Sexual activity occurs between 1.5 and 7 years of age, with an average of 2–3 litters per year, with 3–4 kittens per litter. Adolescent queens (less than 1 year of age) and queens greater than 8 years of age tend to cycle irregularly and have smaller litters, more abortions, more stillbirths, and more kittens with birth defects.

The estrus cycle of the queen consists of five phases: proestrus, estrus, interestrus, diestrus, and anestrus. Estrus is the phase of sexual receptivity that lasts 4–7 days on average, with a range of 1–12 days. Coital contact does not shorten estrus. As cats are polyestrous and do not ovulate following every estrous period, an interestrus period or anoestrous commonly follows estrus. Interestrus is the interval of sexual inactivity between waves of follicular function in cycling queens. Queens

Figure 4.11   The cat.

typically return to proestrus within 1–3 weeks. The domestic cat is an induced ovulator. Ovulation occurs 24 h after mating. Pregnancy lasts 63 days, and ranges from 60 to 68 days. Abdominal palpation is the most common method for diagnosing pregnancy in the queen. The lifetime is 8–14 years.

Major physiological parameters of cats are given in Table 4.5.

4.8.3 Anatomy

The cat dental formula is 2 (incisors 3/3, canines 1/1, premolars 3/3, molars 1/1) = 30 with 12 small incisors, 4 sharp canine teeth, and sharp cheek teeth. Cats have 7 cervical vertebrae, 13 thoracic vertebrae, 7 lumbar vertebrae, 3 sacral vertebrae, 21 caudal vertebrae, 13 ribs, and their foreleg has 5 toes, whereas posterior has 4 toes. The sharp claws are able to stretch and withdraw from the pads with a few sweat glands between the toes.

The chest is small, abdomen is very large, the stomach is monogastric and the intestinal tract is short relative to that of dogs and other monogastric animals. The cecum is small and the intestinal wall is thick. The epiploon is very developed and connects with the stomach, intestine, spleen, and pancreas, with the function of fixing and protecting. Cats have five hepatic lobes, seven lung lobes, a bicornuate uterus, and four pairs of nipples on the abdomen.

There are countless filiform papillae protrusions on the tongue with thicker cuticles, which is a unique characteristic of the cat. The brain and cerebellum is well developed with good balance of the body. Cats have good eyesight and can adjust their pupil according to the strength of the light. When the light is strong at daytime, the pupil shrinks linearly and the pupil dilates at night. The cirrus of the mouth has sensation.

Cats easily vomit and cough, which is induced by mechanical and chemical stimuli.

4.8.4 Management and Husbandry

Cats are housed commonly in three basic arrangements: single cages, multiple runs within a room, or free ranging in a room. Requirements of the guidelines for space and density restrictions should also be consulted because housing must comply with these regulations and exceptions sought for good cause. Domestic cats develop highly structured interactive social groups. Therefore, individual housing should be avoided unless particular experimental objectives dictate the use of single-cage housing or if caging is needed for short periods to permit collection of specimens, to administer material individually, or to accomplish treatment or observation. If caged, cats should be allowed out of their cages daily to exercise.

Cats require a clean and dry environment, are sensitive to abnormal gas and steam, phenols, disinfectants, and phenothiazine.

As obligatory carnivores, cats are nutritionally and metabolically unique, and their dietary requirements differ considerably from those of most other species. They require diets high in protein and fat but low in carbohydrates. Cats lack the ability to synthesize sufficient quantities of essential nutrients, such as taurine, arginine, vitamin A, niacin, and arachidonic acid, which in the wild were present in tissues of their prey. As their intestinal tract is short relative to that of dogs and other monogastric animals, highly digestible diets are preferred.

Environmental requirements and recommended space for cats are provided in Table 4.6.

4.9 PIGS

4.9.1 Introduction

Pigs (Figure 4.12) share similarities with humans in anatomy, physiology, nutrition, and metabolism and have therefore become significant animal models for human disease research. However, there are some disadvantages such as size, body weight, and cost for husbandry using domestic pigs for research. Thus, miniature pigs are ideal models for biomedical studies. Miniature pigs range from 15 to 60 kg in body weight at sexual maturity and are suitable for most biomedical research projects.

4.9.2 Physiology

Pigs are omnivorous animals whose digestive features are between carnivorous and ruminant animals. Pigs and humans share some similarities in skin structure, including the mechanism of epithelial regeneration, and endocrine and metabolic changes.

Males reach sexual maturity at 3 months of age and females reach estrus at 4 months of age, and can then be used for mating. The average estrus cycle is 21 days, with a range of 17–25 days. Estrus typically lasts 48 h, with a range of 1–3 days. In the early stage of estrus, sows will exhibit signs of vulvar reddening and swelling, mucous discharge, nervousness, and increased activity. Optimal fertilization rates occur when insemination takes place 12 h prior to ovulation.

Figure 4.12 Miniature pigs.

Failure to return to estrus 18–24 days following mating is the first sign of pregnancy. Nonestrus sows are most easily detected by daily exposure to a boar during this time. Behavioral changes are seen in only 50% of sows in the absence of a boar. In the absence of a boar, determination of pregnancy can be based on whether or not the physical and behavioral changes of estrus are observed. Estrus detection has been reported to be 98% accurate and can be used to determine pregnancy status soon after failure of conception or death of a litter. The average duration of pregnancy is 114 days. Due to the precocious puberty, multiplets and multiparous females can have 2 litters per year. If the lactation period of piglets is shortened or hormones are used, they can breed 5 times in 2 years or even 3 times per year.

Major physiological parameters of miniature pigs are given in Table 4.5.

4.9.3 Anatomy

The pig's dental formula is 2 (incisors 3/3, canines 1/1, premolars 4/4, molars 3/3) = 44. The number of cervical vertebrae, thoracic vertebrae, lumbar, and caudal vertebrae in pigs is 17, 14, 14, and 21–23, respectively.

The cardiovascular system is similar to that of humans, especially the coronary anatomy. The blood supply from the coronary artery is right-side dominant and does not have preexisting collateral circulation. This is similar to 90% of the human population. The electrophysiological system is more neurogenic than myogenic, and there are prominent Purkinje fibers.

The stomach has a muscular outpouching, the torus pyloricus, near the pylorus. The cardiac gland is more developed than in other animals and accounts for most of the stomach. The pyloric gland is larger than in other animals. The bile duct and pancreatic duct enter the duodenum separately in the proximal portion. The concentrating ability of the gall bladder is weak, and the amount of bile secreted by the liver is low. The mesentery is thin and friable. The majority of the large intestine is arranged as a spiral colon in the left upper quadrant of the abdomen. This series of centrifugal and centripetal coils includes the cecum and ascending, transverse and majority of the descending colon. Tenia and haustra are present in the cecum and large intestine.

The lymph nodes are inverted with the germinal centers being located in the internal portion of the node. The thymus is located on the ventral midline of the trachea near the thoracic inlet rather than proximal to the larynx.

The penis is fibromuscular with a corkscrew-shaped tip located in a preputial diverticulum near the umbilicus. The penis has a sigmoid flexure. The male accessory glands include the prostate, vesicular glands, and bulbourethral glands. The female reproductive system is bicornuate with long, torturous fallopian tubes. The pancreas is bilobed, and surrounds and encompasses the superior mesenteric vein. The liver is divided into lobules by microscopic fibrous septa.

4.9.4 Management and Husbandry

Individual shipments of pigs are best separated by time and distance, and mixing animals from multiple vendors is poor practice. Pigs should be purchased from vendor herds that are validated as brucellosis-free and pseudorabies-negative. Pigs are best housed in pens rather than cages, which may be constructed of either chain-link fencing, or stainless steel or aluminum bars. Pigs prefer to have contact with other members so they may be housed together in social groups. Pigs readily use an automatic watering system. This system should be checked daily to ensure that the water supply is functional because pigs are susceptible to "salt poisoning," which results in a neurological syndrome when they are deprived of water. Food dishes should be secured to the cage or flooring.

Pigs can be fed with mixed provender or a special type of solid feed according to the experimental requirements. Antibiotics and hormonal additives should not be added into the feed. Nutrient requirements are particularly important for newborn piglets. Nursing piglets require 21 mg of iron for each kilogram of growth, and sow's milk contains approximately 1 mg of iron per liter. Pigs, unlike ruminates, do not require elemental sulfur in their diets when adequate sulfur-containing amino acids are available. Methionine alone can meet the total sulfur-containing amino acid requirement in pigs because cysteine can be synthesized from methionine.

Pigs are commonly vaccinated against erysipelas, cholera, Japanese encephalitis, leptospirosis, transmissible gastroenteritis, and atrophic rhinitis. The breeder should observe the appetite of the swine as well as whether the excrement is normal, especially for symptoms caused by abdominal distension or pain. When constipated, diarrhea and vomiting are exhibited, and the pigs should be treated appropriately.

Environmental requirements and recommended space for miniature pigs are provided in Table 4.6.

4.10 GOATS

4.10.1 Biological Characteristics

Both female and male goats (Figure 4.13) have horns. Goats are active, agile, and docile, and need a clean and dry environment. Goats will usually procreate twice a year or 3 times every two years with 1–3 kids per litter. The age of sexual maturity is 6 months and the best reproductive age is 3–5 years. The estrus cycle is approximately 21 days, with estrus duration lasting 2–3 days, generally in the autumn. Ovulation occurs at 9–19 h after estrus. The duration of pregnancy and lactation is 148 days and 3 months, respectively. The jugular vein of the goat is bulky, making blood collection easy. The blood is widely used for hematology diagnosis, microbiology, and blood culture medium. Goats are good laboratory animals for immunology.

Major physiological parameters of goats are given in Table 4.5.

4.10.2 Feeding and Management

Goats are fed in a single stable. Overall, 3 kg of grass and fresh leaves should be supplied per day, and ewes with lambs are supplemented with 0.3–0.5 kg corn

Figure 4.13 Goats.

fodder. Carrots or sweet potatoes can be fed in the winter and spring. The goat stable should be cleaned and disinfected regularly and excreta should be removed daily.

When goats are confined, care should be taken to provide adequate, but draft-free, ventilation. Ammonia buildup and other waste gases may induce respiratory problems. In cold weather, the ventilation should be increased even at the expense of lower temperatures. Even adult goats and younger cattle are quite comfortable in the cold, even subzero temperatures, if provided with adequate amounts of dry dust-free bedding and draft protection. Goats and other ruminants such as sheep are social and herding animals, thus they should be housed in groups or at least within eyesight and hearing of other animals.

Environmental requirements and recommended space for goats are provided in Table 4.6.

4.11 SHEEP

4.11.1 Biological Characteristics

Sheep (Figure 4.14) are even more docile than goats and are gregarious, but their flexibility and endurance are poor. Sheep need a dry environment because humidity can cause foot rot and parasites. Sheep dislike heat and timely shearing should be done in summer. They like eating grass and have a good digestion with a high consumption of feed. Sheep are sharp-tongued, have thin lips, and the jaw incisors incline out of the mouth. Sheep are ruminants with four stomachs. Their pancreas secretes constantly whether eating or not, but the capacity of the gallbladder is poor.

Figure 4.14 Sheep.

Sheep are seasonally polyestrous; most breeds will express estrous in the fall while some may cycle in both the fall and the spring. Sexual maturity of sheep occurs between 7 and 8 months of age, and the sexual cycle is 17 days. The average duration of estrus is 24 h. Ovulation in sheep occurs between 12 and 41 days and pregnancy lasts 150 days. Lactation lasts 4 months, with only 1–2 lambs per litter. Sheep are often selected for studying areas such as ruminant physiology and nutrition. Sheep are also widely used as models for basic and applied fetal and reproductive research.

Major physiological parameters of sheep are given in Table 4.5.

4.11.2 Feeding and Management

Sheep, because of their wool, are remarkably tolerant to both hot and cold extremes. Newborn lambs and recently shorn adults are susceptible to hypothermia and sunburn. Therefore, in outside housing areas, sheep should be provided with shelters to minimize exposure to sun and inclement weather. Many methods can be used for individual identification such as ear tags or ear cuts. Shearing is done during June each year and there is an additional shearing of coarse wool during September or October. Fodder is mixed with coarse green vegetation.

Both sheep and goats are sensitive to changes in light cycle especially reproductive parameters, photoperiod must be taken into account when they are housed indoors.

Environmental requirements and recommended space for sheep are provided in Table 4.6.

4.12 NONHUMAN PRIMATES

Rhesus monkeys (*Macaca mulatta*) (Figure 4.15a) are the most common nonhuman primates used in biomedical research. Rhesus monkeys are a species that includes

(a)   (b)

Figure 4.15   The rhesus monkey (a) and cynomolgus monkey (b).

the following subspecies: *M. m. mulatta, M. m. sancti-johannis, M. m. lasiotus, M. m. tcheliensis, M. m. villosus,* and *M. m. mcmahoni.* The second-most used nonhuman primates are cynomolgus monkeys (*Macaca fascicularis*) (Figure 4.15b).

Monkeys were the laboratory animal models used to investigate, develop, and produce the polio vaccine. Rhesus monkeys are currently the models of choice for human immunodeficiency virus infection/acquired immune deficiency syndrome (HIV/AIDS) vaccine development and study. In addition, cynomolgus monkeys, marmosets, and chimpanzees are also used for biomedical experiments. Here, we will briefly note the characteristics and laboratory management of macaques.

4.12.1 General Characteristics

Macaques are of a medium size, diurnal animals characterized by having brown to gray fur, lighter undersides, and deep eye sockets. They also have a medium-length, nonprehensile tail covered in long, dense fur.

Macaques are tropical and subtropical animals, and they live gregariously in the rain forest or grassland near water. In the wild, Macaques exist primarily in male-dominated, multi-male–multi-female groups ranging in size from 10 to 50 members. Females adhere to a strict hierarchical class system with the dominant male changing groups every few years. The leader of the group of monkeys is the Monkey King, being the most ferocious and strongest monkey.

In the wild, Macaques are mainly frugivorous. Their diet has been observed to be 65%–70% fruit, supplemented with leaves, shoots, roots, bark, fungi, and small invertebrates. They have cheek pouches that allow storage of food for mastication at a later time. They are smart and agile because they have a well-developed brain. They are very curious and have a developed intelligence and neural control. Their most common threat behavior is a wide-open mouth with staring eyes.

4.12.2 Anatomy

Macaques generally share the characteristics of other mammals in that they have claws, a collarbone, and placenta. Macaques have a developed cecum. They have a total of 32 teeth, with a dental formula of 2 (incisors 2/2, canines 1/1, premolars 2/2, molars 3/3) = 32. Lung lobes are unpaired, the right lung is composed of 3–4 lung lobes, while the left is made up of 2–3 lobes. The sexual area of the female is near the genitals and the entire buttocks. At preovulation, ovulation is especially significant by swelling and redness that subsides before menstruation, which is obvious in adolescent and young monkeys.

4.12.3 Physiology

Sexual maturity occurs at 3 years in males and at 2 years in females. Pregnancy lasts 165 days with one infant in each litter each year, rarely two pups. Newborn monkeys weigh approximately 0.40–0.55 kg. Young monkeys will not suck milk the first day after birth. They will latch on to the belly or back of females, often with the help of other females. Seven weeks after birth, the infants can play alone without their mother. Lactation lasts for more than 6 months. The life span is 20–30 years.

Major physiological parameters of nonhuman primates are provided in Table 4.7.

4.12.4 Management and Husbandry

There are two main methods for feeding macaques, caged and house-fed. Cages are for quarantine acclimation to isolate groups and for acute experimental groups. Population and chronic experimental groups may feed in cot.

Food varies from staple cereals and vegetables, but also some animal proteins such as eggs, fish, and milk. The diet should contain enough vitamin C and minerals. Food should be cooked or processed into pellet biscuits. The amount of feed needs to be adjusted for growth. Vegetables and feed must be of high quality, and drinking water needs must be ensured throughout the day.

New monkeys must be individually quarantined for at least 1 month or more. A tuberculin test must be carried out, and internal and external parasites removed. Particular attention should also be paid for other zoonosis inspections.

Feeding in groups can be done without too much management, but the reproduction rate is not high. In a rearing cage, one male often is fed together with 3–12 females. At 11–17 days, the female should be transferred to the male when swelling of the sexual area is obvious. Following the observation of mating, the monkeys should be housed separately. Pregnancy diagnosis should be done by observing for

Table 4.7   Physiological Parameters of Several Common Nonhuman Primates

	Marmoset	Cynomolgus Monkey	Rhesus Monkey	Chimpanzee
Adult Weight (g)				
Male	0.4–0.6	2.5–6	4–9	35–45
Female	0.4–0.5	4–8	6–11	45–60
Life span (year)	10–16	15–25	20–30	40–50
Heart rate (time/min)	–	100–150	100–150	85–90
Respiratory rate (time/min)	–	40–65	40–65	30–60
Body temperature (°C)	–	37–40	36–40	36–39
Chromosome number (2n)	–	–	42	48
Body surface area (cm^2)	–	–	–	–
Water intake (mL/100 g/day)	–	–	–	–
Adolescence (year)				
Female	0.8–1	3–4	3–4	6–8
Male	0.8–1	3–4	3–4	8–10
Breeding Season (year)				
Female	1.5–2	4–5	3–4	9–10
Male	1.5–2	4–5	4–5	10–12
Estrous cycle (day)	27–29	31	29	32–38
Menstrual duration (day)	none	4	3	–
Gestation (day)	142–146	161	155–170	210–250
Litter size	2–3	1	1	1
Newborn weight (g)	25–35	300–400	450–500	1500
Weight at weaning (g)	80–120	800–1200	1000–1500	–
Weaning age (month)	3–6	12–16	12–16	36
Blood Parameters				
Blood volume (mL/kg)	70	–	50–90	62–65
Hemoglobin (g/100 mL)	–	–	11–12.5	10–14
Hematocrit (vol%)	–	–	39–43	38–43
White blood cells($\times$1000/mm^3)	–	–	7–13	10–14
Blood glucose (mg/100 mL)	–	–	60–160	80–95

the signs of menstruation, nipple changes, hormone tests, or ultrasound. Pregnant monkeys are best housed singly, in a room with indoor and outdoor spaces. Females give birth during the night usually without human care, unless dystocia occurs during childbirth. Young monkeys start to feed at 3 months of age. At 6–7 months of age, the young monkeys can be completely fed the same as adult monkeys and gruel or milk sugar should also be added to the diet. The room temperature should be maintained at approximately 20°C.

Environmental requirements and recommended space for nonhuman primates are provided in Table 4.8.

Table 4.8 Feeding Environmental Parameters of Several Common Nonhuman Primates

	Marmoset	Cynomolgus Monkey	Rhesus Monkey	Chimpanzee
Environmental Requirements				
Temperature (°C)	18–29	18–29	18–29	18–29
Relative humidity (%)	30–70	30–70	30–70	30–70
Ventilation rate (time/h)	10–15	10–15	10–15	10–15
Light/dark cycle (h)	10/14 or 12/12	10/14 or 12/12	10/14 or 12/12	10/14 or 12/12
Minimal Feeding Space				
Floor area for single (m²)	<1 kg:0.149	<1 kg:0.149	<1 kg:0.149	<20 kg:0.929
	1–3 kg:0.279	1–3 kg:0.279	1–3 kg:0.279	20–35 kg:1.394
	3–10 kg:0.399	3–10 kg:0.399	3–10 kg:0.399	>35 kg:2.323
	10–15 kg:0.557	10–15 kg:0.557	10–15 kg:0.557	
	15–25 kg:0.743	15–25 kg:0.743	15–25 kg:0.743	
	25–30 kg:0.929	25–30 kg:0.929	25–30 kg:0.929	
	>30 kg:1.394	>30 kg:1.394	>30 kg:1.394	
Floor area (m²)	0.25	0.9	1.1	–
Floor area for group (m²)	0.25	0.7	0.9	–
Height (cm)	<1 kg:50.80	<1 kg:50.80	<1 kg:50.80	<20 kg:139.70
	1–3 kg:76.20	1–3 kg:76.20	1–3 kg:76.20	20–35 kg:152.40
	3–10 kg:76.20	3–10 kg:76.20	3–10 kg:76.20	>35 kg:213.36
	10–15 kg:81.28	10–15 kg:81.28	10–15 kg:81.28	
	15–25 kg:91.44	15–25 kg:91.44	15–25 kg:91.44	
	25–30 kg:116.84	25–30 kg:116.84	25–30 kg:116.84	
	>30 kg:116.84	>30 kg:116.84	>30 kg:116.84	

4.13 BIRDS

4.13.1 Introduction

Approximately 8%–10% of the vertebrates used annually for research are birds. In the United Kingdom, 7% of all animals used in biomedical research studies, as well as in 139 thousand procedures in 2014, were birds. The chicken (*Gallus domesticus*) (Figure 4.16) is the most widely used in biomedical research, whereas the pigeon (*Columba livia*), the dove (*Streptopelia risoria*), and Japanese quail (*Coturnix japonica*) are also commonly used. As laboratory animals, the chicken was used by Pasteur in 1789 for research on chicken cholera.

Figure 4.16 Chicken.

4.13.2 Anatomy and Physiology

Birds have specific physiological and anatomical characteristics when compared with mammals. For example, their skin is devoid of sweat and sebaceous glands. To facilitate preening, birds possess either powdery, downy feathers, and/or two preening glands. Birds have no diaphragm; the lungs are connected to the rib cage and they possess a system of air sacs, which are essential structures in the process of respiration.

The abdominal cavity is rather small and includes the caudal part of the gizzard, the intestine, and the spleen. Birds have no urinary bladder; urine is transported via the ureters, which empty into the cloaca.

Erythrocytes of birds are oval and nucleated. Platelets are absent and "replaced" by nucleated thrombocytes. In birds, the female is heterogametic (two different sex chromosomes (ZW)), whereas the male is homogametic sex (ZZ).

Compared with mammals, birds have a high metabolic rate. This means that a continual intake of food must be guaranteed in order to meet their energetic and nutritional requirements. The absorption, metabolism, and elimination of nutritional elements and waste are also performed at a high rate. These aspects need to be considered when using birds as laboratory animals.

The chicken, the pigeon, and the Japanese quail are herbivorous. Food is mixed with the saliva in the mouth, which contains amylase, and then it is passed down to the crop where it is predigested. Further digestion takes place in the proventriculus, where it is mixed with gastric juices, which contain pepsin and hydrochloric acid. Grinding of seeds and food takes place in the gizzard. Nutrients are subsequently digested and absorbed in the relatively short small intestine, and a certain amount of cellulose will be fermented in the two caeca by the intestinal microflora.

4.13.3 Laboratory Management and Husbandry of Chickens

4.13.3.1 Housing

Chickens are frequently kept in wire cages, usually stacked in two or more levels. This, however, is considered as a poor housing system in which their behavioral needs cannot be exercised. It is possible to house them individually, but also in pairs or groups. Chickens, when grouped together, are better able to meet their behavioral needs, but may demonstrate a marked social hierarchy, which can lead to the violent plucking of each other's feathers and even cannibalism. There are minimum measurements for cages depending upon the weight of the animals and the number of animals per cage. The wire mesh floors should have a maximum mesh width of 10 mm when they are to be used for chicks, and 25 mm width when constructed for young and adult animals. The thickness of the wire must be at least 2 mm in all cases. Laying hens should be provided with a tilted wire mesh floor with maximum 14% tilt. When carrying out experiments of a relatively long duration involving young chicks, the fast rate of growth of the animals must be taken into account. This is necessary to minimize or exclude reallocation and regrouping of the animals during the experiment.

4.13.3.2 Feeding and Management

Chicken feeding and management is divided into several stages: incubation, hatching, brooding, and adult chickens. Hatched chicks at 3 weeks need to have the temperature and relative humidity stable (temperature 38°C, relative humidity of 85%). Eighteen days after placing the eggs into the hatching tray, chicks hatch inside the shell after breaking into the brood chamber after 21 days. Different feeds are given at different stages of development during the brooding period: 0–6 weeks of age are fed chick feed, 10–18 weeks of age are fed large-young animal feed, and after 18 weeks are fed chicken feed. Adult chickens need artificially controlled light for reproduction.

4.13.3.3 Reproduction

After copulation, several eggs will be fertilized, and therefore there is no need for copulation to occur daily. The eggs may be collected 2–4 times a day and, after disinfection, are incubated at 37°C with a relative humidity of 60%. During the first 18 days, the eggs should be turned over 3–5 times per day.

4.13.3.4 Handling

Young chicks should be picked up in one hand with the thumb and index finger placed gently around the neck. When removing a fully grown hen from its cage, both hands will be necessary, with the fingers spread widely and the thumbs placed dorsally over the wings. Hands need to be positioned around the body of the hen in such a way that the bird's head is facing the handler. The hen can then be held under the handler's arm, facing backward, while the handler fixes the legs with one hand. Restraint is possible by holding the wings together on its back. Thereafter, the hen can be placed on its side.

4.14 FISH

4.14.1 Introduction

Fish is a broad, encompassing term for the most diverse and largest taxonomic grouping of vertebrates. In common usage, the term includes all of the members of four taxonomic classes, Myxini, Cephalaspidomorphi, Elasmobranchiomorphi, and Osteichthyes. More than 2400 species are known from the United States and Canada alone. The Osteichthyes alone, the class of the bony fishes, comprises more distinct species than all other mammals, birds, reptiles, and amphibians combined. This massive biodiversity is not a quirk of taxonomic whim. The anatomical and physiological differences among species of fish are every bit as marked as the differences among mammalian species. This characteristic of fish offers tremendous opportunity for the researcher seeking a particular anatomic, physiologic, or disease model, but it also presents serious challenges for laboratory animal managers and veterinarians seeking to maintain the health and well-being of the animals selected. Although certain areas of medicine and health management transcend the species differences of fish, it is important to keep in mind the adage, "A fish is not a fish." What it means is that all fish species are not equivalent in their basic biologic and husbandry needs. It is important to know what fish species you are dealing with when making a diagnosis or designing a health protocol.

The zebrafish (Figure 4.17), *Brachydanio rerio* (also referred to as *Danio rerio* and the zebra danio), is currently emerging as an increasingly popular model of

Figure 4.17 The zebrafish.

vertebrate embryonic development, gene function analysis, and mutagenesis. The zebrafish has gradually become the lower vertebrate model of choice because the fundamental molecular mechanisms of embryonic development are similar in all vertebrates. Mutagenesis screening allows researchers to investigate uncharted areas of the genome without prior knowledge of the function of specific genes. Mice are utilized in more conventional laboratory techniques where knockout mutations of known genes of interest are created in order to study the effects of the gene in the resultant phenotype. Through the creation of mutant phenotypes via chemical mutagenesis, the functions of many genes associated with pigmentation, muscular, cardiovascular, and central nervous system development have been investigated extensively in zebrafish.

4.14.2 Laboratory Management and Husbandry

4.14.2.1 Water System

There are recirculating systems and flow-through systems available for water sources to raise fish in small or large tanks. Three major types of filtration are used in aquatic design: mechanical, biological, and chemical. In a well-established reservoir system, reseeding may require no more than placing the biofilter into the system. It may be beneficial to keep the disinfected biofilter downstream of and in close contact with well-seeded biofilters. It will take a minimum of 3 weeks to reseed the disinfected biofilter to a level that will be of any value in the research tank. This can be hastened somewhat by keeping the biofilter reserve system warmer; however, it is difficult to say whether this is a long-term benefit or not. The bacteria working in the biofilters are susceptible to environmental changes, and a large drop in fixation efficiency can occur with just the careful transfer of a biofilter from the reserve system to a tank. This loss is greater if the environmental conditions, including the water temperature, of the research tank and the reservoir system are widely disparate.

It is important to feed biofilters while they are in the reservoir system. This can be accomplished in a number of ways. Some aquarists maintain fish or invertebrates in the reservoir system to provide waste material to feed the bacteria. This minimizes the work involved in operating the reservoir system, but has the disadvantage of having potential reservoirs of infection in the reservoir system itself. Another approach is to introduce ammonia into the system on a periodic or continuous basis. In marine systems, ammonium chloride solutions are often used for this purpose.

4.14.2.2 Water Management

The most critical issue in water management, after certification and monitoring of the quality of the source, is the water change procedure. Flow-through systems are constantly undergoing water changes; water is removed and new water enters the system continuously. The positioning of inflow and outflow pipes in these systems will have a major impact on the efficacy of this exchange. Ideally, the flow dynamics and convection patterns of the tank would be such that only old water leaves the system and all new water remains, but this is rarely achieved. Some degree of mixing of newly added water to old water nearly always occurs, reducing the effective water change. It is important to avoid significant shunting of new water to the outflow, and it is advantageous to have a pattern of mixing in the tank that avoids "dead spots," or volumes of water that are never or only very slowly exchanged. In recirculating tanks, the challenge of optimizing water mixing is not the primary concern. What is critical is to remove a volume of water prior to introducing the new water.

A frequent failure in recirculating systems is due to the misconception that "topping up," the practice of adding water to compensate for evaporative losses, is equivalent to conducting water changes. Toxic compounds do not evaporate at the same rate as water, and most tend to accumulate in the system water if topping up is allowed to substitute for a true water change. The required rate of water changes is dependent on the configuration of the tanks and the system, as well as the bioload of fish being maintained. As a general rule, 0.75%–1% per day is effective as a routine water exchange in a wide variety of systems. This approach of frequent small changes has the advantage of reducing the potential impact of temperature, pH, ionic, or other shocks that can occur if improperly conditioned water is used in larger volume exchanges.

4.14.2.3 Feeding

The problem of feeding is as complex as is the provision of suitable space and water. The breadth of laboratory fish species diversity contributes to this difficulty. Relatively little is known about the natural diets of many commonly kept fish. Although diets that maintain and even support growth and reproduction are known for common laboratory fish species, these empirically derived diets are usually relatively unrefined. Trace nutrient balance is rarely considered, and quality control, including component selection for these diets, is minimal. Protein sources can vary markedly from lot to lot, as can processing and storage procedures. These problems are all compounded by the relative plasticity of fish growth and development in adapting to nutrition availability.

As fish can survive for relatively long periods without food, not feeding fish is sometimes seen as a way to circumvent the variability in food lots, difficulties in documenting individual intake in group-housed fish or time constraints in acclimating fish to new diets. As a result, a large amount of knowledge of fish physiology and a considerable amount of disease model research are based on a catabolic animal. Fish entering a new system should be allowed time to acclimate to new diets before experiments are initiated. In experiments that depend on accurate assessment of individual food intake for interpretation of the results, fish should be housed and fed in a manner that makes this assessment feasible. Feeding patterns should try to mimic the natural feeding patterns of the fish.

4.14.2.4 Social Grouping Enrichment

Covers and substrates must allow a fish to be comfortable in the tank, but not hinder capture or observation. Plastic piping can often be used to provide hiding places, and clear piping is often accepted by a fish, especially if the tank lighting is kept subdued except during observation. Animals that need to burrow in sand can be easily observed through clear glass gravel placed in patches, in removable containers or covering the entire bottom of the tank. Most laboratory fish are held in species isolation, which can simplify this issue. However, multiple-species housing in a primary enclosure is common in fish management, and interspecific and intraspecific interactions among individuals must be managed appropriately.

4.15 AMPHIBIANS

4.15.1 Introduction

Amphibians are unique among vertebrate species in that they represent the transition between ancestral aquatic life-forms and more recently evolved terrestrial existence. The word *amphibian* is derived from the Greek word "*amphibios*," which means "double life." This "double life" accurately describes the aquatic larval stage and postmetamorphic terrestrial lifestyle of many amphibians. The class amphibia is represented by approximately 4300 species contained in three orders: Gymnophiona, Caudata, and Anura.

4.15.2 Anatomy and Physiology

4.15.2.1 Integumentary System

The skin of most amphibians is smooth, moist, and glandular. Two primary types of skin glands are present in amphibians: mucous glands and granular glands. Mucous glands secrete a slimy protective layer, which prevents mechanical damage to the skin, facilitates retention of body fluids, and provides a barrier against pathogens. Granular glands synthesize and secrete a variety of compounds that protect against predators, as well as chemicals that have antibacterial and antifungal properties. Granular glands are usually found on the head and shoulders, but can be scattered over the body. For example, the parotoid gland of toads (*Bufo*) (Figure 4.18a), located on the head behind the eyes, is a raised cluster of granular glands. Other

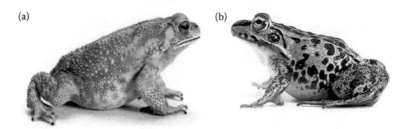

(a)   (b)

Figure 4.18   The toad (a) and frog (b).

chemical compounds secreted by granular glands of several species include phero-mones used in courtship and mating.

Amphibians shed their skin in cycles, which may range from days to weeks. The skin commonly splits mid-dorsally, and the animal uses its limbs to climb out of its skin. Shed skins are commonly eaten. Some amphibian species have a specialized area of permeable skin in the abdominal region, which is pressed against wet sub-strates and used to absorb water from the environment for rehydration.

4.15.2.2 Musculoskeletal System

The amphibian skeleton has undergone several modifications. The salaman-der skeleton is largely cartilaginous. Ribs are absent or greatly reduced in most frogs (Figure 4.18b). Anuran adaptations for jumping include fusion of post-sacral vertebrae into an elongated bone, the urostyle, which articulates with the sacral vertebra and the ilium, and the tibia and fibula fuse into a single, strong bone, the tibiofibular.

4.15.2.3 Respiratory System

Larval amphibians breathe primarily through gills. Adults can retain and use gills, lose gills and develop lungs, breathe with both gills and lungs, or have neither. Skin is the primary respiratory surface in most amphibians. In species that use lungs for respiration, air is forced in and out of the lungs by movement of the buccopharyn-geal floor. Lungs lack alveoli and are very fragile and easily ruptured. In many frog species, the trachea is short and bifurcation occurs close to the glottis; this anatomic feature must be taken into account when performing endotracheal intubation.

4.15.2.4 Cardiovascular System

Most adult amphibians have a three-chambered heart, consisting of paired atria and a single ventricle while larval amphibians have a two-chambered one. Patterns of blood flow and mixing of oxygenated and deoxygenated blood vary among species, depending on degree of pulmonary respiration, physiological state, and anatomic structures. The hepatic portal veins drain blood from the rear half of the amphibian's body; this may impact the pharmacokinetics of drugs with hepatic excretion.

4.15.2.5 Lymphatic System

The lymphatic system of amphibians drains directly into the venous system. At venous junctions, lymph lobes contract and force lymph into the veins. Large sinuses, collection sites for lymph, are found throughout the amphibian's body. In frogs, a pair of these sinuses lies subcutaneously over the sacral area, lateral to the midline. Substances injected into these dorsal lymph sacs will be transported to the venous circulation.

4.15.2.6 Gastrointestinal System

Adult amphibians are carnivorous with a relatively short gastrointestinal tract. The tongue is well developed in all amphibian species and is important for apprehending food. *Xenopus* directs food into the mouth with the front legs. Melanin is commonly found in the amphibian liver and other abdominal organs, and pronounced pigmentation is not unusual. Vomiting is a common defensive mechanism in amphibians, and it is not unusual for some frog species to eject part of the stomach during regurgitation.

4.15.2.7 Excretory System

Frogs have mesonephric kidneys and lack the ability to concentrate urine in excess of plasma levels. Aquatic amphibians excrete ammonia and terrestrial amphibians excrete urea. Most amphibians have a bladder to conserve water more or less. Many frogs, when frightened by predators, will release urine to deter and escape quickly.

4.15.2.8 Nervous System

The cerebral cortical structure in amphibians is not similar to that of higher vertebrates, and the function of the different areas is still controversial. Amphibians have 10 cranial nerves. The hypoglossal nerve (cranial nerve XII) is formed by branches of the first two spinal nerves. A lateral line system is well developed in larval amphibians and is retained by adults of many aquatic species. The lateral line system is recognizable as a linear arrangement of sense cells called neuromasts on the head and along the body. Neuromasts detect changes in water pressure and currents, and function in locating prey. Amphibians can detect higher-frequency sound transmitted through the air to the tympanic membrane, but low-frequency vibration is transmitted through the forelimbs and the cranium to the ear. The amphibian eye has two types of rods, red and green, which are responsible for color sensitivity. Cones detect only the presence of light. A vomeronasal (Jacobson's) organ is responsible for odor detection.

4.15.3 Management and Husbandry

4.15.3.1 Enclosures

Glass aquaria work very well as primary enclosures for amphibians, especially aquatic species. Plastic shoe boxes and sweater boxes also provide appropriate housing and have the advantage of being stackable. Larger aquatic frogs and salamanders are frequently housed in stainless steel, fiberglass tanks. All cages should be constructed of impermeable, easily sanitized material, and should ideally be able to withstand multiple wash. Cages or boxes should be of adequate height to meet behavioral needs of climbing and jumping for some species. Fitted lids or covers are required for most terrestrial and many aquatic species to prevent escape. Commercially available research fish housing units can be adapted for *Xenopus* or other amphibian species.

4.15.3.2 Water

Fresh, dechlorinated water is preferred for amphibians. Many amphibian species are quite sensitive to chlorine and will die from exposure to chlorinated water. Allowing open containers of water for 24–48 h, aerating the water, adding sodium thiosulfate and passing tap water through activated carbon filters are four methods of dechlorination. Maintaining the correct pH of water in the tank or cage is very important. If the preferred pH of a given species is not known, we recommend starting with a pH of 6.8–7.1 (neutral to slightly acidic), then adjusting to a more basic pH if the animal appears irritated. Other parameters that can affect amphibian health are dissolved oxygen and ammonia.

4.15.3.3 Temperature

Many amphibians stay beneath the leaf litter of forest floors or submerged in cool ponds and fast-moving streams. Tropical species can be maintained at 21–29°C, while amphibians from temperate regions do well at 18–22°C.

4.15.3.4 Lighting

Most amphibians live in cool, dark environments in the wild, thus long and direct exposure to bright light should be avoided. Light cycles of 12 h light/12 h dark are recommended. However, photoperiod needs to be managed according to the natural habitat of the amphibian.

4.15.3.5 Handling

Amphibians should be handled carefully to avoid disrupting their protective mucous layer or causing excess secretion of toxins. Gloves free of powder should be worn and moistened with dechlorinated water. Aquatic species can be transferred or held in glass containers to protect the sensitive gills. Small terrestrial amphibians can be manually restrained with one hand. Large salamanders should be firmly but gently grasped behind the head and around the pectoral girdle with one hand, and around the pelvic girdle with the other. Frogs and toads can be held around the pectoral girdle with the strong hind legs restrained to prevent kicking and slipping out.

4.16 REPTILES

4.16.1 Introduction

Reptiles are the first class of vertebrates to evolve an amniotic shelled egg; therefore, they no longer require an aquatic environment for reproduction. Members of class *Reptilia* are derived from two lineages, *Anapsida* and *Diapsida*. Turtles are anapsids, unmistakable due to the presence of a bony shell covering the body. Diapsids include the saurians (crocodilians and, according to many taxonomists, birds) and the lepidosaurians (tuataras, lizards, and snakes).

4.16.2 Anatomy and Physiology

4.16.2.1 Integumentary System

The body of reptiles is covered primarily by scales. Snakes (Figure 4.19a) and lizards (Figure 4.19b) have an epidermal structure with α-keratin on the inside and β-keratin on the outside. Crocodilians (Figure 4.19c), some lizards, and turtles (Figure 4.19d) have osteoderms (bony plates in the dermis) on the dorsal and lateral surfaces of the animals. There are a few skin glands with a variety of functions in reptiles. Many turtles have musk glands in the inguinal and axillary regions. Male tortoises and both sexes of crocodilians have glands in the mandibular area; crocodilians also have cloacal glands. Snakes and some lizards have paired scent glands that empty through the cloaca and seem to function in defense and sexual

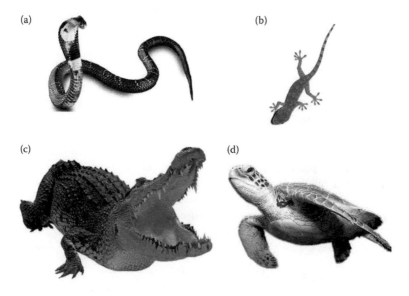

(a)   (b)

(c)   (d)

Figure 4.19   The snake (a), house lizard (b), crocodile (c), turtle (d).

recognition. Salt glands are found in tongue, orbit, nasal passage of some species such as sea snakes, crocodiles, and turtles to remove excess salt from the body.

4.16.2.2 Musculoskeletal System

Reptiles have undergone many musculoskeletal adaptations during evolution. Crocodilians close their mouths with incredible crushing force due to the bony and muscular arrangement, and the refinement of cranial bones in snakes enables them to swallow prey much larger than their own heads.

4.16.2.3 Respiratory System

The primary respiratory organ in reptiles is the lungs. Most reptiles have paired lungs; many snakes have an elongated right lung and either no or a small vestigial left lung. Turtles and crocodilians have compartmentalized lungs. In turtles, movement of the head and limbs causes air to be forced in and out of the lungs. Most reptiles do not routinely vocalize.

4.16.2.4 Cardiovascular System

The reptiles except crocodilians have a three-chambered heart, consisting of paired atria and a single ventricle. Crocodilians have essentially a four-chambered heart, with the foramen of Panizza being the sole intraventricular connection. The functioning of this septum, along with paired aortas arising from opposite separate ventricles, results in shunting of blood to the cephalic and coronary circulation during anoxic events such as diving.

4.16.2.5 Gastrointestinal System

Crocodilians, snakes, and lizards have teeth while turtles have a horny beak to bite food. The esophagus of snakes is thin-walled and distensible to hold large prey. Crocodilian stomachs are round, muscular, and thick-walled, and often contain gastroliths to aid in digestion. Unlike that in mammals, the pancreas in snakes is discrete and compact. The synthesis of glucagon, and insulin secretion and development in the alligator pancreas is identical to that in mammals and birds. Some turtle species have pigmented cell aggregations in the spleen and liver.

The small intestine and colon tend to have larger volumes in herbivorous species of reptiles. In some snake species that eat infrequently, energy is conserved through atrophy of the small intestine and related organs between meals. In all reptiles, products of the digestive, excretory, and reproductive tracts empty into the cloaca.

4.16.2.6 Excretory System

Reptiles have paired kidneys. Aquatic species tend to excrete ammonia, semi-aquatic species excrete urea, and terrestrial species excrete primarily uric acid. A renal portal system that drains venous blood from the caudal half of the reptile's body directly through the kidneys has been reported. Except for snakes and crocodilians, many species of turtles and lizards have a urinary bladder.

4.16.2.7 Nervous System

The reptilian brain has cerebral hemispheres and 12 cranial nerves. The spinal cord extends to the tip of the tail and contains locomotor control centers, thereby allowing animals to respond at the spinal level. Most reptiles except snakes and some lizards have lidded eyes and nictitating membranes.

Many lizard species have a parietal or "third" eye, which contains photoreceptors that may permit enhanced detection of dawn and dusk. Other lizards have dermal photoreceptors that may function in regulation of behavior. Snakes lack external ears. Vibration is transmitted through the body to the quadrate bone, and then to the columella and inner ear.

4.16.3 Laboratory Management and Husbandry

4.16.3.1 Enclosures

Reptiles can be maintained in a variety of primary enclosures. Glass aquaria and plastic brands are the most commonly used type of housing. These cages are unbreakable, and with perforations for ventilation made in the sides, are stackable. Approaching many reptiles directly from the front evokes more defensive responses than reaching in quietly from above. Stainless steel cages can be adapted for larger species of reptiles.

Most reptiles are good at escaping from the enclosures or cages, thus they must be housed in cages with secure lids. Absorbent beddings such as hardwood shavings can be used for reptiles to burrow. Hardwood chips are sometimes used, but may inadvertently be ingested during feeding. Sand and soil should be heat treated or washed prior to use. Most aquatic species do not require a substrate in the aquarium or tank.

Arboreal reptiles should be provided with branches or dowels on which to climb. Species that bask should be provided with basking platforms. This is particularly important for many aquatic reptiles to allow normal drying and behavioral thermoregulation.

4.16.3.2 Water

When aquatic turtles are being housed, water should be at least as deep as the width of the shell. Turtles and crocodilians may generate large amounts of waste in the enclosures. Therefore, static systems with frequent complete water changes or flow-through systems are preferred over recirculating systems.

4.16.3.3 Temperature

Reptiles are cold-blooded animals and unable to maintain their body temperature at a consistent level. Warm temperatures are necessary for normal physiological processes such as digestion, growth, reproduction, and immune function. Heat pads or strips placed under part of the cage will result in a temperature gradient. Reptiles should never be allowed to directly contact with any heat source, or life-threatening thermal burns may result.

4.16.3.4 Lighting

Many reptile species need exposure to ultraviolet light in the appropriate UVB spectrum (290–320 nm) in order to endogenously produce vitamin D3. Several species of lizards and turtles will develop metabolic bone disease if deprived of ultraviolet light. Artificial ultraviolet light sources should be replaced approximately every 6 months. Most reptiles do well with a 12 h light/12 h dark cycle. Breeding animals should have their light cycles adjusted accordingly.

4.16.3.5 Handling

When handling any species of reptile, it is important to support the animal's body as much as possible. Reptiles should not be picked up or restrained by the tail. Many species of lizards have tail autotomy. Lizards should have both the pectoral and pelvic girdles supported, with the tail gently held to prevent slapping. Very aggressive individuals may have to be restrained behind the head to prevent defensive biting. Many lizards also have long, sharp claws that can scratch the handler. Tape can be wrapped around the animal's snout to prevent biting. Many turtles can be restrained by holding the sides of the shell. Many snakes are more comfortable if allowed to move about in the restrainer's hands. Snakes should never be held behind the head unless absolutely necessary. Grabbing a snake too tightly behind the head can damage tissues, restrict breathing, and elicit a much more panicked escape response.

BIBLIOGRAPHY

Festing MFW. Origins and characteristics of inbred strains of mice, 11th listing. *Mouse Genome* 1993;91:393–50.
Fox JG et al. *Laboratory Animal Medicine*, Second Edition. Elsevier Academic Press, San Diego, 2002.
Ji W. Reproduction and Breeding of Rhesus Monkey, Science Press, Beijing, China, 2013.
Hedrich HJ. *The Laboratory Mouse*. Elsevier Academic Press, Boston, MA, 2012.
Liu E. *Animal Models of Human Diseases*, Second Edition. People's Health Publishing House, Beijing, China, 2014.
Liu E, Yin H, Gu W. *Medical Laboratory Animals*. Science Press, Beijing, China, 2008.

Manning PJ, Ringler DH, Newcomer CE. *The Biology of the Laboratory Rabbits.* Academic Press Inc., San Diego, 1994.

National Research Council Laboratory Animal Management. *Rodents.* National Academy Press, Washington, 1996.

Sharp PE, LaRegina MC. *The Laboratory Rat.* CRC Press, New York, 1998.

Suckow MA, Stevens KA, Wilson RP. *The Laboratory Rabbit, Guinea Pig, Hamster, and Other Rodents.* Elsevier Academic Press, Boston, MA, 2012.

Suchow MA, Weisbroth SH, Franklin CL. *The Laboratory Rat.* Elsevier Academic Press, San Diego, 2006.

Van Zutphen LFM, Baumans V, Beynen AC. *Principles of Laboratory Animal Science,* Second Edition. Elsevier Science Publishers, Amsterdam, Netherlands, 2001.

Zhao S et al. Applications of transgenic rabbits in biomedical research—based on the literature search. *World Rabbit Science* 2010;18:159–67.

CHAPTER 5

Principles of Creating and Using Animal Models for Studying Human Diseases

Sihai Zhao and Qi Yu

CONTENTS

5.1 WHY WE NEED ANIMAL MODELS?

It has been found that it is difficult for biomedical studies with humans as the experimental subject to promote the development of biomedicine. There are space and time limitations for clinical experiments, and many human studies are restricted by ethical standards and methodology. However, these disadvantages in humans do not often exist in animal models that are widely used in biomedical experiments. These advantages also make animal models attractive to scientists and have played a unique role in biomedical research. The superiority of animal models is demonstrated by the following aspects.

5.1.1 To Avoid the Risks of Human Experimentation

As any test is accompanied by the potential of injury, biomedical experimentation should not be directly performed on humans from a humanitarian perspective. Research was conducted in the human body or on the researchers themselves in ancient China due to the underdevelopment of animal experimentation. For example, Shennong, the famous founder of Chinese traditional medicine, 5000 years ago, tasted different plants to identify their potential medicinal value. In modern society, some biomedical scientists were also reported to take their own new medicine to identify the effectiveness and detect potential adverse effects. Tu Youyou, a Chinese Nobel Laureate (2015), and her two colleagues were reported to have taken preclinical artemisinin to identify its safety. Barry J. Marshall, Australian Nobel Laureate, ingested the *Helicobacter pylori* culture liquid to identify the causal relationship between *H. pylori* and gastritis. Such practices should not be advocated in the scientific community. A large number of laboratory animals are used in modern biomedical research to mimic and reflect the situations of humans if they were in the same experiment. Thirty million mice were used in the United States in 1971, 7 million laboratory animals in Britain in 1975, and approximately 5 million laboratory animals every year in the 1990s in China. In some instances, it is impossible to perform similar experiments in humans.

In the 1930s, scientists found that the hypothalamus played a regulatory role in the endocrine system, but there was no corresponding hormone discovered in the following 40 years. It was not until the 1970s when two groups of scientists extracted only few milligrams of released hormone from the hypothalamus of more than 100,000 sheep and pigs, which made clear the regulatory mechanism of the hormone.

In addition, trauma, poisoning, and tumors are difficult or even impossible to study by only relying on clinical research; for example, in acute and chronic respiratory disease research the process of environmental pollution on the human body cannot be repeated, nor can radiation damage in the body be repeated on humans. As domesticated animals of humans, laboratory animals can be designed under the experimental conditions, and can be observed and studied repeatedly. Therefore, application of animal models not only overcomes the ethics and social restrictions of human research, but also allows some methodological approaches that cannot be applied to human beings. Animal tissues and organs can be treated and animal lives can be sacrificed for research purposes.

5.1.2 Rare Diseases Can Be Easily Studied in Animals

For radiation exposure, gas poisoning, and lethal infectious diseases, it is difficult to collect clinical patients, whereas these diseases can be successfully induced in laboratory animals for research purposes in a laboratory. Control of the animal living environment and easy observation of symptoms in the laboratory make animal experimentations more reflective of the experimental principles and purpose than human trials.

5.1.3 Chronic Diseases and Slow Progressive Diseases Can Be Easily Studied in Animals

It is difficult for researchers to observe diseases characterized by a long incubation period, long duration, or low incidence. Some disease incidences are very low in clinical practice, such as myasthenia gravis, which makes a problem for scientists to pool enough samples in their research. Similar situations also exist for chronic diseases, such as cancer, as most of them have a slow progression that may take several or more than 10 years. That is very long time for a research study. The animal model can overcome these flaws by shortening the process. It is easy to induce these kinds of disease in animals without wasting time, and they can be repeated for study purposes.

5.1.4 Strict Control of Experimental Conditions with Animal Use

In general, many factors play different roles in the onset and progression of disease, which makes it difficult to clarify the complex mechanism of some diseases. Even patients with exactly the same disease, because of the differences in age, gender, physical condition, and other factors, which all have influence on the development of diseases, may have different clinical features. However, when a study is planned, we often like to focus on the factor that we are interested in and minimize the effects of others. Today, such a situation dreamed by scientists exists in laboratory animal experimentation by standardization of the animals' genetics, microbiological status, living environment, and nutrition.

5.1.5 Convenient Experimental Manipulation and Sample Collection

By using animal models as an "epitome" of human diseases, all valuable samples, such as blood, tissues, and organs, are easily collected for experimentation, which is difficult to do in clinical practice. In addition, laboratory animals are more advantageous for miniaturization of the experiment to be more easily managed daily such as with rodent study.

5.1.6 Animal Models Contribute to More Comprehensive Understanding of Disease

Only relying on clinical research has some limitations. It is known that many pathogens can also cause several animal infections beside humans, and different species have their own characteristics. By the comparative study of zoonoses in humans and animals, effects in different species caused by the same pathogen can be fully understood. Another advantage of animal models is the ability to observe the influence of environmental or genetic factors on different stages of disease management, including its onset, development, therapeutic effects, and prognosis, which is important for the comprehensive understanding of the nature of the disease.

5.2 THE CLASSIFICATION OF ANIMAL MODELS

Much physiological and biochemical knowledge of the human body was derived from animal studies, in which the animals always appeared as the substitute of humans. Most animal models are established for the study of occurrence, development, and treatment of human diseases. According to the methods that the animal models are developed by, they can be classified as the following: induced animal models, spontaneous animal models, negative animal models, and orphan animal models. The induced and spontaneous animal models are the most important in scientific research. The rapid development in genetic engineering brings numerous genetically modified (GM) animal models for scientists to choose from for their research, and these models also play more and more important roles in modern studies. Some scientists suggest listing the GM models out of the induced animal models as another new group besides the above four.

5.2.1 Induced Animal Models

Induced animal models refer to animals in which their tissues, organs, or entire body progress to a certain artificial disease, resembling the human disease by exhibiting similar metabolic disorders or morphological and pathological changes that are induced by physical, chemical, or biological techniques. We think, GM animal models are included in induced animal models. For example, atherosclerosis can be induced in rabbits and apolipoprotein (apo) E deficient mice by a high cholesterol diet. Chemical carcinogenic agents, radiation, and cancer-causing viruses can be used to induce tumors in animals. Since the 1980s, with the development of GM technology, a huge number of GM animal models were created and play increasingly important roles in biomedical research.

Due to the convenience of induction and short replication time, commonly used induced animal models are suitable substitutes of humans for modern biomedical research purposes because the conditions when inducing disease can be easily and strictly controlled, especially in drug screening. However, there are some differences between induced animal models and naturally occurring disease models. Therefore, both the drawbacks and advantages should be considered when designing an induced animal model for human disease.

5.2.2 Spontaneous Animal Models

Spontaneous animal models refer to animals possessing a certain disease under natural conditions without any artificial treatment. These models mainly developed from animals with gene mutations, which include artificially selected mutant strains and inbred strains with some diseases. It is possible to obtain spontaneous animal models either from inbred strains (genetically uniform) or from random-bred populations (heterogeneous) where a high percentage of the animals are affected by the disease. Many spontaneous animal models play important roles in the study of human disease, such as the spontaneously hypertensive rat (SHR) is an animal model of essential hypertension, used to study cardiovascular disease. It is the most studied model of hypertension measured based on number of publications. The SHR strain was obtained during the 1960s by Okamoto and colleagues, who started breeding Wistar-Kyoto rats with high blood pressure.

As for spontaneous animal models, the similarity in the occurrence and development in natural conditions with the corresponding human disease give them high application values. Some differences still exist between induced disease and naturally occurring disease, and even some human diseases still cannot be induced in animals by current techniques; therefore, recent scientists pay more attention when selecting and breeding spontaneous disease animal models from available species and strains. Some large censuses were even performed in animals, such as mice, dogs, and cats, to find mutants and then breed them to keep the mutation to develop spontaneous disease models for study purposes. Numerous genetic disease models are established by this method. Many mouse and rat spontaneous models are developed and applied in nearly all biomedical fields. The use of these models is growing in genetic disease, metabolic disease, immunodeficiency disease, endocrine disease, and even cancer. Although spontaneous models develop rapidly, compared with induced models, the number and convenience of spontaneous models requires improvement.

In conclusion, both the induced and spontaneous models have their advantages but also face some shortcomings. Which models should be used mainly depends on the aims of the study, and only suitable animal models will bring relative valuable results and contribute to correct conclusions that may be helpful for the future final solution of human disease.

5.2.3 Negative Animal Models

Negative animal models refer to some animal strains in which a certain disease does not develop. This term may also be given to those animals that are insensitive to a certain stimulus that would usually have an effect on other species or strains. However, the underlying mechanism of insensitivity can be studied by using these kinds of animal models. For example, most mammals are susceptible to schistosomiasis, except the *Microtus fortis* living east of the Dongting Lake (located in Hunan, China), which are resistant to schistosomiasis, thus it may be used as a valuable tool to the study of the mechanism of anti-schistosomiasis.

5.2.4 Orphan Animal Models

Orphan animal models refer to models in which a certain disease is initially recognized and studied in an animal species, with the knowledge that a human counterpart could be identical at a later stage. Papilloma virus in malignant epithelial tumors and Marek's disease virus as a lymphoproliferative agent are two typical examples of orphan animal models.

5.3 THE SELECTION OF ANIMAL MODELS

The selection of an animal model depends on the aims of research to be performed, as well as on more practical aspects associated with the project, the research team, and experimental facilities. The selection is usually made after fully considering these aspects. Only researchers who fully understand the disease to be studied and know both advantages and disadvantages of the candidate animal models can make a suitable or wise choice of which models match the requirements of the project. As the structures and functions are very similar to humans, mammals have been widely used in biomedical research. The small size, short lifetime, easy manipulation, convenient breeding and housing of mice, rats, and guinea pigs have made them popular with scientists, and most universal animal models are mainly of these species. Choosing a good model when scrutinizing a biological phenomenon or major human disease is important. How to make a good choice requires taking full consideration of the following items.

5.3.1 Similarity between Animals and Humans

There are many spontaneous or induced disease animal models available for researchers that can partly or even fully reflect human diseases. The more the characteristics of the animal model resemble those in the corresponding human disease, the better the animal model in research. Whether the disease features of the model are similar with humans is the first item to take into account when choosing an animal model. In addition, the following also should be considered: If possible, choose animals whose biological characteristics and anatomic physiological characteristics are similar to humans. Generally, the higher degree of evolution of the animal, the more similar to human beings in function, metabolism, and the reaction to disease causing factors. That is why the current widely used animals are mammals in biomedical research. Age is also a factor that should be considered. The life span of animals and humans are very different, hence the study of diseases that have strict age requirements should use the animals in corresponding growth and development stages with humans. In surgery teaching or some surgery strategy studies, medium-sized or large animals, such as dogs and pigs, are generally used for operations as they are closer to humans than small animals.

5.3.2 Priority of Standardized Laboratory Animals

The key point of the study of human diseases by using animal models is how to make animal experiments more accurate and reliable to achieve precise results, and then draw a relative correct conclusion. Therefore, choosing standardized laboratory animals that are bred and maintained under strict genetic monitoring, microbiological, nutritional, and environmental hygiene control is important. Therefore, the influence from bacteria, viruses, parasites, and potential disease, as well as the unwanted effects from individual variation may be excluded.

Standardized laboratory animals have a clear genetic background, are specific pathogen free (SPF), and have stable properties. Laboratory animal quality certification is a sign of standardization as well as commercial product quality certification. All animal suppliers should show quality certification when their animals are provided to academic researchers. In general, hybrid animals bred by random mating and maintained in open conditions not free of bacteria, viruses, or parasites are not suitable in biomedical research.

More information on standardization of laboratory animals can be found in Chapter 3.

5.3.3 The Advantages of Inbred Strains

Inbred strains of animals are frequently used in laboratories for experiments where for reproducibility of conclusions for all the test animals should be as similar as possible. Mating of brother–sister pairs for a minimum of 20 generations results in inbred strains that are roughly 99% genetically identical. Many inbred strains have been inbred for many more generations and are in effect isogenic. For example, the inbred strain of C57BL/6 is the most widely used "genetic background" for GM mice used as models of human disease. They are the most widely used and best-selling mouse strain due to the availability of congenic strains, easy breeding, and robustness. Information on inbred mice can be found on the website (http://www.mgu.har.mrc.ac.uk/mutabase/; http://www.gsf.de/isg/groups/enu-mouse.html). Outbred animals are also often used in research. Outbreeding increases genetic diversity, thereby reducing the probability of an individual being subject to disease or reducing genetic abnormalities. High hetero-zygosity of outbred animals is similar to that of the human population, therefore, out-bred animals have an important value in the studies of human genetics, drug screening, and toxicology research. In biomedical research area, the vast majority of researchers selected inbred strains in the use of mice, and selected outbred if the use of rats.

Although the above rules show what should be considered when an animal is chosen, background research should be done before a decision is made based on scientific, practical, and ethical considerations. Full understanding of the human dis-ease that you are interested in and full understanding of the animal models avail-able can help to make a reasonable decision in selecting an animal model, bringing relative, correct results and conclusions after the experiment. There are also some cases in which suitable animal models are not readily available for research and the researchers may have to develop new models, which is very time-consuming work. In this case, the newly developed models have to be validated to demonstrate that they can indeed act as a model for a certain human disease.

5.4 CREATING ANIMAL MODELS

With the advance of science and technology, the human disease animal models play increasingly important roles in biomedical research. The methods to establish ani-mal models mainly include the following: developing spontaneous animal models by finding and maintaining mutations; inducing a certain disease by chemical, biological, or physical methods; and creating GM models through genetic engineering methods. As Krogh's principle states, "for such a large number of problems there will be some animal of choice, or a few such animals, on which it can be most conveniently studied."

Here, some examples are given to show the typical induced or developed process of animal models.*

5.4.1 Induced Animal Models

5.4.1.1 Atherosclerosis

The ideal animal model of human atherosclerosis should have several important features: It should be easy to acquire and maintain at a reasonable cost, be easy to

---

* Krogh A. The Progress of Physiology. *Am J Physiol.* 1929;90(2):243–251.

Table 5.1   Comparison of Lipoprotein Metabolism Characteristics between Mice, Rabbits, and Humans

	Mouse	Rabbit	Human
Lipoprotein profile	HDL-rich	LDL-rich	LDL-rich
CETP	No	Yes	Yes
Hepatic apoB editing	Yes	No	No
apoB 48	Chylomicron VLDL	Chylomicron	Chylomicron
Hepatic LDL receptor	Usually high	Downregulated	Downregulated
Dietary cholesterol	Resistant	Sensitive	Sensitive
Atherosclerosis	Resistant	Susceptible	Susceptible

Note: CETP—cholesteryl ester transfer protein; HDL—high-density lipoprotein; LDL—low-density lipoprotein; VLDL—very-low-density lipoprotein.

handle, and be of the proper size to allow for all anticipated experimental manipulations; The animal should reproduce in a laboratory setting and have clear genetic information; Finally, the animal model should share with humans the most important aspects of the disease process. The induced atherosclerotic lesions should develop slowly over the animal's lifetime with clinical sequel in later middle to old age. Compared with the most widely used transgenic model, the mouse, rabbits have different lipoprotein metabolism features that are similar to those of humans, as summarized in Table 5.1. Although there is no species that satisfies all these requirements, cholesterol-fed rabbits are the first developed and most generally used model for the study of atherosclerosis. A 0.2%–2% high cholesterol diet is usually used for 8–16 weeks to induce atherosclerosis in rabbits (Figure 5.1). In our laboratory, the

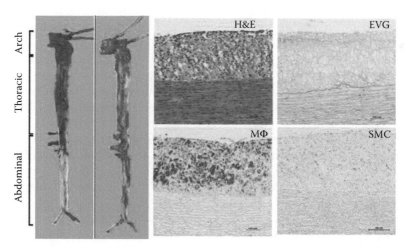

Figure 5.1   The aortic atherosclerotic lesions in rabbit after 16 weeks 0.3% high cholesterol diet feeding. Atherosclerotic lesions can be stained in red by Sudan IV. The aortic sections are stained with H&E, hematoxylin-eosin; EVG, Verhoeff–Van Gieson; or immunohistochemically stained with antibodies against macrophage (Mφ) and smooth muscle cell (SMC).

recommended concentration of cholesterol in the diet does not usually exceed 0.6%. In ApoE-deficient (ApoE$^{-/-}$) mice, an approximately 1.25% high cholesterol diet is used to accelerate the process of atherosclerosis. There are also other species, such as chickens and pigs, which are reported to have induced atherosclerosis by a high cholesterol diet.

5.4.1.2 Diabetes

Both type 1 and type 2 diabetes mellitus are multifactorial diseases in which a very complex genetic background interacts with environmental factors contributing to the disease development. Streptozotocin (STZ) or other diabetogenic agents (e.g., alloxan) with β-cell toxicity abilities have also been used for producing chemically induced type 1 diabetes models when administered in a large dose or in repeated low doses for several days (30–40 mg/kg body weight). STZ and alloxan accumulate in the pancreatic β-cells via the glucose transporter 2 and destroy them through reactive oxygen species and free radical mechanisms. More specifically, STZ is a nitrosourea-related antibiotic and antineoplastic drug, which is produced by *Streptomyces achromogenes*, and due to its alkylating properties, causes alkylation and thus fragmentation of DNA, modifies biological macromolecules and finally destroys β-cells, causing insulin-dependent-like diabetes.

5.4.1.3 Carcinogen-Induced Cancer Models

Decades of studies on the laboratory mouse have led to the knowledge that certain mouse strains are prone to developing certain cancers, and treatment strategies can be tested on mouse models. Some chemical agents may have broad-spectrum effects in mouse models. For example, cadmium has been shown to induce a range of cancers in animals. Arsenic has multiple impacts that can be observed in animal models. These may offer many leads on the general mechanisms of cellular alterations, as well as insights into tissue-specific features as well. Radiation-induced cancers offer another route to understanding the mechanisms of cancer. Ultraviolet radiation can be used to study skin cancer development in mice. Ionizing radiation is known to cause leukemic changes in mice. Intriguing mouse strain-specific features of radiation effects may lead to increased understanding of the process of cancer development. Physical agonists may play a role in some cancers. Asbestos fibers may elicit an effect on cells that may initiate cancerous alterations in multiple tissues, and can be studied in mice.

There are many ways to induce human disease animal models for the mechanistic or strategy research of a certain disease. Generally, researchers can choose different methods for different study purposes. The main rule for the scientist to follow is that they should choose the model that truly exhibits the features of the human disease that they study. Although there are many methods to induce a certain disease, even in same species, which one will be used mainly depends on which aspect of the disease will be focused on.

5.4.2 Spontaneous Animal Models

5.4.2.1 Nude Mice

Nude mice are derived from a strain with a genetic mutation that causes a deteriorated or absent thymus, resulting in an inhibited immune system due to a greatly reduced number of T cells. The phenotype, or main outward appearance of the mouse, is a lack of body hair, which gives it the "nude" nickname. The genetic basis of the nude mouse mutation is a disruption of the *Foxn1* gene. Nude mice were first discovered by Dr. Grist in 1962. Without a thymus, nude mice cannot generate mature T lymphocytes. Therefore, they are unable to mount many types of adaptive immune responses including antibody formation that requires CD4$^+$ helper T cells; cell-mediated immune responses, which require CD4$^+$ and/ or CD8$^+$ T cells; delayed-type hypersensitivity responses (require CD4$^+$ T cells); killing of virus-infected or malignant cells (requires CD8$^+$ cytotoxic T cells); graft rejection (requires both CD4$^+$ and CD8$^+$ T cells). Due to the above characteristics, nude mice have served in the laboratory to gain insights into the immune system, leukemia, solid tumors, and other forms of immune deficiency. Moreover, the absence of functioning T cells prevents nude mice from rejecting not only allografts, but they cannot even reject xenografts; that is, grafts of tissue from another species. These factors make nude mice a very suitable model for cancer research. Both cancer cell or cancer tissue grafts can grow on nude mice, which can then be used as models for cancer research. Most strains of nude mice are slightly "leaky" and do have a few T cells, especially as they age. For this reason, nude mice are less popular in research today as knockout mice with more complete defects in the immune system have been constructed (e.g., RAG1 and RAG2 knockout mice).

5.4.2.2 Spontaneous Type 2 Diabetes Obese Rodent Models

The ob/ob mouse, db/db mouse, and Zucker fa/fa rat are examples of type 2 diabetes models with a monogenic background. These diabetic models develop obesity due to mutations in the leptin gene (ob/ob) or leptin receptors (db/db and fa/fa), which may finally lead to the emergence of diabetes. The ob/ob (currently named as Lepob) genotype has been observed in the C57BL/6J mouse strain and this model is characterized by hyperphagia and low-energy expense, and thus becomes obese approximately at the age of 4 weeks. The ob/ob mouse is characterized by mild hyperglycemia due to compensatory hyperinsulinemia, which is observed at the age of 3–4 weeks together with hyperphagia, obesity, and insulin resistance. However, diabetes becomes very severe and lethal when the ob/ob genotype is expressed in the C57BL/KS strain. On the other hand, the db/db mouse also becomes hyperphagic, obese (approximately at the age of 4 weeks), hyperinsulinemic (approximately at the age of 2 weeks), and insulin resistant, but later (4–8 weeks) develops hyperglycemia, due to β-cell failure and does not live longer than 8–10 months. The Zucker (fa/fa) fatty (obese) rat develops the same pathophysiological characteristics with the db/db mouse and is mainly used as a model of human obesity accompanied with hyperlipidemia and hypertension. However, selective inbreeding of fa/fa rat for hyperglycemia gave birth to the Zucker diabetic fatty (ZDF) rat strain, which develops severe diabetes (only in males) at about 8 weeks after birth, due to enhanced apoptosis of β-cells, which are not able to compensate for the insulin resistance, as in the fa/fa rat, and becomes insulinopenic at approximately 14 weeks of age.

5.4.2.3 *Watanabe Heritable Hyperlipidemic Rabbit*

Dr. Yoshio Watanabe discovered a male Japanese white rabbit in 1973 that showed hyperlipidemia despite feeding on a normal standard diet and named it the Watanabe heritable hyperlipidemic (WHHL) rabbit. Similar to human familial hypercholesterolemia, the low-density lipoprotein (LDL) receptor's function is genetically reduced in WHHL rabbits and they exhibit hypercholesterolemia. This defect arises from an in-frame deletion of 12 nucleotides that eliminates four amino acids from the cysteine-rich ligand-binding domains of the LDL receptor. From this mutant gene, although the precursor of LDL receptor proteins are synthesized normally, the maturation of the LDL receptor protein is delayed and is not transported to the cell surface at a normal rate. In WHHL rabbits, the average plasma cholesterol levels are approximately 1100 mg/dL when they are below 6 months old, approximately 900 mg/dL at 12 months old, and approximately 800 mg/dL at 18 months old. The average plasma triglyceride levels are between 150 and 300 mg/dL. Approximately 70% of the cholesterol is distributed in LDL fractions, and only a few percent in high-density lipoprotein (HDL) fractions. In WHHL rabbits, aortic atherosclerosis is observed grossly from 2 months old despite feeding them a normal standard rabbit chow. The lesions develop first at the orifices of the branches. At 6 months of age, aortic lesions expand to approximately 40% of the aortic surface. At 12 months of age, the lesions cover approximately 70% of the aortic surface. Above 18 months of age, the aortic lesions cover most of the aorta. In lipid metabolism related disease research, such as atherosclerosis, WHHL rabbits play an important role. WHHL rabbits have contributed to the studies of cholesterol metabolism and atherosclerosis *in vivo*. For example, WHHL rabbits contributed to the Nobel Prize study by Goldstein JL and Brown MS, who were allocated WHHL rabbits in 1980, which they used to verify their hypothesis of the LDL receptor pathway.

There are enormous amounts of induced or spontaneous animal models that are widely used in biomedical research. Here, we just listed few of them to show their application as examples. We hope these examples can bring some ideas to the readers when facing the problem of how to choose the best model for their study or test.

5.5 GM ANIMAL MODELS

A GM animal is any animal whose genetic material has been altered using genetic engineering techniques. Genetic modification involves the mutation, insertion, or deletion of genes. GM animals currently being developed can be placed into several broad classes based on the intended purpose of the genetic modification: to develop animal models for these diseases; to produce industrial or consumer products; to produce products intended for human therapeutic use; to enhance production or food quality traits; and so on.

The following is a brief introduction of several GM animal production methods, mainly transgenic and knockout animals.

5.5.1 Transgenic Animals

The term "transgenic animals" refers to a class of animals that have an exogenous gene integrated into their genomes. The exogenous gene that is integrated into the animal genome is known as a transgene, which can be inherited by the offspring. The emergence of transgenic animals has shocked the world. It is considered a revolutionary milestone in genetic research and marks the beginning of a new era where humans have gained the capability to alter and modify the genetic information in animal genomes at the whole animal level.

5.5.1.1 Introduction

In 1974, Jaenisch and Mintz successfully microinjected, for the first time, simian virus 40 (SV40) DNA into the blastocoel cavity of mouse embryos using a microinjection technique and detected SV40 DNA in the offspring. During the 1980s, scientists began to conduct studies via pronuclear injection of zygotes. In 1980, Gordon et al. successfully generated transgenic mice using the pronuclear microinjection method. In 1982, Palmiter et al. created the ground-breaking "super mouse" by injecting a rat growth hormone gene into the pronuclei of mouse zygotes. The purpose of establishing transgenic animal technology is not simply to prove that humans have the ability to conduct genetic manipulation, modify genetic materials, or alter cellular behaviors *in vitro* to utilize the capability of "transforming individuals" at the *in vivo* level. Rather, the goals of developing transgenic animal technology include theoretical and applied perspectives. From the theoretical perspective it includes the study of gene functions and gene effects on the organisms; the study of gene expression and gene regulatory patterns; and the exploration of the pathways and mechanisms of gene transfer. From the perspective of application, studies focusing on "transformation of individuals" will generate new strains that possess advantages, such as high yield, high quality, low cost, and stress resistance; and studies focusing on "utilization of individuals" employ transgenic animal models to conduct biomedical, pharmaceutical, and veterinary research. For decades, transgenic animal technology has rapidly developed in the abovementioned directions and has obtained significant achievements.

5.5.1.2 Transgenic Animal Technology

Transgenic animal technology is an advanced gene expression approach. The process of producing transgenic animals can be divided into the following three steps: preparation of constructs carrying the gene of interest (GOI) (upstream), gene transfer (midstream), and validation and construction of animal models (downstream).

The upstream step focuses on genetic modification and vector construction. The exogenous gene construct consists of the flanking sequence containing gene regulatory elements, expressible structural gene sequences, and transcription termination signals. To facilitate detection, reporter genes or reporter sequences are also introduced into the exogenous gene construct. In addition, the native promoter of the GOI is deleted. A strong promoter that may even contain an enhancer sequence is linked to the GOI to form a fusion gene, resulting in high GOI expression.

The midstream step includes gene transfer and embryo transfer. To introduce the GOI, physical, chemical, and biological approaches are employed to deliver the successfully constructed vector system carrying the exogenous gene into the cells. The recipient cell and embryo transfer are critical for generating transgenic animals, which determine the selection at the cellular level and the transfer of exogenous genes. The recipient cells (typically early embryonic cells) are transplanted into the oviducts or uterus of the recipient animals, which will eventually develop into a new animal.

The downstream step refers to the detection of gene integration and expression. Gene integration and expression in transgenic animals are examined at the chromosome, gene, transcription, and protein levels.

- **At the DNA level:** Only a very small portion of the exogenously introduced DNA is integrated into the host genome, which can be detected by Southern blot, *in situ* hybridization, and polymerase chain reaction (PCR).
- **At the RNA level:** Hybridization-based Northern blot analysis is a commonly used method. However, the application of Northern blots is limited. Northern blots cannot be employed to analyze the expression of an exogenous gene if the gene is expressed at an extremely low level or an endogenous homologous gene is expressed. Reverse transcription polymerase chain reaction (RT-PCR) is a highly sensitive, specific method for detecting the expression of transgenes.
- **At the protein level:** Western blot analysis is an effective method. However, the presence of endogenous homologous gene products may interfere with the results of Western blots. Therefore, antibody specificity is extremely important.

The upstream and downstream steps involve a variety of molecular biology techniques, which will not be further described in detail (please refer to the relevant literature). The present chapter introduces specific techniques and approaches employed to generate transgenic animals. Based on the concepts and systems of transgenic animal technology, it can be concluded that gene transfer into reproductive cells or early embryonic cells is the core technology for producing transgenic animals. The gene transfer techniques are described below. The main methods for the construction of transgenic animals include microinjection, retroviral infection, the embryonic stem (ES) cell method, and the sperm carrier method. Zygote microinjection is one of the main methods of creating transgenic animals. In the present chapter, we use the microinjection-based production of transgenic mice as an example to introduce the methods of creating transgenic animals. We will also present several other transgenic techniques and describe the applications of transgenic animals.

5.5.1.3 Creation of Transgenic Animals by Microinjection

Among the transgenic techniques, microinjection was developed earliest and is currently the most widely used and effective method. The advantages of microinjection include the following: an acceptable rate of gene transfer; ability to directly transfer exogenous genes without involving prokaryotic vector DNA fragments; unrestricted length of the exogenous genes (up to 100 kb); capability of generating inbred strain animals regularly; and a relatively short experimental period.

However, certain shortcomings have limited the application of the microinjection technique. Microinjection requires expensive and sophisticated equipment, a complicated operational procedure and specialized technical personnel; When an exogenous gene is introduced by microinjection, the copy number of the exogenous gene cannot be controlled, and the exogenous gene is often present in multiple copies and may occasionally reach several hundred copies; Microinjection often leads to host genetic mutations, such as large DNA fragment deletion and genetic recombination near the site of exogenous gene insertion, which may cause severe physiological defects in the animals.

Nevertheless, as microinjection allows for the direct manipulation of genes and renders a high rate of gene integration, it remains an extremely important method for generating transgenic animals. The technical route of microinjection is illustrated in Figure 5.2.

Microinjection is a technique that utilizes micromanipulators to inject exogenous genes into the zygotes of recipient animals. The exogenous gene may integrate into the chromosome of recipient cells, resulting in the development of transgenic animals. The microinjection-based production of transgenic animals involves multiple steps. First, a GOI construct is prepared. Based on experimental requirements, DNA fragments of the GOI suitable for microinjection are obtained using a variety of genetic engineering techniques. The subsequent experimental procedures are then performed. Starting from hormone injection to the birth of transgenic mice, the whole experimentation lasts approximately 1 month (Figure 5.3).

- **Estrus synchronization and superovulation:** On the first day of the experimentation, female donor mice are injected with pregnant mare serum gonadotropin (PMSG), which induces a synchronized estrus in the female donor mice. After an interval of 46–48 h (i.e., on the third day of the experimentation), the donor mice are injected with human chorionic gonadotropin (hCG) to induce superovulation. In the afternoon of the day of hCG injection, the donor mice and male mice are placed in the same cage for mating.
- **Preparation of the recipient mice:** To prepare recipient mice (pseudopregnancy) for embryo transfer, female mice in estrus are selected and caged with spermaduct ligated male mice in the afternoon of the third day of the experimentation. Usually, male mice are vasoligated in advance prior to the experimentation. Only ligated male mice that have failed at least twice to impregnate female mice in the mating test are used for the preparation of recipient mice.

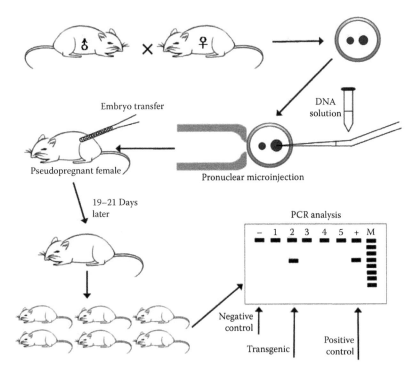

Figure 5.2 Flowchart illustrating the production of transgenic mice by the microinjection method.

- **Collection of zygotes:** In the morning of the fourth day of experimentation, donor and recipient mice are examined for the presence of vaginal plugs. Mice that have visible vaginal plugs are considered pregnant. The female donor mice are sacrificed. After exposure of the abdominal cavity, oviducts and a small portion of the uterus are collected. The ampulla, which is the swollen segment of the oviduct, is placed under a microscope and carefully sliced open. Clusters of mouse eggs are automatically released. The foam cells surrounding the egg cells are digested with hyaluronidase. Morphologically normal zygotes are collected and cultured in modified M_{16} medium at $37°C$ and 5% CO_2 until microinjection.

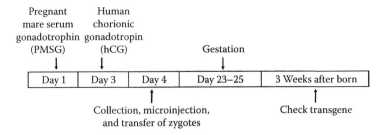

Figure 5.3 The timeline for generating transgenic mice.

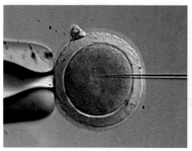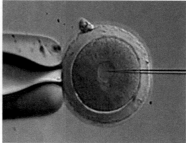

Figure 5.4 Microinjection of mouse zygote.

- **Microinjection:** Injection of an exogenous gene into the zygotes of female mice is performed under an inverted differential interference contrast microscope at 200-fold magnification using a robotic arm. A zygote is held with a holding pipette. The exogenous DNA solution is sucked into a microinjection syringe and injected into the male pronucleus of the zygote (Figure 5.4). After injection, zygotes are cultured briefly in modified M_{16} medium at 37°C and 5% CO_2. Zygotes that maintain morphological integrity are selected for transplantation.
- **Embryo transfer:** Pseudopregnant recipient mice are anesthetized. A small incision is made in the back of a recipient mouse, through which the ovaries and oviducts are located. The oviduct is exteriorized on one side through the small cut. The fat pad surrounding the oviduct is clamped with extra-fine vascular clamps to hold the exteriorized oviduct in place. Approximately 15–20 microinjected zygotes are loaded into a transfer pipette in the following order: mineral oil-air bubble-M_2 medium-air bubble-zygote-air bubble-M_2 medium (Figure 5.5). Under a dissecting microscope, the zygotes are transplanted into the oviduct of the recipient mouse. Subsequently, the oviduct is carefully pushed back into the abdominal cavity, and the incision is sutured. The same procedure is conducted to transplant zygotes into the oviduct on the other side.
- **Identification of transgenic mice:** Two to three weeks after birth, tail tissues are collected from the neonatal mice. Genomic DNA is extracted and dissolved in Tris-EDTA (TE) buffer. PCR or Southern blot analysis is conducted to determine whether the exogenous gene is integrated into the genome of the neonatal mice.

By virtue of the single-celled zygotes, the microinjection technique achieves gene transfer at cell level. The zygotes are allowed to develop under appropriate conditions. After appropriate selection and mating, homozygous transgenic animals

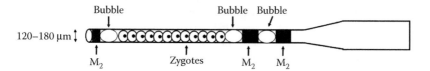

Figure 5.5 Loading of zygotes into a transfer pipette for transfer into an oviduct.

are generated. This technology is now well established. People have accumulated an enormous amount of experience in zygote isolation and culture, microinjection (including improvement of instruments), embryo transfer, and *in vivo* development, and have achieved great success. Therefore, microinjection remains the most widely used effective method in transgenic animal research.

5.5.1.4 Improved Techniques Based on Microinjection

To overcome the shortcomings of the classic zygote microinjection technique, such as the low rate of gene integration and the inability to achieve site-specific integration, a number of improved techniques have been developed in the past 30 years, and significant breakthroughs have been made.

- Utilizing a replication-defective retrovirus as a vector
  This method takes advantage of the fact that the long terminal repeat (LTR) region of retroviral DNA possesses promoter activity. Exogenous genes are inserted downstream of the LTR to construct recombinant retroviral vectors, which are packaged into high-titer viral particles. The particles are used to directly infect the zygotes or are microinjected into the blastocoel cavity. The retroviral DNA carrying the exogenous gene can be integrated into the host chromosome.
  Transfection of early embryos with retroviral vectors often results in the production of chimeras. Eggs remain in the metaphase II (MII) phase for a long period of time, during which the nuclear envelope disintegrates. The breakdown of the nuclear envelope facilitates gene integration. Based on the above characteristics, retroviral vectors carrying exogenous genes have been injected into the zona pellucida of MII phase eggs. Subsequently, *in vitro* fertilization is conducted. The resulting blastulas are randomly selected for embryo transfer. Such approaches have achieved great success. The improved approach displays several advantages: the method greatly improves the rate of gene integration, retroviral vectors are integrated randomly at multiple sites and the overall integration rate exceeds 30%; it is of moderate technical difficulty and low cost. However, the method has its own limitations: retroviral vectors have limited capacity for exogenous genes, and such vectors only accommodate exogenous genes less than 10 kb; multisite integration results in genetic differences among the offspring; and genes in the retroviral vectors may affect expression of the exogenous gene.
- Utilizing sperm as a vector
  Sperm that possess fertilizing capability are incubated with exogenous DNA and used for *in vitro* fertilization. Subsequently, embryo transfer is conducted, allowing the expression of exogenous genes. In 1989, a researcher reported a successful transgenic study using live sperm as a vector. However, there was controversy surrounding the report. In 1999, Perry et al. modified and improved the method based on the fact that the injection of sperm head or dead sperm (damaged membrane) into the cytoplasm of egg cells successfully produces offspring. In the study conducted by Perry et al., mouse sperm were first treated with detergent and then subjected to freezing-and-thawing to compromise the sperm membrane. Subsequently, the sperm were incubated with exogenous genes such that exogenous genes were in direct contact with the sperm DNA or sperm surface. After incubation, the sperm were microinjected into MII phase mouse eggs, which prevented the degradation of the exogenous genes in the cytoplasm. Finally, the blastocysts were transferred

into the uterus of pseudopregnant female mice, thereby producing transgenic mice. Sperm-mediated gene transfer has many advantages. The needle used in the intra-cytoplasmic sperm injection (ICSI) of eggs has a significantly larger diameter (approximately 100-fold) compared to that required for the intranuclear injection of eggs. This method is suitable to manipulate large-sized exogenous genetic material (such as artificial chromosomes). *In vitro* construction of recombinant retroviruses is not required. In addition, the method is simple to perform and renders a high integration rate. Therefore, sperm-mediated gene transfer is likely to become popular.

- Somatic cell nuclear transfer

  In 1997, the English biotechnology company PPL Therapeutics and the Roslin Institute jointly created the first clone sheep "Dolly" using somatic cell nuclear transfer (SCNT) technology. The researchers then cotransduced sheep fetal fibroblasts with human coagulation factor IX and neomycin resistance genes. The fibroblasts were first selected with G418 and then examined using DNA hybridization to identify cells that had integrated both genes in the chromosome. Using sheep fetal fibroblasts carrying the two genes as a nuclear donor, researchers obtained three transgenic sheep (another one died shortly after birth). The transgenic sheep were named "Polly," "Molly," "Holly," and "Olly," which were as world famous as "Dolly." Cibelli et al. created three calves carrying the exogenous marker gene LacZ via SCNT technology. The nuclear donor utilized by Cibelli et al. was also fetal fibroblasts. In transgenic animals created using the SCNT method, the exogenous gene-integration rate reaches 100%. However, the SCNT method has a number of disadvantages such as high technical difficulty, low success rate, and unsatisfactory fetal survival rate.

In addition to the methods described above, a number of other methods have been explored, including embryo stem (ES) cell-mediated gene transfer, teratoma cell-mediated gene transfer, receptor-mediated gene transfer, high-efficiency micro-projectile bombardment, the needle-pricking method, laser-mediated gene transfer, protoplast-mediated gene transfer, and calcium phosphate co-precipitation.

5.5.2 Gene Targeting

Gene targeting is a technique that employs site-specific homologous DNA recombination to alter the structure of a particular gene. Gene targeting allows the *in vivo* study of gene functions. Animals produced using gene targeting and ES cell technology that have a target gene inactivated or deleted from a specific genomic locus are known as gene knockout animals.

ES cell gene targeting is an efficient technology to study gene function. From the 1980s to the early 1990s, gene targeting technology in the context of mouse ES cells was greatly developed and reached a mature stage. Mouse ES cells after gene targeting can be transferred by microinjection into the blastocoel cavity, which is then transplanted into pseudopregnant female mice to create germline chimeric mice. After appropriate mating, an inbred strain of mice derived from ES cells is obtained. Currently, this technology is widely applied in several fields of biomedicine.

The prerequisites for gene targeting include the following:

5.5.2.1 *ES Cells*

ES cells used in gene targeting are derived from the inner cell mass of early mouse embryos, namely the blastocyst on the fourth or fifth day of mouse zygotic development. ES cells display unique characteristics. These cells can be cultured *in vitro* while retaining developmental totipotency. The morphological character-istics of ES cells during *in vitro* adherent growth include large nucleus, little cyto-plasm, compact cell arrangement, and growth in colonies. When ES cells are in a poorly differentiated state, many functional genes are not expressed. Such ES cells only express a number of genes that are involved in maintaining cell prolifera-tion and controlling cell differentiation. However, *in vitro* cultured ES cells display a tendency to differentiate toward multiple different cell types in the process of proliferation.

The key issue facing the *in vitro* culture of ES cells is maintaining cell prolifera-tion and a normal karyotype while inhibiting cell differentiation. When ES cells are re-transplanted into mouse embryos after *in vitro* genetic manipulation, the cells will develop into different embryonic tissues, eventually leading to the generation of chimeric mice. If the genetically manipulated ES cells develop into mouse germ cells, knockout or knock-in mice can be generated through mating.

5.5.2.2 *Targeting Vectors*

Targeting vectors contain two types of selectable markers: a positive selectable marker (the neomycin resistance gene, *neo*) and a negative selectable marker (herpes simplex virus thymidine kinase gene, *HSV-tk*). Cells undergoing homologous recom-bination can be screened by virtue of the two types of selectable markers (Figure 5.6).

The *neo* positive selectable marker: the *neo* gene is inserted into the exogenous DNA intended for targeting. When homologous recombination occurs between the exogenous DNA and its homologous sequence on the chromosome of the cell, the *neo* gene is also inserted into the chromosome. Therefore, the ES cells that have undergone homologous recombination are able to grow in culture medium contain-ing G418 (geneticin).

The *HSV-tk* negative selectable marker: the *HSV-tk* gene is a thymidine kinase (TK) gene derived from herpes simplex virus. The product of the *HSV-tk* gene can degrade mononucleotide analogs to produce toxic metabolites. The *HSV-tk* gene is inserted into the vector sequence outside of the exogenous gene. When homologous recombination occurs between the exogenous DNA and its homologous sequence on the chromosome of the ES cell, the vector sequence (containing *HSV-tk* gene) will not be integrated into the chromosome. In the event that the ES cells are able to grow in medium containing the mononucleotide analog, the vector DNA is also inserted into the chromosome through recombination. On the contrary, it can be concluded that the vector sequence is not inserted into the genome if the mononucleotide ana-log-containing medium exerts a cytotoxic effect on the ES cells.

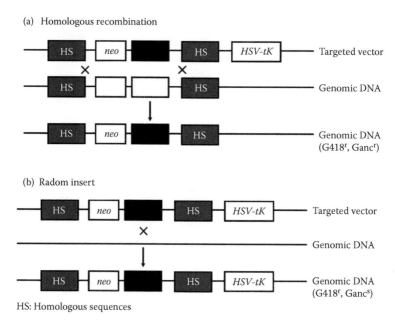

(a) Homologous recombination

(b) Radom insert

HS: Homologous sequences

Figure 5.6 Schematic diagram of homologous recombination: (a) homologous recombination and (b) random insert.

5.5.2.3 *The Basic Procedure of Gene Knockout*

Due to homologous recombination between DNA molecules, a specific endogenous gene of ES cells is destroyed, resulting in loss of function of the specific gene. Subsequently, a mouse model that has lost a specific gene is obtained through ES cell transfer (Figure 5.7).

The basic gene knockout procedure includes construction of targeting vectors, *in vitro* culture of ES cells, transfection of ES cells with recombinant vectors, identification of ES cells successfully transfected with the recombinant vectors, transfer of embryos carrying the transfected ES cells, and crossbreeding of the chimeras.

- **Construction of targeting vectors:** The frequency of homologous recombination between DNA molecules is very low (10^3–10^7). When designing gene targeting strategies, enhancement of homologous recombination frequency and introduction of a selectable system are the key factors for successful experimentations. The construction of vectors using DNA fragments of the homologous genes may enhance the frequency of homologous recombination by 20-fold. As the length of homology arms increases, the recombination frequency increases accordingly. In general, high recombination efficiency can be achieved if the length of the homologous sequence in each homology arm exceeds 250 bp. Therefore, the first step toward constructing a vector is to obtain gene fragments from the same strain as the ES cells. The gene fragments are inserted into the vectors as the homologous fragments. In addition, selectable marker genes are also inserted into the vectors (Figure 5.6).

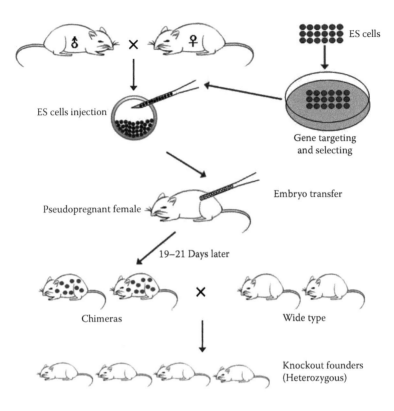

Figure 5.7   Schematic diagram of a gene knockout procedure.

- **The basic process of constructing a targeting vector is as follows:** A DNA fragment homologous to the target gene (the gene to be knocked out) is obtained and cloned into a regular plasmid vector; a large portion of the DNA sequence homologous to the target gene is cleaved from the recombinant plasmid, leaving part of the homologous sequence at both ends of the linearized plasmid vector; the *neo* gene is cloned into the linear plasmid containing the sequence homologous to the target gene such that the *neo* gene is located in the middle of the residual homologous sequence; and the recombinant plasmid vector is linearized outside of the sequence homologous to the target gene. Subsequently, the *HSV-tk* gene is cloned into the linear vector. The residual fragment homologous to the target gene, the *neo* gene located inside of the homologous fragment and the *HSV-tk* gene located outside of the homologous fragment constitute the targeting vector.
- **Introduction of the targeting vector into ES cells:** The targeting vector is introduced into ES cells. Replacement recombination occurs between the DNA sequence homologous to the target gene that is located in the targeting vector and the gene to be knocked out that is located on the chromosome of ES cells. The target gene in the ES cell genome is replaced by the *neo* gene in the vector. As a result, ES cells without the target gene (gene knockout cells) are developed.

- **Injection of the gene knockout ES cells into blastocysts:** The knockout ES cells are injected into blastocysts. The injected ES cells and the original cells in the blastocysts constitute the inner cell mass of the blastocysts.

- **Implantation of the blastocysts into the uterus of pseudopregnant mice:** The blastocysts containing the gene knockout ES cells are transplanted into the uterine cavity of pseudopregnant mice such that the ES cells have the opportunity to develop into mice or certain types of tissues. Such blastocysts contain both the knockout ES cells and the normal ES cells originally present in the blastocysts. Therefore, the offspring developed from such blastocysts include the mice originating from the knockout ES cells and the mice originating from the normal embryo cells.

- **Crossbreeding of chimeras:** The offspring mice are screened to obtain gene knockout chimeric mice. It is generally believed that homologous gene recombination only occurs on one chromosome. Once a gene located on a chromosome is replaced by a homologous sequence, its allele located on the other chromosome will no longer be replaced. Therefore, homologous recombination only gives rise to chimeras. Subsequently, crossbreeding is performed. According to Mendel's principles of inheritance, the offspring has a 25% chance of being homozygous. After mating the chimeric mice with normal mice, inbred strains of mice that have specific target genes knocked out, that is, knockout mice, are generated.

5.5.3 Tissue-Specific Gene Knockout

Traditional gene targeting strategies all result in complete gene knockouts, regardless of whether the replacement or insertion strategy is used. The introduced exogenous selectable gene fragments may cause irreversible interference in the genome, thus affecting the function of genomic DNA. If a target gene is essential for embryonic development, gene knockouts at an early embryonic stage will lead to death of the embryo, rendering it impossible to conduct in-depth studies on the target gene. To resolve the above issues, conditional gene knockout techniques, such as the Cre–loxP and flippase (FLP)-flippase recognition target (FRT) systems, have been developed. As a result, tissue-specific gene knockouts become feasible, and time- and tissue-specific gene targeting is ultimately achieved. This type of targeting technique limits the inactivation of target genes to a specific time or within a specific type of tissue/cell, and thus is of great scientific value.

The Cre–loxP system was first proposed by Gu et al. in 1993. The system consists of two components: Cre recombinase and loxP sites. Cre recombinase is a site-specific recombinase encoded by the *Cre* gene of the *Escherichia coli* bacteriophage P1. LoxP is composed of two 13-bp inverted repeats separated by an 8-bp spacer region. Cre recombinase mediates recombination between the 34-bp repeat units, allowing excision of the DNA fragment flanked by two directly repeated loxP sites and one of the loxP sites while maintaining the other loxP site. The Cre–loxP system can be operated in two ways. (1) When constructing a targeting vector, a marker gene is inserted inside the target gene. The loxP sequences are placed on both sides of the marker gene and are oriented in the same direction. Subsequently, at the cell level, the Cre recombinase expression plasmid is transfected into the target cells. Cre recombinase is expressed, which excises the resistance marker gene through recognition of the loxP sites. (2) At the individual level, heterozygous-targeted mice are crossed with Cre transgenic mice. Conditional knockout mice, in which the exogenous marker gene has been deleted, can be obtained through screening the offspring.

As shown in Figure 5.8, loxP sites are introduced into the targeting vector on both sides of the target gene and the marker gene, which allows homologous recombination between the vector and the genome of ES cells. ES cells that have undergone homologous recombination are obtained. Subsequently, these ES cells are transiently transfected with the *Cre* gene, allowing the transient expression of Cre recombinase. Under the action of Cre recombinase, recombination may occur in the following three ways. (1) Recombination occurs between loxP1 and loxP2. The marker gene becomes lost, whereas the target gene and the loxP sites flanking the target gene on both sides are preserved, which is the desired result. (2) Recombination occurs between loxP2 and loxP3. As a result, the marker gene is preserved. As the product of the *HSV-tk* gene is able to degrade mononucleotide analogs to produce toxic metabolites, the addition of mononucleotide analogs to the culture medium will eliminate the cells undergoing this type of recombination. (3) Recombination occurs between loxP1 and loxP3, which results in the removal of both the marker gene and the target gene. The ES cells that have undergone the first type of recombination are screened out and used to construct the gene-targeted mice. To obtain tissue-specific knockout mice, homozygous gene-targeted mice are then crossed with preprepared transgenic mice that only express Cre recombinase in specific tissues.

To date, nearly thousands of Cre transgenic mouse strains have been reported or are being investigated. Vectors have been constructed by linking Cre to a variety of specific promoters (tissue-, site-, time-, or developmental stage-specific). Subsequently, traditional transgenic techniques or gene knockout techniques are employed to produce the corresponding transgenic mice. Once the gene encoding Cre recombinase is placed under the control of an inducible promoter, any gene located between the loxP sites can be excised through the induction of Cre recombinase expression. Thus, inactivation of a specific gene at a specific time can be achieved. However, at present, only a limited number of transgenic mouse strains are available that allow the tissue-specific expression of Cre recombinase. The application of the Cre–loxP system will also depend on the discovery of novel tissue-specific marker genes and further study of the artificially regulated gene expression system.

The yeast-derived FLP-frt system functions in the same way as the Cre–loxP system. Yeast expresses an enzyme called flippase (FLP). The mode of action of FLP is similar to that of Cre. However, the recognition site for FLP is "*frt*," and FLP can excise a DNA fragment flanked by *frt* sites. The FLP-frt system can also be used to achieve tissue-specific gene knockouts.

5.5.4 Emerging New Gene Editing Technologies

Traditional gene targeting technologies rely on homologous recombination induced by random double-strand breaks (DSB), and the probability of DSB formation

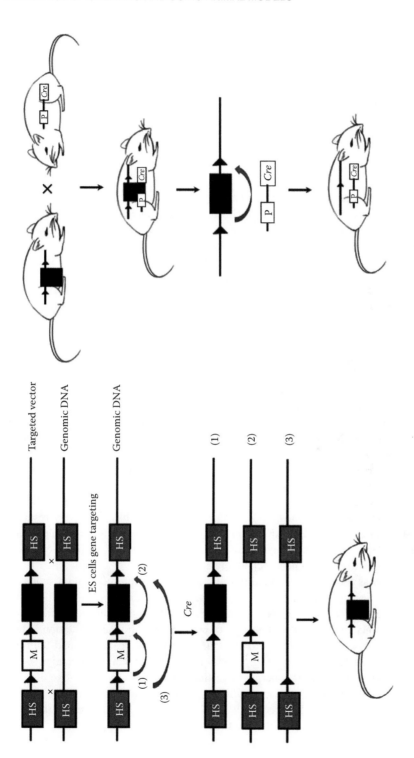

Figure 5.8 Schematic diagram of tissue-specific gene knockout.

largely determines the efficiency of homologous recombination. As molecular biology techniques advance, scientists have begun to seek methods capable of actively creating DSBs in an attempt to achieve more precise alterations and modifications of animal genomes. Recently, several nuclease-based techniques have emerged. These techniques have significantly improved the low efficiency of traditional homologous recombination. Therefore, these techniques are important for the development of future genetic engineering tools.

5.5.4.1 Zinc Finger Nucleases

Each zinc finger nuclease (ZFN) can be divided into two parts. The first part is the DNA recognition domain, which comprises a tandem array of Cys2-His2 zinc finger proteins (ZFPs). Each ZFP recognizes and binds to a specific 3-bp DNA sequence. The other part is the catalytic domain, which consists of the nonspecific endonuclease FokI. The DNA recognition domain of ZFN recognizes a specific DNA sequence, thereby directing the catalytic domain to the target site. The endonuclease activity of ZFN then cleaves the target DNA, which gives rise to DSBs and induces DNA damage repair. Cells may repair DNA damage through the mechanism of nonhomologous end-joining (NHEJ). NHEJ-mediated DNA repair lacks accuracy and is extremely error-prone (insertion/deletion). As NHEJ causes frame shift mutations; it can be employed to achieve gene knockout animal.

As described above, the ZFN technique may achieve gene mutations through NHEJ. In addition, because the appearance of large numbers of DSBs greatly enhances the efficiency of homologous recombination, zinc finger nucleases (ZFNs) can be used to improve the efficiency of gene knockin and gene repairs. Simultaneous introduction of ZFNs and exogenous DNA into cells induces homologous recombination-based site-specific repair through the production of DSBs, resulting in the formation of point mutations capable of causing certain diseases such as sickle cell anemia.

The lack of ES cells in large animals results in a low efficiency of gene targeting. The emergence and application of the ZFN technique has effectively resolved this issue. For example, ZFN expression vectors have been introduced into pig fetal fibroblasts. Upon screening, clones positive for the mutation are obtained. SCNT is then performed to create gene knockout pigs.

However, as the ZFN technique initially adopted a 9-bp recognition sequence and there are issues of patent protection surrounding the ZFN technique, the selection of target sequences for ZFN is greatly restricted. Significant differences exist between distinct ZFNs in the efficiencies of target recognition and mutation induction. Moreover, technological monopoly caused by patent protection and the relative high cost of ZFN design and synthesis have become major challenges faced by researchers. The ensuing emergence of the transcription activator-like effector nuclease (TALEN) and clustered regularly interspaced short palindromic repeats (CRISPR)/CRISPR associated protein 9 (Cas9) techniques have overcome the technical defects of ZFN and have extended the range of nuclease application.

5.5.4.2 *Transcription Activator-Like Effector Nucleases*

The core region of transcription activator-like effector nucleases (TALENs) is the TALEN DNA recognition domain. TALENs are discovered specific DNA-binding proteins derived from *Xanthomonas* (a genus of plant pathogenic bacteria). The function of TALENs in nature is the direct regulation of host gene expression. Upon injection into host cells by the bacterial type III secretion system, TALENs enter the nucleus and bind to the effector-specific sequences in host gene promoters, thereby activating gene transcription and achieving bacterial invasion. TALENs comprise a set of specific effector proteins, including the N- and C-terminus that confer positioning and activating functions, as well as a central domain responsible for specific DNA recognition and binding. The transcription activator-like (TAL) effector DNA recognition domain consists of a tandem array of repetitive structural units. The number of repetitive structural units ranges from 5 to 30, with an average of 17.5 units. Each repetitive unit is composed of 34 amino acids, among which, 32 amino acids are conserved. However, the 12th and 13th amino acids are hypervariable among different units. Therefore, these two amino acids are called repeat variable diresidues (RVDs), and are responsible for base recognition and binding. The RVDs belong to the core recognition domain of TALENs (Figure 5.9).

The recognition of DNA bases by TAL effectors is determined by the RVDs. The RVDs NI (Asn Ile), HD (His Asp), NN (Asn Asn)/NK (Asn Lys), and NG (Asn Gly) recognize adenine (A), cytosine (C), guanine (G), and thymine (T), respectively. Recent studies have found that NK is more efficient in recognizing G compared with NN. In addition, naturally occurring TALs prefer to recognize sequences whose first 5′ base is T.

Upon understanding the highly specific base-recognition mechanism of TAL effectors, researchers began to consider developing TAL effector-based techniques and widely applying the techniques toward the GM animals. Therefore, TALENs are developed by fusing TAL effectors with the FokI catalytic domain.

The genetically engineered TALENs contain an N-terminal nuclear localization sequence, which is connected to a portion of the N-terminal sequence of TALENs. The middle portion of TALENs is the DNA recognition domain composed of 15–24 RVDs. TALENs also contain the partial C-terminal sequence followed by the FokI

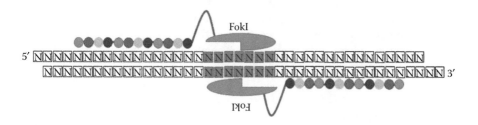

Figure 5.9 Schematic diagram of TALEN-mediated gene targeting.

catalytic domain. Through recognition of the sequences located upstream and down-stream of the target site, two TALEN molecules are directed to the target site. The TALENs then create DSBs at the target site via FokI enzyme digestion. DSBs induce NHEJ, resulting in gene insertion or deletion mutations. Simultaneous introduction of exogenous DNA sequences and TALENs into cells may lead to site-specific gene insertion through highly efficient homologous recombination, which is known as knock-in.

TALENs provide a faster and easier way to achieve target gene insertion/deletion mutations. The TALEN technique possesses a huge advantage over traditional methods, especially in the creation of gene knockout animals. Large segment dele-tions, insertions, or translocations that cannot be readily achieved using traditional homologous recombination-mediated gene targeting have been achieved though the introduction of multiple TALENs. Due to TALEN technology, laboratory animals and large animals (such as pigs) with a target gene knocked out have successfully been created.

The longer recognition sequence, better design, and open-source platform of TALENs give them a lower occurrence rate of off-target effects compared with ZFNs. However, because of the complex structure of the DNA recognition domain, TALENs are more difficult to construct than ZFNs. The emergence of several TALEN construction methods (such as Golden Gate cloning) has rendered it easier to design, construct, and utilize TALENs compared with ZFNs. Moreover, ZFNs fail to exhibit an apparent advantage over TALENs in terms of the efficiency of mutation induction. Therefore, the application of TALENs is becoming increasingly extensive.

5.5.4.3 CRISPR/Cas9

The discovery of CRISPR dates back more than 30 years ago. CRISPRs are bacterial genomic sequences that account for approximately 40% of the bacte-rial genome. The CRISPR locus is composed of an array of successively arranged short direct repeats, separated by short spacer sequences. The length of the repeats ranges from 21 to 47 bp (average length 32 bp). For a given CRISPR, the length and sequence of the repeats are known, whereas the spacers are completely different. Comparative genomics has shown that these distinct spacers share high homology with bacteriophage and plasmid DNA. Further studies demonstrated that CRISPR loci give rise to CRISPR-derived RNA (crRNA), which binds to CRISPR-associated protein (CRISPR/Cas9) to form a complex with substrate recognition and catalytic activity. The complex recognizes exogenous DNA sequences and removes muta-tions, thereby achieving defensive effects. Therefore, CRISPR/Cas9 is also known as the CRISPR defense system.

Researchers have developed a set of tools for genetic modification and gene edit-ing based on the type II CRISPR/Cas9 prokaryotic adaptive immune system. The CRISPR/Cas9 system includes crRNA and trans-activating RNA (tracrRNA) with transcriptional activation activity. CrRNA and tracrRNA form a double-stranded RNA, which assembles with Cas9. The RNA recognizes target DNA sequences and directs Cas9 to the genomic loci of the target DNA. Cas9 then cleaves both the

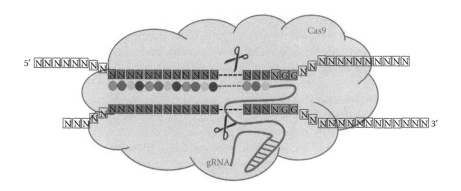

Figure 5.10 Schematic diagram of CRISPR/Cas9-mediated gene targeting.

complementary DNA strand and the noncomplementary DNA strand. Researchers have used the synthesized guide RNA (gRNA) to substitute for the complementary crRNA/tracrRNA, which functions together with Cas9 to create DSBs. Using the CRISPR/Cas9 system, researchers have successfully introduced gene mutations into mammalian cells. In human and many animal cells, the joint action of Cas9 and gRNA has successfully achieved specific DNA recognition, generated DSBs and induced NHEJ (Figure 5.10).

The CRISPR/Cas9 technique is the fourth technique established, after ES cell targeting, and the ZFN and TALEN techniques, that can be used for site-specific construction of gene knockout animals. The CRISPR/Cas9-, ZFN-, and TALEN-based methods have been compared for their efficiency in the induction of gene mutations. It has been found that there are no significant differences between the three methods in the efficiency of gene deletion. However, compared with the other two methods, the CRISPR/Cas9 system appears to be simpler and more convenient to operate. In addition, the CRISPR/Cas9 system possesses many other characteristics; it is efficient, less time-consuming, simple, economic, and highly potent in germline transfer. Therefore, the CRISPR/Cas9 system has broad prospects for application in the construction of animal models.

The CRISPR/Cas9 system recognizes target genes through RNA:DNA hybridization. Due to the firmness of RNA:DNA binding and the short length of the core sequence recognized by the CRISPR/Cas9 system, off-target effects can readily occur. However, the design and construction of CRISPR/Cas9 gRNA is relatively simple, whereas the construction of TALEN plasmids involves tedious steps. Therefore, many researchers still choose the easily operable CRISPR/Cas9 system despite its off-target effects.

Currently, the TALEN and CRISPR/Cas9 techniques have been widely applied in the preparation and study of many model animals besides the commonly used mice, including rats, rabbits, pigs, dogs, and nonhuman primates. In addition, simultaneous gene knockouts at multiple sites have been achieved using the TALEN or CRISPR/Cas9 techniques, as well as simultaneous introduction of multiple mutations at distinct gene loci.

At present, site-specific gene integration (namely gene knockin) has also been achieved using a combination of gene targeting and transgenic techniques, which allows for the site-specific integration of exogenous genes into the host genome through homologous recombination. The technical process employed for gene "knock-in," which will not be introduced in detail in the present chapter, is similar to that of gene knockout. Special mechanisms, such as RNA interference (RNAi), exist in cells, which regulate gene expression at the RNA level. Such mechanisms can be exploited to artificially reduce gene expression levels. Gene knockdown animal models are actually transgenic animals expressing RNAi molecules. Genes encoding RNAi molecules are integrated randomly or site-specifically into the host genome, which function to reduce the expression levels of certain genes through RNAi. The above two animal models, together with gene knockout animal models and transgenic animal models, constitute the commonly referred to GM animal models.

5.6 HUMANIZED ANIMAL MODEL

A humanized animal model is an animal carrying functioning human genes, cells, tissues, and/or organs. Now, nearly all humanized animals that were used as human disease models in biological and biomedical research for human therapeutics models are developed from mice. Immunodeficient mouse, such as nude mouse, severe combined immunodeficiency (SCID), NOG (NOD/Shi-*scid*/IL-2Rγ^{null}) mouse, and NSG (NOD.Cg-*Prkdc*^{scid} *Il2rg*^{tm1Wjl}/SzJ) mice are often used as recipients for human cells or tissues, because they can relatively easily accept heterologous cells due to lack of host immunity. Several humanized immunodeficient mouse models is summarized in Table 5.2. Humanized animal models could be applied in many promising biomedical research for human therapeutics including cancer, infectious diseases, regenerative medicine, and hematology.

For genes humanized animal model, such as the transgenic mice with a diverse human T-cell antigen receptor (TCR) repertoire was a good sample, which was developed by Li LP and his colleagues and can be used to identify pathogenic and therapeutic human TCRs. This transgenic mice was generated with the entire human TCR alpha beta gene loci (1.1 and 0.7 Mb), whose T cells express a diverse human TCR repertoire that compensates for mouse TCR deficiency. A human major histocompatibility class I transgene increases the generation of CD8$^+$ T cells with human compared to mouse TCRs. Functional CD8$^+$ T cells against several human tumor antigens were induced, and those against the Melan-A melanoma antigen used similar TCRs to those that have been detected in T cell clones from individuals with autoimmune vitiligo or melanoma. For Biologics Evaluation, Humanization of CTLA-4 (cytotoxic T-lymphocyte antigen) knockin mice was developed and used for *in vivo* testing of monoclonal antibodies specificity and efficacy. With the development of gene editing technologies in recent years, more and more genes humanized animal models were created for specific purposes in biomedical research.

Humanized cells or tissues mice were also developed by scientists. Over the last decades, incrementally improved xenograft mouse models, supporting the

Table 5.2   Humanized Immunodeficient Mouse Models

Strain Name	Mutated Gene	Advantage	Disadvantage
Nude	$Foxn1^{nu}$	No T cells	NK activity high, Very low engraftment of human cells
SCID	$Prkdc^{scid}$	No functional T cell and B cell	NK activity high, Low engraftment of human cells
NOD/SCID	$Prkdc^{scid}$	No functional T cell and B cell, Lowered NK level, Promoted engraftment of human cells and tissues	Short lifespan, NK activity till present
NSB	$Prdkc^{scid}$ $B2\,m^{tm1Unc\text{-}J}$	No functional T cells and B cells, NK activity very low, Promoted engraftment of human cells and tissues	Short lifespan
NSG	$Prkdc^{scid}$ $Il2rg^{tm1Wjl}$	No functional T cells and B cells, No NK cells, Long lifespan, High engraftment of human cells and tissues	No human MHC, No human cytokines
NOG	$Prkdc^{scid}$ $Il2rg^{tm1Sug}$	Similar to NSG	Similar to NSG
RG	$Rag2^{tm1Fwa}$ $Il2rg^{tm1Sug}$	Similar to NSG	Similar to NSG

engraftment and development of a human hemato-lymphoid system, have been developed and now represent an important research tool in the field. The use of NOG/NSG mice has greatly improved human hematopoiesis, as shown by the development of multiple human cell lineages, including B and T lymphocytes, NK cells, myeloid DC, plasmacytoid DC, macrophages, and erythroblasts. Healthy or neoplastic hematopoietic cells are transplanted into immunocompromised mice to develop humanized mouse models. The models help to learn and understand the physiology and pathophysiology of human hematopoiesis. The knowledge gained with these models can then be translated to humans.

Certain viruses are specific to humans as they require human cells for infection, replication, and pathogenesis, which are absent in regular animal models. Humanized liver models were very successful models for the study of infectious or other chronic liver diseases. Humanized liver models, in which the mouse liver cells are replaced with human liver cells, which is useful for evaluating drug metabolism in the human liver. Generally, in immunodeficiency mice (e.g., SCID or NOG mice), a urokinase-type plasminogen activator transgene (Alb-uPA) was used to induce the self-hepatocyte death. Human hepatocytes were transplanted into the inferior splenic pole and demonstrated that human hepatocytes could be engrafted over 50%–80% in the liver of these mice. Recently, a novel NOG substrain that expresses the herpes simplex virus type 1 TK transgene under the control of a mouse albumin promoter were developed. Administration of ganciclovir, which is nontoxic to human and mouse tissues, ablated TK-expressing liver parenchymal cells. Herpes simplex virus type 1 TK NOG mice allowed high engraftment of human hepatocytes (over 80%) and did not develop systemic morbidity (liver disease, renal disease, and bleeding diathesis) as seen in other uPA-dependent models. These humanized liver models facilitate studies of drug metabolism, toxicology, and the virology of hepatitis viruses.

It can be said that the humanized models are very valuable tools to mimic human diseases and also break the bottlenecks of studying human-specific pathogens in small animals (e.g., hepatitis B and C virus). These animal models might also be very useful to test not only approved but also new experimental drugs and to find novel treatment options as well.

BIBLIOGRAPHY

Bedell VM, Wang Y, Campbell JM et al. In vivo genome editing using a high-efficiency TALEN system. *Nature* 2012;491:114–8.

Behringer R, Gertsenstein M, Nagy KV et al. *Manipulating the Mouse Embryo: A Laboratory Manual*, Fourth Edition. Cold Spring Harbor Laboratory Press, New York, 2014.

Brosius FC 3rd, Alpers CE, Bottinger EP et al. Mouse models of diabetic nephropathy. *Journal of American Society of Nephrology* 2009;2:2503–12.

Crabtree WN, Soloway MS, Matheny RB Jr et al. Metastatic characteristics of four FANFT-induced murine bladder tumors. *Urology* 1983;22:529–31.

Doyle A, McGarry MP, Lee NA et al. The construction of transgenic and gene knockout/knockin mouse models of human disease. *Transgenic Research* 2012;21:327–49.

Dutt A, Wong KK. Mouse models of lung cancer. *Clinical Cancer Research* 2006; 12:4396s–402s.

Ernst W. Humanized mice in infectious diseases. *Comparative Immunology, Microbiology and Infectious Diseases* 2016;49:29–38.

Gordon JW, Scangos GA, Plotkin DJ et al. Genetic transformation of mouse embryos by microinjection of purified DNA. *Proceedings of the National Academy Science USA* 1980;77:7380–4.

Gu H, Zou YR, Rajewsky K. Independent control of immunoglobulin switch recombination at individual switch regions evidenced through Cre-loxP-mediated gene targeting. *Cell* 1993;73:1155–64.

Hasegawa M, Kawai K, Mitsui T et al. The reconstituted "humanized liver" in TK-NOG mice is mature and functional. *Biochemical and Biophysical Research Communications* 2011;405:405–10.

Houdebine LM, Fan J. *Rabbit Biotechnology*. Springer, Netherlands, 2009.

Ito R, Takahashi T, Katano I et al. Current advances in humanized mouse models. *Cellular and Molecular Immunology* 2012;9:208–14.

Kang JH, Mori T, Niidome T et al. A syngeneic hepatocellular carcinoma model rapidly and simply prepared using a hydrodynamics-based procedure. *Veterinary Journal* 2009;181:336–9.

Li LP, Lampert JC, Chen X et al. Transgenic mice with a diverse human T cell antigen receptor repertoire. *Nature Medicine* 2010;16:1029–34.

Liu E. *Animal Models of Human Diseases*, Second Edition. People's Health Publishing House, Beijing, China, 2014.

Morton JJ, Bird G, Refaeli Y, Jimeno A. Humanized mouse xenograft models: Narrowing the tumor-microenvironment gap. *Cancer Research* 2016;76:6153–8.

Palmiter RD, Brinster RL, Hammer RE et al. Dramatic growth of mice that develop from eggs microinjected with metallothionein-growth hormone fusion genes. *Nature* 1982;300:611–5.

Quaife CJ, Kelly EJ, Masters BA et al. Ectopic expression of metalothionein-III causes pancreatic acinar cell necrosis in transgenic mice. *Toxicology and Applied Pharmacology* 1998;148:148–57.

Quinn BA, Xiao F, Bickel L et al. Development of a syngeneic mouse model of epithelial ovarian cancer. *Journal of Ovarian Research* 2010;3:24.

Ross PJ, Cibelli JB. Bovine somatic cell nuclear transfer. *Methods in Molecular Biology* 2010;636:155–77.

Theocharides AP, Rongvaux A, Fritsch K et al. Humanized hemato-lymphoid system mice. *Haematologica* 2016;101:5–19.

Ting AY, Kimler BF, Fabian CJ et al. Characterization of a preclinical model of simultaneous breast and ovarian cancer progression. *Carcinogenesis* 2007;28:130–5.

Utomo AR, Nikitin AY, Lee WH. Temporal, spatial, and cell type-specific control of Cre-mediated DNA recombination in transgenic mice. *Nature Biotechnology* 1999;17(11):1091–6.

Wang H, Yang H, Shivalila CS et al. One-step generation of mice carrying mutations in multiple genes by CRISPR/Cas-mediated genome engineering. *Cell* 2013;153:910–8.

Wang YW, Sun GD, Sun J et al. Spontaneous type 2 diabetic rodent models. *Journal of Diabetes Research* 2013;2013:401723.

White JK, Gerdin AK, Karp NA et al. Genome-wide generation and systematic phenotyping of knockout mice reveals new roles for many genes. *Cell* 2013;154:452–64.

Wilmut I, Schnieke AE, McWhir J et al. Viable offspring derived from fetal and adult mammalian cells. *Nature* 1997;385:810–3.

Zhang XH, Takenaka I. Morphological changes in the basement membrane during progression from non-invasive to invasive rat bladder cancer induced by N-butyl-N-(4-hyhydroxybutyl) nitrosamine. *British Journal of Urology* 1997;79:378–82.

CHAPTER 6

Practical Techniques for Animal Experimentation

Enqi Liu and Jianglin Fan

CONTENTS

6.1 BASIC EXPERIMENTAL TECHNIQUES

6.1.1 Handling and Restraining of Animals

Researchers should be mentally prepared before handling or restraining animals. First, when making contact, researchers should choose the correct method of handling or restraining in mind the safety of the animal; second, animals should be handled or restrained as early as possible, and a bold but cautious approach is required; finally, self-defense knowledge is indispensable in dangerous circumstances.

Generally speaking, it is advisable to wait until the animals go completely quiet rather than handling them when they are excited. When handling, instead of suddenly touching the animals, researchers are supposed to give animals a hint in advance. For example, when handling rats and hamsters, researchers may first tap the cage gently so as to give the animals a signal, then move gradually closer to handle.

Researchers should choose one or two people to use their hands or designate a holder to restrain in advance in accordance with the different administrative site or blood collecting methods of laboratory animals.

Researchers should wear gloves in case of bitten fingers. Thick work pants are suggested when restraining between the thighs. To avoid being bitten by some animals, such as rats, timely restraining is required.

6.1.1.1 Mice

Use the finger pads of the right-hand thumb and forefinger to grab the middle of the tail of a mouse to pick it up (Figure 6.1a). If the researcher just wants to move the animal, two hands should be used to hold it (Figure 6.1b).

- **Hand restraining:** After putting the lifted mouse on the lid of the feeding cage, the mouse will struggle forward and the researcher should use the thumb and forefinger to grab the skin of the neck, back, and central back so that the mouse cannot move its head (Figure 6.1c).

However, grabbing tightly will lead to asphyxia or cervical dislocation; conversely, grabbing loosely increases the risk of being bitten. Invert the left hand grasping the skin of the back and neck to hold the mouse body in a straight line. Meanwhile, the little finger of the left hand presses the tail (Figure 6.1d).

- **Holder restraining:** After anesthetization, the mouse is hogtied by a 20–30 cm thread. As is shown in Figure 6.1e, a 15–20 cm plank with 5 nails wedged on the edges is prepared, and one end of each thread that hogties the mouse is restrained to the nail. One thread hooked to the upper teeth is also restrained on a nail to immobilize the mouse completely. For intravenous (i.v.) administration, containers of appropriate size and weight (e.g., beakers) are inverted, and the mouse is put inside with the tail outside. Such

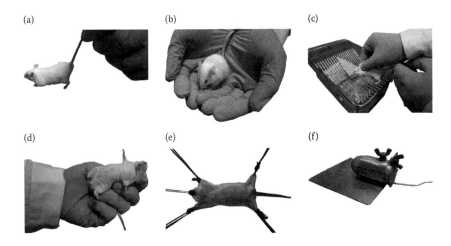

Figure 6.1   (a–f) Handling and restraining a mouse.

containers can stop the mouse from moving by pinning the tail. In addition (Figure 6.1f), special holders can be used to restrain mice.

6.1.1.2 Rats and Hamsters

Rats and hamsters within 4–5 weeks can be seized by the tail like mice. For older animals, being picked up by the tail is unsuitable due to easy skin detachment on the tail.

Using the left hand, grab the central back to chest, press down, and then pick up the animal with the index finger on the neck and back, and thumb and other three fingers on the chest. Use the index finger and middle finger to separate the two forelimbs and hold them up. The right hand restrains the hind legs (Figure 6.2a).

During administration, researchers should use their left-hand thumb and forefinger to grab the animal's skin in the neck and back, and the remaining fingers hold the back firmly (Figure 6.2b).

Like mice, rats and hamsters can also be restrained using wood, thread, or special holders.

6.1.1.3 Guinea Pigs

When handling young guinea pigs, researchers should hold them between their hands. Mature guinea pigs, however, can be seized with the left hand. Guinea pigs will not bite.

For restraining, use the left index finger and middle finger on either side of the guinea pig's neck and back, put the thumb and ring finger on the chest, and then use fingers to hold its forelimbs (Figure 6.3). Next, inverting the left hand, use the right-hand thumb and forefinger to hold the right hindlimbs, and with the middle finger and ring finger hold the left-hind legs to pull the body into a straight line.

(a)   (b)

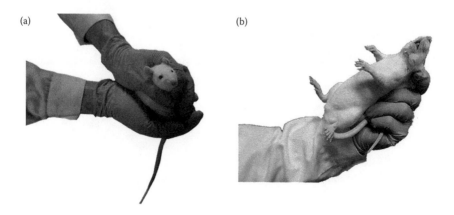

Figure 6.2 (a, b) Handling and restraining a rat.

Figure 6.3 Handling and restraining a guinea pig.

If sitting in a chair, hold the guinea pig with its hind legs held in the lap using the right hand, and then pin down the animal with the legs instead of the hands.

6.1.1.4 Rabbits

With one hand grasping the skin of the neck and back, hold the hip of the rabbit out of the cage with the other hand (Figure 6.4a). When moving the animals, continue to hold the neck and back skin of the rabbit (Figure 6.4b). When moving rabbits, do not grab the ears as this may cause ear cartilage injury.

For oral administration, sit on a chair with one hand holding the nape and the other handling the two hind legs held in the thigh. Pin the rabbit's lower body with

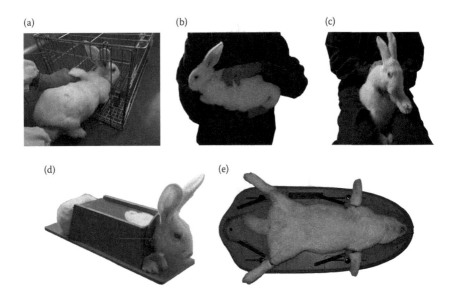

Figure 6.4 (a–e) Handling and restraining a rabbit.

the thigh and use the free hand to handle the two forelimbs. Hold the nape and ears with the same hand so that the head cannot move (Figure 6.4c).

There are several holders used for rabbits. Metal semicircle buckets or square box-type rabbit restraining can be used (Figure 6.4d) when administered through auricular veins or for blood collection and drug pyrogen testing. Special restraining (Figure 6.4e) for carotid artery blood collection and operation is available.

Semicircle bucket or square box-type rabbit restraining puts the rabbit into a cylinder with only their heads protruding forward, and thus rabbits are restrained by twisting restraining (Figure 6.4d). Special restraining for rabbits is to let the rabbit remain in a supine position, use gauze to bind limbs on both sides of the restrained rod, and then restrain the head by metallic pillory and mouth ring (Figure 6.4e).

One simple way of restraining is to wrap the rabbits in old work clothes with only two ears showing for direct manipulation.

6.1.2 Gender Identification

6.1.2.1 Mice, Rats, and Hamsters

We can judge the gender of new born animals by observing the distance between the genital opening and anus. A short distance indicates a female and a long distance indicates a male (Figure 6.5). The male genitalia (penis) is bigger than that (clitoris) of females, but judgment of this requires some experience.

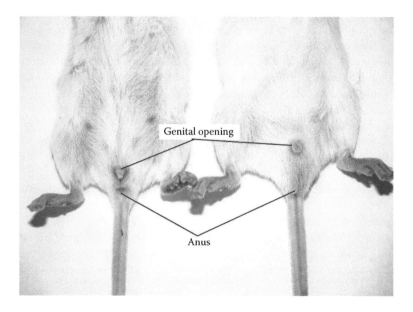

Figure 6.5 Gender identification of mice, rats, and hamsters. A short distance between genital opening and anus indicates a female (right) and a long distance indicates a male (left).

It is quite easy to identify the gender of adult animals. Females have a vaginal opening, while males have scrotum swelling and a penis.

6.1.2.2 Guinea Pigs

Guinea pigs have a long pregnancy period, and the newborns can open their eyes, and have a coat and permanent teeth. The gender of newborns is also easy to distinguish through the genitals. For female guinea pigs, their genitalia, clitoris protuberant, is comparatively small. Use the thumb to hold this protrusion and crease the folds of the labia majora to see the vaginal opening. Guinea pig vaginal openings have closed, locking membranes (except in estrus). Male guinea pig external genital organs consist of a small bulge, which is covered with the foreskin of the penis. Use the thumb to gently hold the base of the small bump and the glans penis is easily identified.

6.1.2.3 Rabbits

Identification of the gender of newborn rabbits is more difficult than that of rodents, but can be mastered after professional guidance and training. They can be differentiated by measuring the distance between the urethra and anus openings, and the form of urethral opening. Compared with female rabbits, the distance between openings of male rabbits is 1.5–2 times in length. Press the lower abdomen with the finger that is located near the urethral opening. If the urethral opening points in the direction of the anus, and the distance between the anus and urethral opening is not obvious, it is a female rabbit. If the urethral opening and the anus are in the opposite direction, and the distance between the anus and urethral opening is obviously elongated, it is a male rabbit. In addition, identifying the shape of the urethral opening is crucial for determining gender. The urethral opening of female rabbits is a long slit, while that of male rabbits is cylindrical.

For adult rabbits, it is easy to differentiate the gender by judging whether vaginal opening or swelling scrotal and penis exist or not.

6.1.3 Animal Identification and Marking

For animal experimentation, individual animals must be identified or marked so as to differentiate them. Animal marks and numbers are divided into two types: permanent markers that can be recognized for life and short-term markers.

The ways to mark animals are based on the following principles: harmless to animals, simple operation, and long-term recognition. There is no certain rule to identify and number animals, as long the same method is adopted within one unit.

Another way that is equal to marking is to treat cage numbers as individual animal markers instead of animal tags. This way is applicable when one animal is kept in one cage.

6.1.3.1 Mice, Rats, and Hamsters

Short-term markers: Light-colored or white animals can be marked with a biological stain that does not easily fade. For example, the biological stain that is a solute of picric acid in 80%–90% saturated ethanol solution is used on the back of mice. This method will keep for 2–3 months. Dyes, such as basic fuchsin and methylene blue, are also acceptable.

As shown in Figure 6.6, dyes are added to the mouse's head, back, and front and hind legs. The number of labeled mice is 10. In addition, when using short-term markers, fur can be shaved and then that area marked if the animals were colored. For newborn mice, the four toes of the forelimbs and five toes of the hind legs can be cut.

Permanent markers: Under light anesthesia, specialized tools can be used to pierce the ears or scissors are used to cut a triangular shape at the edge of the ears. This method can be used to mark approximately 100 animals.

Long-term individual identification may be achieved by using recently developed methods to implant micro-integrated circuits under an animal's neck and back. Special reading instruments can be used to identify the microchips achieving individual recognition for a long period. At present, this method has been used for many kinds of animals, including mice.

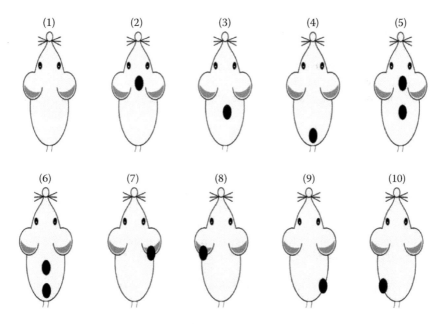

Figure 6.6 Mouse markers and numbers.

Figure 6.7   Rabbit marker with aluminum earring.

6.1.3.2 *Guinea Pigs and Rabbits*

For medium-sized animals that were kept separately, rearing cage marks replace individual marks. However, it is better to adopt an individual tag at the same time.

Short-term markers: For white animals, dye may be used to mark as mentioned above.

Permanent markers: This method requires special equipment. There are two commonly used methods. The first method is to use a manual stamp or electric ink-adding apparatus to print numbers on the inner ear where the vessels are not very dense, and identify them. The second method is to use a special ear clamp to put an aluminum earring with a number and markers on the animals' ears (Figure 6.7).

6.1.4 Fur Removal

There are three methods to remove animal fur: plucking, mechanical methods, and chemical methods. Mechanical methods are subdivided into the shearing method and shaving method. The former needs curved tip scissors; the latter requires a razor, electric razor, or special animal shears (similar to mechanical scissors). Chemical methods require depilatory cream.

6.1.4.1 *Plucking*

This method is simple and practical. It is commonly used for i.v. injection in the hindlimbs, injection in the rabbit's ear margin vein or for blood collection.

After restraining the animal, use the thumb and index finger to remove hair at the desired site.

6.1.4.2 Shearing

Shearing is the most commonly used method in acute animal study. After restraining animals, first moisten the area to be sheared so as to keep the local skin taut, then use curved tip scissors to cut the fur in the desired location. It is advisable to roughly cut longer fur and then carefully cut at the root. However, when cutting the fur, do not lift the animal's skin by hand as this may easily break the skin and affect the next experiment. The cut fur should be put into beakers with water or cleaned with a vacuum to remove loose fur.

6.1.4.3 Shaving

This method is generally used for chronic surgery in large animals. A warm, soapy brush is first used to fully moisten the fur that needs to be shaved, and a razor is used to shave in the direction of fur growth.

If using an electric razor or special animal shears, shave against the direction of fur growth.

6.1.4.4 Depilatories

These are commonly used in sterile operations for large animals, local skin irritation studies, observing an animal's local blood circulation, or other forms of pathological changes.

The major components of commonly used animal chemical depilatories are sodium sulfide (Na_2S), calcium sulfide (CaS), and barium sulfide (BaS).

The following can be used for an animal depilatory: 8% sodium sulfide aqueous solution (8 g of sodium sulfide is dissolved in 100 mL of water) or mix 8 g of sodium sulfide, 7 g of starch, 4 g of glucose, 5 g of glycerol, 1 g of borax, and 75 mL of water (100 g in total) to make a paste.

It is advisable to use scissors to cut the fur short in the desired areas before removal. This is to save on the amount of depilatory used. Then, use a cotton ball or a piece of gauze to coat a thin layer of depilatory on desired areas. After 2–3 min, wash away the removed fur with warm water. Next, dry gauze is used to dry the water and finally an oily coating is spread over the area. Skin hyperemia and inflammation rarely occur after depilatory use.

However, it is better to not use the animals for experiments the same day of fur removal. After 24 h, if there is no inflammation or other adverse reactions, only then can they be used for formal experimentation.

To avoid the depilatory soaking into the roots, stimulating the skin, causing skin inflammation or other changes, do not wash the animals before fur removal.

6.2 ADMINISTRATION TO ANIMALS

Compared with conventional administration routes, laboratory animal administration routes can be divided into three types: transdermal administration, enteral administration, and parenteral administration.

6.2.1 Transdermal Administration

This administration route is simple and the used drugs are usually in liquid or ointment form. However, there are disadvantages. This method is comparatively rough, because there are great differences in drug permeability for different animals or different skin parts on the same animal, and after lapping and friction, the drugs may be removed from the skin. Moreover, if the drug irritates the skin or mucous membranes, it may cause the animal discomfort, which is unfavorable in scientific studies.

6.2.2 Enteral Administration

This route refers to drugs sent through the mouth to the gastrointestinal system or use of suppositories to send drugs to the rectum via the anus. The anal administration route is not applicable for some small laboratory animals. The easiest way is to add drugs into the animal feed or drinking water, thus, animals are able to take the drugs along with their normal diet. However, for some poor tasting drugs, this method does not work. Animals may refuse to eat food or drink water with the drug. In addition, in order to take via water, the drugs themselves must be soluble and stable in water. It is also difficult to measure accurately the dose of drugs that each animal has taken.

The most popular way to perform forced oral administration is with a stomach tube. This practice allows for exact dosing and accurate recording of the time of symptom onset and the corresponding course. However, compared with adding the drug, forced oral administration with a stomach tube will consume extra time for tube insertion as well as cause a certain degree of mechanical and psychological impact on the animals. To reduce these negative influences, it is necessary to perfect forced oral administration skills.

For mice, the suitable diameter of gastric tubes is 0.8 mm; rats, 1–2 mm; guinea pigs, 1.5–2 mm; rabbits and cats, 3–5 mm; and dogs, 5–7 mm.

When giving drugs to small laboratory animals (such as mice, rats, hamsters, and guinea pigs) by oral administration, we recommend using a special gavage needle (Figure 6.8). The gavage needle is easily available. The administration protocol is as follows: First, confirm the position of the stomach where the needle needs to be placed (Figure 6.9a and b). Next, after restraining the animal with the hand, put the front end of gavage needle into the animal's mouth, follow along the roof of the palate, and insert the needle into the throat. Do not force the needle. Hold the syringe gently, and carry out the procedure quickly and quietly. When the front end of the gavage needle goes into the throat, there will be resistance. The front end of

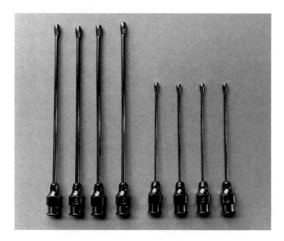

Figure 6.8 Reusable stainless steel oral gavage needles.

the gavage needle should be moved and gently inserted parallel to the animal's longitudinal axis. After entering the esophagus without resistance, the gavage needle is inserted into the needed position and the solution is slowly injected. After injection, gently remove the gavage needle (Figure 6.9c).

For rabbits, rubber stomach tubes, mouth gags, and injectors are needed. The specific operation methods are as follows: after restraining the rabbit, use the thumb and middle finger to squeeze the rabbit's cheeks to the jaw tightly, and then press the mouth gag into its mouth to make the rabbit bite the mouth gag; the tongue is placed at the bottom of the mouth gag, slightly sticking out. The mouth gag is put close to the corner of the mouth, and the cloth tied at the two ends of the mouth gag

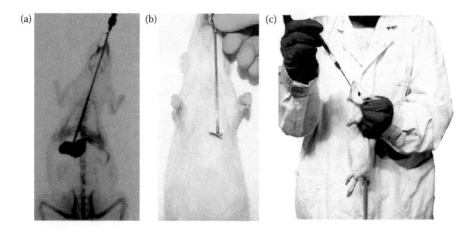

Figure 6.9 (a–c) The rat oral administration by gavage.

cross the back of head and are tied beside the base of the ears. Next, the mouth tag is restrained. The gastric tube should be soaked in saline in advance so the tube will insert easily and not damage the esophagus. To insert the tube, an injector is attached to the opposite end of the gastric tube, the inner barrel is removed, and the drug is injected after confirming that no air is present. To avoid residual liquid in the gastric tube, 5 mL of saline solution is injected and the tube pulled out. However, this approach requires that operators have experience because it is important to prevent the catheter from being inserted into the trachea. If the catheter is inserted into the trachea, the animals will cough and the catheter will touch the tracheal cartilage ring causing air to be inhaled into the pipe. This will not occur if the catheter is inserted into the esophagus correctly.

For large animals (e.g., dogs and pigs), oral administration requires special surgical instruments. It is best to use a nasal catheter for administration to primates, cats, and horses.

6.2.3 Parenteral Administration

Parenteral administration refers to the methods other than the two methods above. The main way is by injection.

The most commonly used administration routes include the following: intracutaneous (i.c.) or intradermal (i.d.), to inject drugs into the inside of skin; subcutaneous (s.c.), which is absorbed relatively slowly; intramuscular (i.m.), this method is often used in gluteus and back muscles, and while it has the advantage of speedier absorption, it can cause pain; intraperitoneal (i.p.), injection into the abdomen. The peritoneal cavity has the ability to absorb drugs, which promotes the speed of reabsorption; i.v., the quickest and most accurate method as drugs are injected into the veins. Tables 6.1 and 6.2 show the administration routes, methods, and maximum dose for common laboratory mammals, birds, reptiles, and amphibians.

Here are several kinds of commonly used administration methods for laboratory animals:

6.2.3.1 Caudal Vein Injection in Mice and Rats

Using special restraining to restrain the animal with its tail outside the container (Figure 6.10), the tail is rotated and oriented, and a glass container is used to pin the tail. The lateral veins in the tail will expand under the weight of the glass container. For injection, use the thumb to restrain the end of the tail and forefinger to hold the tail in place. Before injection, use alcohol cotton balls to disinfect the tail and expand the blood vessels. The weight of container will cause the vessels to fully expand when restrained. Choose a site close to the end of the expansion with an injection angle of approximately 30°. Aiming at the central blood vessels and in the direction of vessels, gently lift the tip of the needle to insert into the vessel; do not pull the needle and injector, use the left thumb and middle finger to restrain it in place. Confirm whether the blood returns or not. After puncturing the blood vessel, inject the drugs slowly. If the needle has not completely punctured intravascularly,

Table 6.1 Commonly Used Administration Route and the Maximum Dosage for Mammals (mL)

	Mice 20–25 g	Hamsters 25–30 g	Rats 250 g	Guinea Pigs 350 g	Rabbits 2.5 kg	Cats 4 kg	Dogs 20 kg	Pigs 50 kg	Primates 10 kg	Sheep 60 kg	Horses 500 kg
Oral administration	Use a blunt lavage needle along the direction of the longitudinal axis and then insert it into the esophagus				Stomach tube	Nasal tube	Stomach tube	Stomach tube	Nasal tube	Stomach tube	Nasal tube
Maximum (mL)	0.5	0.5–1.0	1.0		7.5	10.0	20.0	100	30	100	100
Diameter (mm)	1.0		2.0		5.0						
Intradermal injection				Back or abdominal skin							
Maximum	For every injection, volume of injection for every animal is 0.05–0.1 mL each time										
Needle						26 G					
Subcutaneous injection		Neck, back		Neck	Neck, back	Neck, back, chest			Neck		Neck, back, chest
Maximum (mL)	0.5–1.0		1.0–5.0	1.0–2.0	1.5–5.0	2.0	10.0				
Needle	26 G			25 G	21 G	23 G	21 G	19 G	23 G	19 G	
i.m. injection			Gluteus				Gluteus, dorsal muscles, and pectorals				
Maximum (mL)	0.05		0.1		0.2	0.2	0.2				
Needle	26 G	26 G	25 G	25 G	21 G	23 G	21–23 G	19 G	23 G	19 G	
Intraperitoneal injection		Next to the midline and near navel				!		Edge of pelvis and near navel	!	!	!
Maximum (mL)	1.0		10.0		20.0						
Needle	25 G		24 G	21 G				20 G	25 G	20 G	

(Continued)

Table 6.1 (Continued) Commonly Used Administration Route and the Maximum Dosage for Mammals (mL)

	Mice 20–25 g	Hamsters 25–30 g	Rats 250 g	Guinea Pigs 350 g	Rabbits 2.5 kg	Cats 4 kg	Dogs 20 kg	Pigs 50 kg	Primates 10 kg	Sheep 60 kg	Horses 500 kg
Intravenous injection	Vena caudalis	Sublingual vein, penile venous	Vena caudalis, hindlimb vein, jugular vein, penile venous	Foreleg vein, hindlimb vein, sublingual vein, penile venous	Ear vein	Foreleg vein, hindlimb vein	Foreleg vein, hindlimb vein	Ear vein, jugular vein	Foreleg vein, hindlimb vein	Jugular vein	Jugular vein
Maximum	0.2		0.5		1–5	2–5	10–15				
Needle	25 G	27 G	23–25 G	26–27 G	21–23 G	21–24 G	21–24 G	16–24 G	21–25 G	16–19 G	16–19 G

Note: 1. In the table, hamster refers to the Syrian hamster.
2. "Diameter" refers to outer diameter of gastric needles or stomach tube, which is used for parenteral administration.
3. "Needles" refers to the needle size. "G" is gauge, in addition, 19 G = 1.00 mm, 20 G = 0.90 mm, 21 G = 0.80 mm, 22 G = 0.70 mm, 23 G = 0.60 mm, 24 G = 0.55 mm, 25 G = 0.50 mm, 26 G = 0.45 mm, and 27 G = 0.40 mm.
4. "!" means this administration route cannot be used.

Table 6.2 Injection Sites for Birds, Reptiles, and Amphibians

	Birds	Snakes	Turtles	Frogs
Subcutaneous (s.c.)	Neck	–	Limbs	–
Intramuscular (i.m.)	Pectorals and hindlimb muscles	Dorsal muscles	Hindlimb muscles	–
Intraperitoneal (i.p.)	The middle of the sternum and cloaca	2/3 from heart to cloaca	The border of the tail and body	Midline
Intravenous (i.v.)	Hindlimb veins and wing veins	Injection inside the heart	Peritoneovenous, jugular vein (anesthesia)	–
Dorsal lymph sacs	–	–	–	+

there will be resistance and local swelling. Pull the injector out immediately when the injection is done, and use fingers, absorbent cotton, or gauze to press hard on the injection site to stop the bleeding.

6.2.3.2 *Ear Margin Vein Administration to Rabbits*

After restraining the rabbit, place the tip of the rabbit's ear in the hand, and use the thumb and little finger, and index and middle fingers to hold the inside and outside of the ear, respectively. Gently pull the ear, using the little finger, middle finger, and ring finger to close the contralateral blood vessels; after disinfection using alcohol, put the needle in the ear along the direction of blood vessels. Confirm whether

Figure 6.10 The rat caudal vein injection.

the needle is in the blood vessel, and use the left-hand thumb, index, and middle fingers to restrain the ear and needle together to inspect whether the blood returns or not. Slowly inject the drug, and use sterilized gauze or cotton wool for compression hemostasis after the completion of local injection.

6.2.3.3 Intraperitoneal Injection

After restraining the animal in the hand to display the abdomen, choose a place at the left or right side deviating from the midline and disinfect with alcohol. When puncturing the skin, keep the needle parallel with the animal's skin. Insert the needle approximately 5 mm into the skin. Free movement indicates the needle is inserted subcutaneously. Next, angle the needle 45°and insert the needle through the abdominal muscle into the abdominal cavity without resistance, then slowly inject the drug (Figure 6.11).

6.2.3.4 Subcutaneous Injection

Disinfect the skin at the injection site, back, and neck skin (from above, a triangle is formed when the hindquarters are pulled and restrained). In a vertical direction from the head, puncture the skin and move the needle 5–10 mm along the body axis. Turn the tip (if skin is punctured, it will easily turn) to confirm the needle has punctured the skin, and then inject the drug. After the injection, pull out the needle slowly and press slightly at the injection site with a finger to avoid leakage of liquid from the injection site (Figure 6.12).

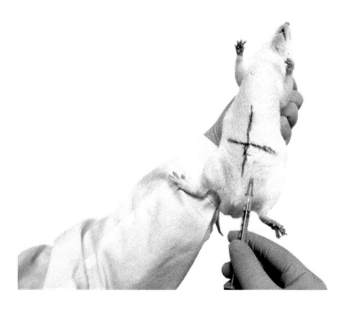

Figure 6.11   The rat intraperitoneal injection.

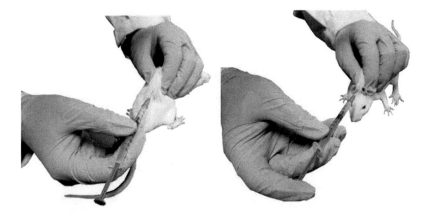

Figure 6.12 The rat subcutaneous injection.

6.2.3.5 Intradermal Injection

Before injection, use a razor to cut the fur from the injection site and surrounding area. For guinea pigs and rabbits, barium sulfide and hair removal cream are used to remove remaining fur after cutting (performed more than one day after hair removal). Use the same method to restrain as subcutaneous injection and disinfect with alcohol. Keep the cross section of the needle upward, and stab parallel with the direction of the skin. Keep the needle just under the skin surface. High resistance is felt during injection. If apophysis appear at the injection site and the pores stand out, the needle has completely punctured into the skin. Remove the needle 5 s after finishing the injection as the drug will leak out if the needle is removed immediately.

6.2.3.6 Intramuscular Injection

When the needle is inserted subcutaneously in the hip through the skin, the needle can penetrate deeper into the skin and the tip of the needle can no longer move freely. If no blood exists when the injector is pulled back, the drugs can be injected (Figure 6.13).

6.2.3.7 Intracerebral Injection

If a drug must directly affect the brain, it should be injected into the cerebrospinal fluid to avoid blood–brain barrier. Under anesthesia, at the midpoint between the ears and eyes, that is, where the cranium slightly deviates from the midline posterior to the orbital line, vertically insert in a needle 2–3 mm (in mice) deep and slowly inject. Take care as the skull is very hard and brain tissue is very soft.

A set of formal instruments is required to inject drugs into a certain place of the brain. After restraining the head of an anesthetized animal on the instrument, the exact position of injection may be shown through special grid lines.

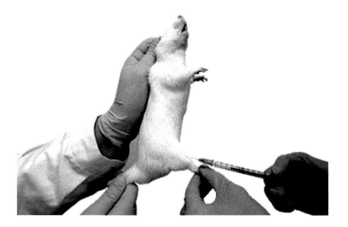

Figure 6.13   The rat intramuscular injection.

According to special research needs, drugs can also be directly injected into certain parts of the body such as articular cavity or trachea.

Injection techniques need to be learned under the guidance of professionals and technical personnel. The technique can then be implemented after training.

For injections, the following points must be considered:

- Use clean, sharp, and sterile needles.
- Choose an injection needle of appropriate size; fine needles will lessen pain and prevent the backflow of liquid, but finer needles do not necessarily better the effects. The size of needles depends on the viscosity of the liquid. Especially fine needles are liable to cause obstruction.
- The amount of injected drugs shall not exceed the maximum capacity that the method has recommended as in Table 6.1.
- Avoid air bubbles in the injection as they may cause embolization.
- Before injection, preheat the injection to room temperature or body temperature. If the injection temperature is too low, it can significantly aggravate pain.
- Be very careful when doing intraperitoneal injection so as to avoid damaging animal internal organs. To avoid damaging the bladder, the injection site should slightly deviate from the midline. The needle position should not be too parallel (to avoid injecting between skin and abdominal wall) or too vertical (it may damage the kidney). The position of needle insertion should be in the animals' abdomen, slightly back and left of the stomach. This reduces the chance of intestinal damage.
- Use short needles to reduce the damage to animals.
- Some injections can strongly stimulate animal tissues (e.g., pH value too high or too low). For these kinds of injections, dilute with normal saline or sterile water before using so as to reduce the stimulation to animals. Dilution is also needed when performing an intraperitoneal injection as excessive stimulation may cause peritonitis and/or intussusception.
- When animals need to be given large amounts of fluid, large animals can use i.v. drips, while small animals may receive through intraperitoneal injection. Oral administration is also acceptable depending on the animal experimental situation.

For all methods, the infusion speed should be slow, especially when giving large amounts of i.v. medication. Furthermore, more attention should be paid to avoid causing pain or shock.

- Through blood circulation, the rate of drug spreading from the injection site to the target tissue depends on the administration route. In general, the fastest route is i.v. injection, while oral administration is the slowest.

6.2.4 Dose Translation from Animal to Human

As new drugs are developed, it is essential to appropriately translate the drug dosage from one animal species to another or to humans. A misunderstanding appears to exist regarding the appropriate method for allometric dose translations, especially when starting new animal or clinical studies. The animal dose should not be extrapolated to a human equivalent dose by a simple conversion based on body weight. For the more appropriate conversion of drug doses from animal studies to human studies, we suggest using the body surface area (BSA) normalization method. BSA correlates well across several mammalian species with several parameters of biology, including oxygen utilization, caloric expenditure, basal metabolism, blood volume, circulating plasma proteins, and renal function.

According to the following formula for dose translation based on BSA and Table 6.3, we can convert a dose for translation from animals to humans, especially for phase I and phase II clinical trials.

$$\frac{\text{A animal (mg/kg)}}{\text{B animal (mg/kg)}} = \frac{\text{B animal } K_m}{\text{A animal } K_m}$$

Table 6.3   Conversion of Animal Doses to Another Equivalent Dose Based on BSA

Species	Weight (kg)	BSA (m²)	K_m factor
Human			
Adult	60	1.6	37
Child	20	0.8	25
Baboon	12	0.6	20
Dog	10	0.5	20
Monkey	3	0.24	12
Rabbit	1.8	0.15	12
Guinea pig	0.4	0.05	8
Rat	0.15	0.025	6
Hamster	0.08	0.02	5
Mouse	0.02	0.007	3

Note: BSA, body surface area; K_m factor is body weight (kg) divided by BSA (m²); to convert dose in mg/kg to dose in mg/m², multiply by K_m value.

6.3 ANIMAL FLUID COLLECTION

6.3.1 Blood Collection

During animal experimentation design, the method of blood collection to be used should be decided. Several methods are available to collect blood samples from different parts of animals. Veins, arteries, orbital arteriovenous, or intracardiac puncture can be used to collect blood samples. The selection of blood collection methods is determined by the purpose of research, such as collection of arterial blood, venous blood, or arteriovenous mixed blood, the duration of blood collection, frequency, and whether the experiment is lethal or not. In some species of laboratory animals (e.g., hamster), enough blood can only be collected under anesthesia. When the research requires repeated blood collection, researchers should consider using the intubation method.

It is advisable to carry out the experiment under anesthesia when collecting blood from small laboratory animals. In this way, we can ensure that animals are not able to move. If the blood collection is performed while the animal is conscious, the researcher should consider that stress from blood collection may lead to the deviations in physiological and biochemical parameters.

Tables 6.4 and 6.5 introduce commonly used blood collection methods for laboratory animals and the respective maximum blood collection volumes. The specific blood collection sites and amounts of mice, rats, guinea pigs, and rabbits are discussed in detail.

6.3.1.1 Venous Puncture Sampling

Vein puncture is the commonly used collection route in animal experiments, and veins near the skin that easily expand after pressure are selected for blood collection. For example, the most commonly used blood vessels are veins in the neck (jugular veins), thigh (femoral veins), cephalic veins in dorsal forelimbs, commonly used rat and mouse tail veins, and the saphenous vein in the back or middle part of the hind legs. For rabbits, auricular veins are the most commonly used blood vessels.

When using vein puncture to collect blood, the fur in the corresponding sites should be shaved and antimicrobials should be used to clean the blood collection sites. After vasodilatation of the blood vessel by pressure, insert the needle into the vein to let blood flow into the test tubes, injectors, or vacuum tubes. Before pulling out the needle, relieve the pressure on the blood vessels. Gently compress the blood collection site to stop bleeding. For example, when performing blood collection via rabbit ear margin veins, remove the fur at the blood-collection site, wipe with alcohol to disinfect, and make vascular filling. Next, dry the skin with dry sterile gauze. At an angle of 30°, aim at the direction opposite to the direction of blood flow and find the expanded veins. Gently lift the needle and insert into the skin in the direction of the vessel, using the required instrument to collect blood. After blood collection, use sterile gauze to compress for 5–10 s to stop the bleeding.

Table 6.4 Commonly Used Blood-Collection Method and the Maximum Blood Volume (mL) in Animals

	Mice	Hamsters	Rats	Guinea Pigs	Rabbits	Cats	Dogs	Primates	Pigs	Sheep	Horses
Jugular vein	+	+	+	+	+	+	+	+	+	+	+
Foreleg vein						+	+	+		+	
Hindlimb vein	+	+	+	+			+	+		+	
Femoral vein	+	+	+	+	+	+	+	+			
Ear vein				+	+						
Caudal vein	+		+						+		
Orbital puncture	+	+	+	+					+		
Cardiac puncture	+	+	+	+	+	+	+	+	+	+	+
Tail tip	+		+								
Maximum (mL)	0.3	0.3	2.0	5.0	15	20	100–500	20–200	200–500	200–600	500–7000

Table 6.5   Blood Collection Site and Blood Capacity of Mice, Rats, Guinea Pigs, and Rabbits

	Blood Collection Site	Blood Capacity (mL)	Measures of Blood Collection
Mice			
Partial blood collection	Caudal vein	0.03–0.05	
	Caudal artery	0.1–0.3	
	Orbital venous plexus	0.05–0.8	Anesthesia
Whole blood collection	Jugular vein	0.5–1.0	Anesthesia, operation
	Carotid artery	0.5–1.0	Anesthesia, operation
	Decollation	0.5–1.0	
	Heart	0.5–0.8	Anesthesia, operation
Rats			
Partial blood collection	Caudal vein	0.3–0.5	
	Caudal artery	0.5–1.0	
	Saphenous vein	0.1–0.3	
	Orbital venous plexus	0.5–5.0	Anesthesia
Whole blood collection	Jugular vein	3.0–5.0	Anesthesia, operation
	Decollation	5.0–10.0	
	Heart	3.0–5.0	Anesthesia, operation
Guinea Pigs			
Partial blood collection	Auricular artery	<0.5	
	Auricular vein	<0.5	
	Heart	3.0–5.0	
Whole blood collection	Heart	5.0–10.0	Anesthesia, operation
	Ventral artery	5.0–10.0	Anesthesia, operation
	Jugular vein	3.0–5.0	Anesthesia, operation
Rabbits			
Partial blood collection	Ear marginal artery	5.0–10.0	
	Ear marginal vein	2.0–5.0	
	Heart	10.0–15.0	
Whole blood collection	Heart	80.0–100.0	Anesthesia, operation
	Carotid artery	80.0–120.0	Anesthesia, operation

6.3.1.2 *Blood Collection from Orbital Vessel Puncture*

The jugular vein of small rodents is too thin for blood sampling. Therefore, the eyeball blood vessels are sometimes chosen for blood collection (such as in mice, rats, gerbils, and guinea pigs). During blood collection, put animals under shallow anesthesia with diethyl ether or other anesthetics, and keep unilateral eyes upward. Tightly hold the back and neck skin of anesthetized animals to expand the eye. Then with a suitable glass tube or Pasteur pipette, gently insert it into the corner of the eye between the eyelid and eye ball. When the needle reaches the depth of the sphenoid bone, slightly rotate the tube or pipette to siphon blood. After blood collection, use sterile gauze to compress the eyeball for 30 s to stop the bleeding. The collection site will heal after 3–7 days. In this way, blood from each eye can be collected alternately many times.

Sterile blood samples cannot be collected in this way because there may be mixing with tissue fluid within the eye socket and gland secretions, which will contaminate blood samples. When collecting blood from one eye many times, it may cause some complications such as bleeding, inflammation, and blindness. In addition, this method may cause undesired effects on vision, thus some countries have already banned the use of this method.

6.3.1.3 Blood Collection by Heart Puncture

After anesthetizing animals, insert a needle into the ventricle directly to collect heart blood. With this method, do not allow the needle to insert into the animal's atrium. The atrium is connected with the pericardium and this can stop the heart, causing death. If the animals are needed alive following blood collection, this point is very important.

Heart blood collection is generally carried out under anesthesia with an open chest. The puncture method can be used for some or all blood collection in place of opening the chest. For guinea pig heart blood, after disinfecting the chest of restrained guinea pigs, use fingers to find the position of the heart throb and position the needle. Insert needle from the determined position over the left ribs and slightly tilt 2 cm. Pull out the needle immediately if the animal appears nervous or upset and wait until the animal calms down before puncturing again.

6.3.1.4 Blood Collection by Arterial Puncture

For obtaining oxygen-rich blood samples, arterial puncture or artery intubation can be used to collect blood. The femoral artery and arteria carotis communis are the preferred sites for blood sample collection. When collecting blood from the carotid artery of rabbits, restrain anesthetized rabbits in a supine position. Remove fur from the middle of the neck and use alcohol to disinfect the skin. Use scissors to cut through the skin (approximately 5–6 cm) at the neck and head junction, and then move the neck muscles with tweezers to the sides to expose the trachea. The vagus nerves of the trachea are white and the carotid artery is pinkish. If the rabbit is to remain alive after blood collection, do not injure the vagus nerves and small vessels, which are responsible for supplying the trachea, when separating the carotid. The distal artery is ligated and put one suture line in proximally. Allow an artery to block the carotid artery in place of a suture. Then use ophthalmology scissors to cut the blood vessels between the ligation suture and proximal line. Next, insert an oblique tip of a plastic catheter into the artery along the direction of the heart for 3–5 cm, and then ligate the heart suture. Finally, put the other end of the plastic catheter in the blood vessel and blood will flow out. For rabbits, ear arteriopuncture is more favorable if the required amount of blood is not large.

6.3.1.5 Blood Collection via Docking and Caudal Vein

For small rodents, such as mice and rats, we can collect a small amount of blood by cutting the tail with scissors. This method can be used for research such as blood smears. After blood collection, use local compression on the wounds to stop bleeding. When using this method for rats, it is recommended to use anesthesia first. If more sample is needed, it can be collected from the tail vein. To expand mouse and rat tail veins, warm the animal's tail. Then, use alcohol to disinfect. Use sterile gauze to dry the alcohol or blood will leak out after wetting the puncture site, which may cause difficulty for blood collection and lead to hemolysis. Use the thumb and index finger to catch the tip of the tail several cm from tail tip and insert the needle into the caudal vein to collect blood samples. This method is a percutaneous vein puncture causing blood to be possibly polluted by tissue fluid, which may affect the results of the experiment. If sterile blood is needed for experiments, it is necessary to anesthetize and take other surgical measures to collect venous blood.

6.3.1.6 Blood Collection from Nonmammals

Blood samples can be collected by cutting a small incision in a bird's crown, wing vein or jugular vein, or by cardiac puncture. For turtles and snakes, blood collection should be carried out under anesthesia, and heart puncture or jugular vein puncture is required to collect blood.

6.3.1.7 Maximum Collected Blood Volume

The blood volume of each animal is relatively constant, accounting for about 8% of the body weight. If blood samples collected are more than 10% of the total blood, animals may have low blood volume (hypovolemia) shock or heart vascular shock. The general principle is blood can be collected every two weeks, and the collected blood volume cannot exceed 8 mL/kg body weight (Table 6.4).

6.3.1.8 Repeated Blood Collection by Cannula Implantation

When collecting blood, it is acceptable to use the same size injector to perform i.v. administration. For large animals, a trocar may be used, which is a wire needle with large hollow inclusion. The study of laboratory animals sometimes requires blood samples to be collected repeatedly. For rabbits or larger animals, venipuncture can be used to collect blood repeatedly. For small rodents, a cannula can be implanted between the jugular and cranial vein cavity or between the femoral vein and caudal vena. After restraining the neck or back subcutaneous ring, the cannula then scratches at the top or back up. Next, use a bolted joint and acrylic glue to restrain in place.

6.3.1.9 Exsanguination

To get the maximum blood sample required for the study, a blade or scissors can be used for decollation or surgical aortic puncture (exsanguination) under anesthesia. In this way, 30 mL of blood per kg of body weight or 50% of the animal's whole blood can be collected. Under anesthesia, through eye removal and cervical dislocation, blood collection from the ophthalmic artery can also be performed.

6.3.2 Collection of Feces and Urine

6.3.2.1 Metabolism Cage

Small rodent excreta (e.g., urine and feces) quantitative determination requires the use of special metabolic cages (Figure 6.14). During experimentation, laboratory animals are fed in the upper part of the metabolic cage and the lower part is a funnel, which is used for separation and collection of animal feces and urine.

A simple method of compulsory urination can be used to collect a small amount of urine from mice and rats, hamsters, gerbils, and guinea pigs. Press on either side of the animals' sacrum gently or press over the bladder to make animals discharge a small amount of urine, and then collect in a container prepared in advance.

6.3.2.2 Catheterization

For some animals, it is acceptable to collect urine through an intraurethral cannula. Insert the catheter into the urethra and push forward into the bladder. It is relatively simple for males because the urethra ends in the penis. Most female

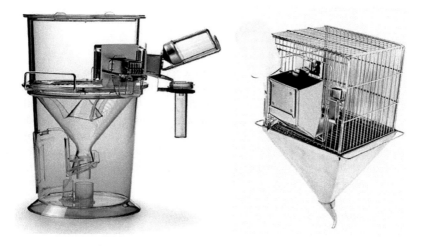

Figure 6.14 The mouse metabolic cage.

mammals' urethral openings are in the vagina and are difficult to see. For mice, rats, hamsters, and guinea pigs, the urethra and vagina are completely separated. If catheterization must be used for females, a more detailed understanding of the animal's anatomy is required in advance. For rabbits, restrain the rabbit on its back and use glycerol to lubricate the catheter. For males, hold the penis with one hand and use the other hand to stroke the penis foreskin and expose the glans crack to open the urethral opening. Next, insert the catheter slowly from the urethral opening. There will be resistance in the urethral sphincter. Do not force the catheter, rather insert into the bladder gently and urine will flow out naturally. For females, the external urethral opening is inside the vaginal vestibule and cannot be seen from the outside. Insert the catheter along the direction of the ventral vagina clitoris in the vaginal vestibule abdominal wall. Some species of animals (e.g., cat) require a sedative for catheter insertion.

The diameter of the catheter used in laboratory animals depends on the type of laboratory animal such as 0.5 mm for guinea pigs and 3 mm for dogs.

6.3.3 Collection of Other Bodily Fluids

Other bodily fluids include cerebrospinal fluid, bile, lymph fluid, and ascitic fluid.

6.3.3.1 Cerebrospinal Fluid

There are two ways to collect cerebrospinal fluid from animals: to puncture the cerebellomedullary cistern, which is located between the skull and the first cervical vertebrae or to puncture the lumbosacral space, which is located between the bottom lumbosacral region and the sacrum. For dura mater puncture, always use a trocar containing maidrin between the two vertebrae. After puncturing, remove the maidrin and suck out the cerebrospinal fluid. This process is carried out under local anesthetic or sedation.

6.3.3.2 Bile

Bile collection requires surgery to open the animal's abdomen and to insert the cannula into the cystic duct. One end of the cystic duct lies in the hepatic hilar region in the liver and the other end opens to the duodenum. Bile collection is generally performed last. After inserting the catheter, bile will no longer flow into the duodenum, which will affect the animal's ability to digest food. If chronic studies are needed without influencing the animal's digestion ability, T intubation can be used to make bile continue to flow into the duodenum and regain enterohepatic circulation in addition to bile collection.

6.3.3.3 Lymph Fluid

Lymph fluid can be collected by inserting the thoracic duct between the abdominal vertebrae and aorta.

6.3.3.4 Ascitic Fluid

Ascites are produced by implanting hybridoma cells into the abdominal cavity of rats and mice, producing monoclonal antibodies that are secreted by the hybridomas. The monoclonal antibodies are then collected and purified. The whole amount of animal ascites collected should not exceed 20% of the body weight. Ascites collection by puncturing should be performed under anesthesia. Moreover, collecting ascites by puncturing will make animals very uncomfortable. Other methods should be used in place of animal ascites to produce monoclonal antibodies.

6.4 ANIMAL SURGICAL OPERATIONS AND IMAGING TECHNOLOGY

The following items should be carried out by implementing surgery in laboratory animals: surgical teaching, new surgical techniques and new materials testing, production of animal models of human diseases (e.g., renal artery stenosis model of hypertension or partial liver resection of regeneration model), etc.

In recent years, microsurgery has been one of the developments in surgical techniques. Microsurgery refers to surgery by microscope and by employing microsurgical techniques, it is possible to perform organ transplantation in mice and rats.

Animal surgical operations need professional technology, which can only be performed under professional guidance and training, and advanced laboratory equipment is also required. In addition, familiarity with anesthesiology and anatomy is indispensable.

A sterile environment in every animal surgical operation is required, and when necessary, antibacterial treatments are performed such as the use of antibiotics to prevent infection. There is no evidence that small rodents have more resistance to infection caused by surgery than other animals. Thus, for these animals, operations performed under asepsis and necessary antibiotics are needed to prevent infection.

Several surgical instruments, ways to prevent bleeding, wound suturing, and other professional knowledge of surgeries will not be discussed here.

If animal blood loss occurs or too much liquid is lost during the surgery, fluid replacement therapy should be used in time. Small animals should be given warm saline subcutaneously to prevent dehydration. Action should be taken to avoid low temperatures. Injuries must be checked once a day. If animals try to tear apart the sutures, use outer garments or cervical collar for protection. In general, after 7–10 days, the sutures can be removed.

6.4.1 Invasive Techniques

6.4.1.1 Ectomy

Ectomy refers to the removal of animal organs or parts of organs, and one of the goals is to make a special animal model. This technique is commonly used in studies on the endocrine and immune system during animal experimentation.

Endocrine system: Includes hypophysectomy (pituitary gland) removal of the control center of the endocrine system; parathyroidectomy (parathyroid gland), these glands are in the cervical region, and it is difficult to remove these glands, respectively; pancreatectomy (pancreas); adrenalectomy (adrenal); gonadectomy (gonad).

Immune system: Includes thymectomy (thymus, mainly located in the chest). As nude mice and nude rats who lack thymus have been found, thymectomy is no longer used; lymph node resection (lymph nodes); resection of spleen (spleen).

Other systematic hepatectomy: Includes hepatic resection (liver), which is commonly used to remove one hepatic lobe, thus referred to as partial liver resection; nephrectomy (kidney); hysterectomy (uterus); removal of part of the brain tissue.

6.4.1.2 Fistulas

Fistulas are artificial channels that are implanted in the animals' body. One end opens to the gastrointestinal region, gall bladder, or urinary bladder, and the other end is connected *in vitro*. By using intestinal fistulae, animal digestion, absorption, and intestinal secretions can be studied.

6.4.1.3 Transplantation

The most common transplantation in laboratory animals is skin grafts among inbred strain animals, which are done to test the animal's genetic background. Currently, lung, heart, liver, kidney, and pancreas transplantation experimentations can be carried out in laboratory animals. With the development of microsurgical operations, organ transplantation in small laboratory animals, such as rats, can be performed. The purpose of these methods is to study tissue and organ rejection and evaluation of rejection drugs.

6.4.1.4 Implantation

This refers to the process by which some materials are implanted into the animal's body. Tumor tissue implantation is the most common form, in particular, implanting human tumor tissue into nude mice to study the characteristics of tumor tissue. Animal skin is preferred for the implant position, but other positions are also acceptable such as rat's renal capsule or liver, or in the hamster's cheek pouch.

6.4.1.5 Shunts

This is a pathway to connect body blood vessels, and it is often used for the connection between the arteries and veins such as the arteriovenous fistula between the common carotid artery and jugular vein. There are also fistula between vein and vein, such as an end-to-side portacaval shunt built between the portal vein and the caudal vena cava, thereby making the intestinal blood flow to the caudal vena cava directly.

6.4.2 Stereotactic Technology

In the study of the brain, stereotactic technology has been widely used. This technology uses single or double electrodes in different areas of the brain. Through these electrodes, brain potentials or electrical stimulation can be detected. Using this method, thin tubes can be inserted in the brain and through this intubation, traces of drugs can be released and whether the drugs excite or suppress certain brain regions can be observed.

It is necessary to understand animals' stereotactic atlas and specific stereo positioning equipment when conducting these studies. According to the skull, the brain can be three dimensionally classified. The animal's brain has three obvious planes: horizontal plane, referring to the center of the outer ear bone and the margin of eye socket (in mice, rats, and guinea pigs it is the edge of the maxillary bone between the incisor); frontal plane, referring to the center of the outer ear bone, and present perpendicular to the horizontal plane; sagittal plane, referring to the plane through the center of the skull and perpendicular to the horizontal plane.

Through the application of three-dimensional positioning equipment and a stereotactic atlas, electrodes and tubes can be inserted into specific areas of the brain.

6.4.3 Perfusions

Perfusion refers to pouring liquid into an animal's body or organs. Perfusion liquid can be used to replace blood flow from animals' bodies or organs. Aside from individual organ perfusion (e.g., perfusion of kidney), it is difficult for perfused animals to survive. Under anesthesia, arteries and veins can be separated from the surrounding tissues, then the arteries intubated, ligated, and restrained, followed by perfusion. Organs can be separated from the body by cutting the two ends of veins and arteries, which are inside organs, and then perfusion can be implemented.

6.4.4 Biotelemetry

Biological telemetry (biotelemetry) is a technique of measuring animal physiological indexes without directly contacting the animal's body or performing traumatic experimentation. Biotelemetry stimulation on animals is low. Heart rate, electrocardiogram (ECG), blood pressure, temperature, and other physical parameters can be measured in the free movement state of laboratory animals. After anesthetizing animals, one implantable transmitter is implanted into the animal's body. This transmitter's outer receiver can send a signal, and these original materials may be converted and stored in a computer through a data collection system.

6.4.5 Imaging Technology

In vivo imaging is the new standard in biomedical research and drug development.

Radiology is often used in animal experimentation such as for anesthesia and cannula insertion. After giving contrast medium, blood flowing through organs can be observed under x-ray irradiation.

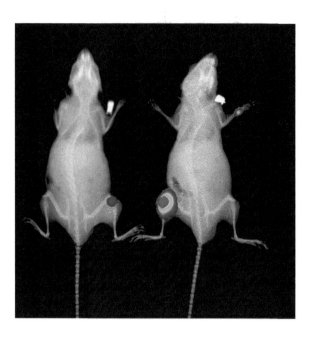

Figure 6.15 The human prostate cancer cell strain PC-3 labeled with luciferase was injected into the nude mice. After 6 weeks, the nude mice were injected intraperitoneally with luciferase substrate, bioluminescence imaging could detect upper and lower limbs tumor bone metastasis.

Scanning technology was used to study organ transplantation and tumor growth. For detection, markers can be combined with transplanted or tumor tissue by injecting a special radioactive marker or fluorescence dye. Using a scanner, features, such as the size of the tumor, can be estimated. Computer tomography (CT), positron emission tomography (PET), PET–CT, nuclear magnetic resonance (NMR), magnetic resonance imaging (MRI), or bioluminescence imaging technology (Figure 6.15) can obtain clear images. In NMR technology, the magnetic differences of different compounds in the tissues become visible, organized images. The detected signals are from protons in the compounds, and the image is clearer.

6.5 ANESTHETIZATION OF LABORATORY ANIMALS

Animal experimentation can lead to pain and distress. For both humanitarian and scientific research reasons, animal pain and suffering in animal experimentation should be reduced or even eliminated.

Scientific studies have confirmed that pain and suffering can cause animal tissues and organs to have obvious physiological reactions, which can be reduced by reducing or eliminating the pain and suffering of animals, thereby improving the accuracy and validity of animal experimentation.

Surgical operations on laboratory animals can be simply divided into two types: survival and nonsurvival. The former refers to those operations in which animals need to be alive after anesthesia or at the end of surgeries; the latter refers to those operations in which animals do not need to survive after anesthesia or at the end of the surgery, and are euthanized. When performing survival surgery, all procedures, including caesarean section, amputation, injectable drugs, or dealing with wounds, shaving, disinfection, sterilization, and aseptic operation of the surgical site need to be strictly implemented. According to experimentation conditions, the animals can be given antibiotics to prevent infection before the surgery. For nonsurvival surgery, although there is no need for such strict aseptic operation, the operation site should be shaved, disinfected, and processed. The researcher should also wear gloves.

Pain caused by surgical experimentation can be avoided through appropriate anesthesia. The researcher should be aware that most anesthetics have certain effects on animal tissues and organs, and the use of anesthetics may affect the experimental scheme. To minimize anesthetics' effect on the experimental results, full consideration should be given to the anesthetic pharmacological effects when choosing an anesthetic.

Postoperative pain and pain caused by nonsurgical operation can be eased through the use of analgesics. To effectively control the pain, it is important to assess the pain grade of animals correctly. Pain is an animal's individual sensory and emotional experience, and as animals cannot communicate with us, it will be very difficult to establish animal pain models. By comparing the similarity of structure and function of human and animal central nervous systems, the mechanism of animal pain can be studied. In addition, scientists now understand the human pain reaction process, and the widespread use of analgesics changes the human response to pain. Based on the similarity of animal and human responses to pain, the conclusions from the human body are also applied to animals. Animals and human beings are the same: harmful stimulus is likely to produce discomfort.

6.5.1 The Concept and Classification of Anesthesia

Researchers must make a detailed plan before the operation, and understand the anatomical structure and physiological characteristics of laboratory animals. In addition, the aseptic concepts and technologies of the operator must also be considered before the operation. In general, when performing the experiment for the first time, veterinarians or animal technicians should be consulted to ensure the experimental steps, drugs, and equipment are correct. During the experimental process or operation, animals must be given pain relief or anesthetics.

Anesthesia is in the form of drugs or other means to suppress the activity of local or central nerve tissue, so as to make a part or the entire body of animals feel completely lose. Anesthesia is a reversible, controlled situation, which may inhibit harmful feelings from the central nervous system (pain) and other stimuli. Anesthetics can be produced by drugs, which may lead to loss of consciousness and loss of pain (general anesthetic) or by medicines that may result in losing feeling in parts of the body (local anesthetic). Many kinds of drugs can be used for anesthetizing animals. Tables 6.5 through 6.7 list the commonly used laboratory animal species and dosage of anesthetic. The choice of anesthetic techniques depends on the animal species, way of implementing the experimental operation, the duration of operation, the experience of the researcher, and the purpose of the experimentation. If the execution is noninvasive or painless, sedation or mild sleep is best. For invasive operations, animals must be restrained, and valid measures implemented to ease the pain of animals. To collect laboratory animal physiological data, there must be a stable state, and the dosage of anesthetic should be as accurate as possible in the experiment. In addition, the anesthetic that causes the least damage should be chosen.

Ideal anesthetics should have the following characteristics: easy to control, ability to keep animals under deep and stable anesthesia, without affecting the animals' physiological functions, safe for humans and animals, and easy to reverse.

At present, the most effective anesthetic techniques have not yet fully achieved the above standards. For this reason, when giving anesthesia, animal anesthesia experts should be consulted to obtain better anesthetic methods. In addition, there are great differences in anesthetic effects between different kinds of animals, and the anesthetic method may be invalid from one species to another. Furthermore, great differences exist in different strains of the same species. Therefore, the animal species, strain, sex, body weight, animal diseases, and other comprehensive factors should be taken into consideration when choosing an anesthetic.

Animal anesthesia can be divided into general anesthesia and local anesthesia.

6.5.1.1 General Anesthesia

There are four types of general anesthesia: loss of consciousness (hypnosis), loss of feeling function (loss the feeling of pain), skeletal muscle relaxation, and reflex activity inhibition (autonomous capability stabilization). General anesthesia is subdivided into injectable anesthetics and inhaled anesthetics, and they can produce the effect of one anesthetic. For example, the subject may inhale the anesthetics halothane or isoflurane, and receive the anesthetics phenobarbital and propofol intravenously. In many animal studies, however, the degree of animal consciousness inhibition, reflex reaction process, and the degree of pain is not a constant requirement. One anesthetic cannot adequately anesthetize different parts alone. In contrast, if different anesthetics are applied to one or more parts to meet the requirements for special experimentation and degree of anesthesia inhibition, the anesthetics that produce balanced anesthesia can be selected. When using balanced anesthesia methods, the dose of anesthetic is generally lower in order to avoid too much damage. The disadvantage of balanced anesthesia is that there is a high potential that the anesthetics will interact with each other, and each anesthetic may have an effect on the animals' physiological activities. Therefore, a full understanding of the pharmacological effects of the anesthetics used in the experiments is required.

Table 6.6 Commonly Used Anesthetics and Dosage for Rodents and Rabbits

	Mice	Rats	Hamsters	Gerbils	Guinea Pigs	Rabbits
Preanesthetic (Anticholinergics)						
Atropine			0.05 mg/kg, s.c.			0.05 mg/kg, i.m.
Glycopyrrolate			0.01 mg/kg, s.c.			0.1 mg/kg, s.c.
Preanesthetic (Sedatives)						
Diazepam	2.5–5 mg/kg, i.p.	2.5 mg/kg, i.p.		5 mg/kg, i.p.		1–2 mg/kg, i.v.
Acepromazine	2–5 mg/kg, s.c.	2.5 mg/kg, s.c.	5 mg/kg, s.c.	3 mg/kg, s.c.	2.5 mg/kg, s.c.	1 mg/kg, s.c.
Fentanyl/fluanisone	0.1–0.3 mg/kg, i.p.	0.3–0.5 mg/kg, i.p.	0.5 mg/kg, i.p.	0.5–1 mg/kg, i.p.	1 mg/kg, i.p.	0.2–0.5 mg/kg, i.m.
Xylazine	5–10 mg/kg, i.p.	1–5 mg/kg, i.p.	5–10 mg/kg, i.p.	2–3 mg/kg, i.p.	5 mg/kg, i.p.	2–5 mg/kg, i.m.
Medetomidine	0.1–0.3, mg/kg, s.c.		0.1 mg/kg, s.c., i.p.	0.1–0.2 mg/kg i.p.	0.3–0.5 mg/kg, i.p.	0.2–0.3 mg/kg, i.m.
Anesthetic (Anesthesia Duration: 5–10 min)						
Alphaxalone/alphadolone	10–15 mg/kg, i.v.	10–12 mg/kg, i.v.	150 mg/kg, i.p.	80–120 mg/kg, i.p.	40 mg/kg, i.p.	6–9 mg/kg, i.v.
Propofol	26 mg/kg, i.v.	10 mg/kg, i.v.	–	–	–	10 mg/kg, i.v.
Thiopentone	30–40 mg/kg, i.v.	30 mg/kg, i.v.	–	–	–	30 mg/kg, i.v.
Medetomidine	10 mg/kg, i.v.	7–10 mg/kg, i.v.	–	–	31 mg/kg, i.p.	10–15 mg/kg, i.v.
Anesthetic (Anesthesia Duration: 20–60 min)						
Ketamine/Acepromazine	100 mg/kg, i.p. 2.5–5 mg/kg, i.p.	75 mg/kg, i.p. 2.5 mg/kg, i.p.	150 mg/kg, i.p. 5 mg/kg, i.p.	75 mg/kg, i.p. 3 mg/kg, i.p.	125 mg/kg, i.p. 5 mg/kg, i.p.	50 mg/kg, i.m. 1 mg/kg, i.m.
Ketamine/Diazepam	100 mg/kg, i.p. 5 mg/kg, i.p.	75 mg/kg, i.p. 8 mg/kg, i.p.	70 mg/kg, i.p. 2 mg/kg, i.p.	50 mg/kg, i.p. 5 mg/kg, i.p.	100 mg/kg, i.p. 5 mg/kg, i.p.	25 mg/kg, i.m. 5 mg/kg, i.m.
Ketamine/Xylazine	100 mg/kg, i.p. 10 mg/kg, i.p.	90 mg/kg, i.p. 10 mg/kg, i.p.	200 mg/kg, i.p. 10 mg/kg, i.p.	50 mg/kg, i.p. 2 mg/kg, i.p.	40 mg/kg, i.p. 5 mg/kg, i.p.	25–35 mg/kg, i.m. 5 mg/kg, i.m.
Ketamine/Medetomidine	75 mg/kg, i.p. 1 mg/kg, i.p.	75 mg/kg, i.p. 0.5 mg/kg, i.p.	100 mg/kg, i.p. 0.25 mg/kg, i.p.	75 mg/kg, i.p. 0.5 mg/kg, i.p.	40 mg/kg, i.p. 0.5 mg/kg, i.p.	25 mg/kg, s.c. 0.2–0.3 mg/kg, s.c.

(Continued)

Table 6.6 (*Continued*) Commonly Used Anesthetics and Dosage for Rodents and Rabbits

	Mice	Rats	Hamsters	Gerbils	Guinea Pigs	Rabbits
Pentobarbital	40–60 mg/kg, i.p.	40–55 mg/kg, i.p.	50 mg/kg, i.p.	60 mg/kg, i.p.	37 mg/kg, i.p.	30–45 mg/kg, i.v.
Anesthetic (Long Duration, Non-Recovery)						
Chloralose	50–100 mg/kg, i.p.	55–65 mg/kg, i.p.	50–100 mg/kg, i.p.		70 mg/kg, i.p.	80–100 mg/kg, i.v.
Urethane	1 g/kg, i.p.	1–2 g/kg, i.p.	1–2 g/kg, i.p.		1.5 g/k, i.p.	1 g/kg, i.p.
Anesthetic (Inhaled Anesthetic)						
Ether	induction concentration 15%–20%, maintenance concentration 5%					
Halothane	induction concentration 4%–5%, maintenance concentration 1%–2%					
Isoflurane	induction concentration 4%, maintenance concentration 1.5%–3%					
Sevoflurane	induction concentration 8%, maintenance concentration 3%–4%					
Methoxyflurane	induction concentration 4%, maintenance concentration 0.5%–1%					

Note: i.p., intraperitoneal injection; i.v., intravenous injection; i.m., intramuscular injection; s.c., subcutaneous injection.

Table 6.7 Commonly Used Anesthetics and Dosage for Dogs, Cats, Ferrets, and Large Animals

	Dogs	Cats	Ferrets	Goats/Sheep	Pigs	Primates
Preanesthetic (Anticholinergics)						
Atropine	0.05 mg/kg, s.c.	0.05 mg/kg, s.c.	–		0.05 mg/kg, s.c.	0.05 mg/kg, s.c.
Glycopyrrolate	0.01 mg/kg, s.c.	0.01 mg/kg, s.c.	0.1 mg/kg, s.c.		0.01 mg/kg, s.c.	0.01 mg/kg, s.c.
Preanesthetic (Sedatives)						
Diazepam	–	–	2 mg/kg, i.m.	2 mg/kg, i.m. 1 mg/kg, i.v.	1 mg/kg, i.m.	1 mg/kg, i.m.
Acepromazine	0.03–0.06, mg/kg, i.m.	0.05–0.1 mg/kg, i.m.	0.2 mg/kg, i.m.	0.1 mg/kg, i.m.	0.2 mg/kg, i.m.	
Fentanyl/fluanisone	–	0.2–0.3 mg/kg, i.m.		0.5 mg/kg, i.m.	–	0.3 mg/kg, i.m.
Xylazine	0.5–1 mg/kg, i.m.			Sheep: 1 mg/kg, i.m. Goats: 0.05 mg/kg, i.m.	–	–
Medetomidine	0.03–0.05 mg/kg, i.m.	0.04–0.08 mg/kg, i.m.	–		–	–
Anesthetic (Anesthesia Duration: 5–10 min)						
Alphaxalone/alphadolone	–	9–12 mg/kg, i.v.	8–12 mg/kg, i.v.	2.2 mg/kg, i.v. then 2 mg/kg, i.v.	6 mg/kg, i.m.	10–12 mg/kg, i.v.
Propofol	5–7.5 mg/kg, i.v.	7.5 mg/kg, i.v.		3–4 mg/kg, i.v.	3 mg/kg, i.v.	
Thiopentone	10–20 mg/kg, i.v.	10–15 mg/kg, i.v.		10–15 mg/kg, i.v.	6–9 mg/kg, i.v.	15–20 mg/kg, i.v.
Medetomidine	4–8 mg/kg, i.v.			4 mg/kg, i.v.	5 mg/kg, i.v.	10 mg/kg, i.v.
Anesthetic (Anesthesia Duration: 20–60 min)						
Ketamine/Diazepam	10 mg/kg, i.v. 0.5 mg/kg, i.v.	–	25 mg/kg, i.m. 2 mg/kg, i.m.	4 mg/kg, i.v. 1 mg/kg, i.v.	10 mg/kg, i.m. 2 mg/kg, i.m.	15 mg/kg, i.m. 1 mg/kg, i.m.
Ketamine/Xylazine	15 mg/kg, i.v. 1 mg/kg, i.v.	15 mg/kg, i.m. 1 mg/kg, s.c.	10 mg/kg, i.m. 0.5 mg/kg, i.m.	4 mg/kg, i.v. Sheep: 1 mg/kg, i.v. Goats: 0.05 mg/kg, i.v.	10 mg/kg, i.m. 1 mg/kg, i.m.	10 mg/kg, i.m. 0.5 mg/kg, i.m.
Ketamine/Medetomidine	5 mg/kg, i.m. 0.03–0.05 mg/kg, i.m.	5–8 mg/kg, i.m. 0.5–0.8 mg/kg, i.m.	–		–	–

(Continued)

Table 6.7 (*Continued*) Commonly Used Anesthetics and Dosage for Dogs, Cats, Ferrets, and Large Animals

	Dogs	Cats	Ferrets	Goats/Sheep	Pigs	Primates
Pentobarbital	20–30 mg/kg, i.v.	25 mg/kg, i.v.	25–30 mg/kg, i.v.	30 mg/kg, i.v.	30 mg/kg, i.v.	5–15 mg/kg, i.v.
Anesthetic (Long Duration, Non-Recovery)						
Chloralose	80–110 mg/kg, i.v.	80–90 mg/kg, i.v.	—	—	—	60 mg/kg, i.v.
Urethane	1 g/kg, i.v.	1.25 g/kg, i.v.	—	—	—	—
Anesthetic (Inhaled Anesthetics)						
Ether	induction concentration 15%–20%, maintenance concentration 5%					
Halothane	induction concentration 4%–5%, maintenance concentration 1%–2%					
Isoflurane	induction concentration 4%, maintenance concentration 1.5%–3%					
Sevoflurane	induction concentration 8%, maintenance concentration 3%–4%					
Methoxyflurane	induction concentration 4%, maintenance concentration 0.5%–1%					

Note: i.p., intraperitoneal injection; i.v., intravenous injection; i.m., intramuscular injection; s.c., subcutaneous injection.

In animal experiments, general anesthesia is often used in animals such as rodents, rabbits, dogs, and cats.

6.5.1.2 Local Anesthesia

Local anesthesia only acts on one part of the body and animals still remain conscious. Local anesthesia is widely used in many ways, and can be used to study the degree of anesthesia in different parts of animals. Local anesthesia includes the following:

- **Surface anesthesia**: Apply or spray the anesthetic directly onto the mucous membranes, or apply it on undamaged skin. In this way, the effect of local anesthesia is produced. It is usually used for small superficial surgeries over a small surface area such as insertion into the ureter or piercing the skin into the superficial blood vessels.
- **Infiltration**: The anesthetic infiltrates into deep tissue and achieve anesthesia. This method is suitable for small surgeries such as skin biopsies.
- **Local nerve block**: Local anesthetic is injected into certain parts of the nerve in the animals' body to control the neural controlling area. It is commonly used in anesthetizing an animals' limbs or tail, and is sometimes also used in surgery.
- **Regional anesthesia**: For large-area applications, anesthetics are injected into the external cavity by epidural. This method is also employed for limb plexus anesthesia. If the anesthetic is injected into the epidural space, the method is called spinal anesthesia. If the anesthetic is injected into the external epidural cavity, then it is called epidural anesthesia. Under correct dosing, spinal and epidural anesthesia are used in large animals; in particular, in surgeries carried out on the lower parts of the body in sheep and cattle such as the legs and abdomen. This kind of anesthesia is sometimes used in smaller mammals (e.g., dogs or rabbits). Plexus anesthesia is often used in the operation of limbs.

The local anesthetics often used in animal experiments are procaine, lidocaine, and bupivacaine. Adding epinephrine or norepinephrine may cause local vasoconstriction, and thus reduce anesthetic absorption and extend the lasting time of the anesthesia effects. However, there may be a cardiovascular response when animals are given catecholamine.

Compared with general anesthesia, the advantage of local anesthesia is that it rarely affects the normal physiological function of animals. Under spinal and epidural anesthesia, anesthetics will suppress the sympathetic nervous system, thus causing vasodilation in the anesthetized area, leading to lower blood pressure and tachycardia. The side effects of local anesthetics depend on the concentration and volume injected. For researchers who do not perform restraining of laboratory animals, the biggest limitation of local anesthesia or regional anesthesia is that the animals are still conscious, but the injection of sedatives can solve this problem. In addition, local anesthesia can take away pain for those animals that are waking due to injection of low doses of soporific or general anesthesia.

Local anesthesia is often given to large animals during operation such as cattle and horses.

6.5.2 Induction and Maintenance of Anesthesia

6.5.2.1 Preparation before Anesthesia

There are necessary preparations needed before the implementation of anesthesia. First, the most appropriate anesthetic must be selected; the instruments or equipment used for injecting the anesthetic need to be checked in advance to ensure normal operation. Furthermore, enough anesthetic and first-aid medicine is required. Animals to be anesthetized should spend 1–2 weeks for adaptation. Before anesthesia, the animals' body weight, and the intake of food and water must be observed and recorded. Animals should be healthy and without any clinical symptoms of diseases. It is better to confirm whether the animals have potential diseases before anesthesia and surgery. For example, in some studies, laboratory tests are often done on animals to confirm whether the animal's health situation is appropriate for anesthesia or not. In addition, for large laboratory animals, hematology and biochemistry tests must be conducted and evaluated, such as detection capacity of hematocrit and hemoglobin, which can be used to determine the animals' intraoperative bleeding, and records of these parameters are helpful for animal anesthesia and surgery.

Large animals must undergo fasting in advance in order to prevent ruminating and aspirating gastric contents. Similarly, in the process of anesthesia recovery, some feeding animals may vomit. To avoid such problems, animals, such as dogs, cats, ferrets, pigs, and nonhuman primates, must fast for 6–12 h in advance. For rodents and rabbits, it is not necessary to fast before anesthesia. Guinea pigs require extra attention because of more saliva secretion.

If the animal is to recover from anesthesia, postoperative nursing work should be done before starting the anesthesia operation. For example, preparation of a suitable recovery box, and ensuring that the recovery box's thermostat can be maintained long enough.

When using a new method of anesthesia for the first time, it is best to perform a preliminary experiment, that is, put one animal under anesthesia to ensure an appropriate anesthesia depth and smooth recovery. There may be different reactions to anesthesia among different stocks and strains of animals, and sometimes this difference is considerably large. Thus, the doses of anesthetic used to implement anesthesia may vary for particular strains of animals.

Before anesthesia, drugs can be given to animals to reduce the side effects of anesthetics and to keep pain at the minimum when performing the anesthesia procedure. Application of anticholinergic drugs, such as atropine or glycopyrronium bromide, can inhibit the respiratory tract and salivary gland secretion, prevent slow heartbeat caused by vagus nerve excitement, and block any unnecessary autonomic response caused by visceral traction during drug or surgical operation. However, this is not suitable for animals with arrhythmia.

Sedatives or tranquilizers can be used to ease animal tension and make animals easy to be controlled. Sedatives can also induce moderate anesthesia, and reduce or directly stop the anesthesia reaction and excitement period. This reduces the dose of anesthetic indirectly and reduces mortality from anesthesia allowing the anesthesia and recovery process to go more smoothly. Most commonly used sedatives and tranquilizers do not lead to loss of pain and other drugs are needed to control the preoperative and postoperative pain.

Drugs for different species before anesthesia are shown in Tables 6.6 through 6.8.

6.5.2.2 Induction and Maintenance of General Anesthesia

General anesthesia can be induced by injecting one or more compounds into the skin, muscle, or intraperitoneal cavity, as well as through inhaling volatile anesthetics. For laboratory animals whose body weight is less than 1 kg, they may inhale anesthesia in an anesthesia box. For larger animals, after restraining, a mask may be used to enable anesthesia inhalation. It may be difficult to restrain the animals during the implementation of anesthesia, and sedatives can be used to reduce stress before anesthesia induction. Alternatively, a short-acting anesthetic can be used to make animals lose consciousness and an inhaled anesthetic to maintain the effect of anesthesia in the later stages.

The advantage of inhaled anesthesia is that by using a standard caliber sprayer, researchers can easily determine the depth of anesthesia. This can also reflect the effects on animals caused by trauma levels generated in surgery. Unless a continuous i.v. drip is used, repeated i.v. anesthesia may lead to unstable anesthesia depth. In addition, anesthetic accumulation can also occur, which may prolong recovery time. The recovery of short-term anesthesia (less than 30 min) is rapid because this method is easy to implement for small rodents and these small rodents recover very fast. Inhaled anesthetics may be a choice when studying these species.

6.5.3 Inhaled Anesthetics

6.5.3.1 Commonly Used Inhaled Anesthetics

- **Isoflurane**: An effective anesthetic to quickly induce anesthesia and enable rapid recovery after surgery. Isoflurane should be used within a standard atomizer, and it produces safe and effective anesthesia effects in most laboratory animals. Isoflurane is not an explosive or flammable substance, but it may stimulate the respiratory tract of animals and slow down the process of anesthesia induction. Isoflurane does not transform within the body and almost all of the isoflurane can be excreted through the breath. It does not consume enzymes in the liver, and in studies on drug metabolism, it can minimize the risk caused by anesthetic effects on metabolism. Although isoflurane will not reduce the reaction of pressure receptors, it leads to lower blood pressure and tachycardia.
- **Halothane**: Like isoflurane, halothane is also an effective anesthetic, which should be used in a standard atomizer. The induction and recovery speed of halothane is slightly slower than that of isoflurane. Halothane may cause local moderate hypotension at the level of anesthesia used when performing operations. Halothane may not only be excreted out of the body through breathing, but may also play an important role in metabolism.
- **Methoxyflurane**: Methoxyflurane is also an effective anesthetic. Compared with isoflurane and halothane, the atomization of methoxyflurane is a safe and simple

Table 6.8 Commonly Used Anesthetics and Dosage for Birds, Amphibians, Reptiles, and Fish

	Ketamine	Pentobarbital	Urethane	Halothane	Isoflurane	Note
Birds	—	—	—	2%–4%	3%–5%	Pigeons: ketamine 30 mg/kg + medetomidine 10 mg/kg, i.m., or pentobarbital 10–20 mg/kg, i.m.
0.1 kg	10–20 mg/kg, i.m.	—	—	—	—	
0.1–0.5 kg	5–10 mg/kg, i.m.	—	—	—	—	
0.5–3 kg	2–5 mg/kg, i.m.	—	—	—	—	
Snakes	20–80 mg/kg, i.p.	15–30 mg/kg, i.p.	—	3.5%–6.5%	4%–6.5%	Medetomidine 8–10 mg/kg, i.p. Thiopentone 8–45 mg/kg, i.p.
Lizards	15–17 mg/kg, i.m.	10–25 mg/kg, i.p.	—	4%–5% induction; 1%–2% maintenance	4%–5% induction 2%–3% maintenance	Induced duration 10 min
Tortoises	60–120 mg/kg, i.m.	10–30 mg/kg, i.p.	—	4% induction 1.5% maintenance	4% induction 2% maintenance	Induced duration 10 min
Frogs	—	30–60 mg/kg, Inject to lymph sack	20 mL/kg, Inject to lymph sack	—	—	10% chloral hydrate, 1–2 mL inject to lymph sack
Fish	—	—	10–90 g/L, water	—	—	Propanidid 0.2–1.5 mg/L, water

Note: i.p., intraperitoneal injection; i.m., intramuscular injection.

process performed in a simple container. If an atomizer is not available, methoxy-flurane liquid can be dumped on absorbent cotton directly and placed inside the inhalation chamber. A metal net or similar device should be used to avoid the absorbent pads with methoxyflurane from contacting the animals directly as methoxy-flurane liquid is irritating. The metabolism of methoxyflurane may happen widely within the body, and the release of inorganic fluoride ions can cause damage to the kidneys, and should therefore be avoided in the study of the kidneys. Some countries have already banned selling methoxyflurane on the market.

- **Enflurane**: Similar to halothane, but enflurane has a slightly quicker induction and recovery than halothane. It is not metabolized in as many places as halothane, and as it has no obvious advantages compared with isoflurane and halothane. It is rarely used for laboratory animals.

- **Sevoflurane and desflurane**: Two kinds of new inhalation anesthetics with fast induction and recovery. The metabolic location of sevoflurane is widespread, but it does not damage liver or kidney function. Desflurane itself does not participate in the metabolism, but as it can lead to the excitement of the sympathetic nervous system, it is seldom used for laboratory animals.

- **Ether**: An anesthetic that is flammable, explosive, and irritating. Ether gas can lead to high saliva secretion, increased bronchial secretions, and sometimes accidental laryngospasms. Ether can increase original respiratory disease in rodents and rabbits. Despite these obvious shortcomings, ether is still used as an anesthetic for small rodents. When used, ether is put into a simple container or placed on skim pads in the anesthesia box. Excessive anesthesia is rare, even when used roughly. However, due to animal welfare, the safety aspects of ether and other factors, many laboratories no longer use ether and tend to choose anesthetics that are more secure, humane, and effective.

- **Nitrous oxide**: Commercial nitrous oxide is stored in a pressurized device as a liquid. To avoid hypoxia, it should be used after mixing with oxygen. The mixture ratio is that nitrous oxide is less than 65% and oxygen is greater than 35%. Nitrous oxide is an effective painkiller, but there are no obvious analgesic effects in experimentation on mammals. It cannot make the animal lose consciousness even if the concentration is higher than 70%. This feature limits its applications for laboratory animals and it is only used in combination with other anesthetics in larger laboratory animal anesthesia such as in dogs and cats.

Most inhaled anesthetics can cause bronchiectasis and lead to the disappearance of hypoxic pulmonary vasoconstriction and reduction of airway mucus cilia function. This results in the mismatch of exhaling and inhaling, thus increasing the probability of lung infection.

Inhaled anesthetics can reduce myocardial contractility, and cause vasodilation and hypotension. Most of the i.v. anesthetics produce the same response in the cardiovascular system, but ketamine can increase myocardial contractility and vasoconstriction.

6.5.3.2 Anesthetic Potency

Inhaled anesthetic-induced loss of consciousness and the required inhaled anesthesia concentration are correlated with the animal species, the implementation of the steps of the anesthesia and the used anesthetic. Minimum alveolar concentration (MAC) refers to the concentration percentage of inhalation anesthetics, that is, at this concentration, 50% of individuals do not show reactions to standard pain stimulus (Table 6.9). For general surgery anesthesia, 1.5 times of the MAC is used. In special cases, some certain species may need twice or more the MAC to meet the operation requirements.

6.5.3.3 The Removal of Inhaled Anesthetics

Long-term exposure to inhaled anesthesia gas is harmful to humans and animals. In the animal operating room, special equipment is needed to remove indoor anesthetic gases effectively, which is necessary for normal operation. Inhaled anesthesia gases not only affect cognitive ability, but also produce teratogenic effects by raising the incidence of spontaneous abortion. In many countries, it has been clearly defined in the relative security law that in the use of inhaled anesthesia gas, anesthetic exhaust must be removed effectively to reduce pollution in the laboratory and operating room.

6.5.4 Anesthetics Used for Injection

Many different anesthetics used for injection can be applied for laboratory animals. Next, we will provide a brief summary and list the reference doses in Table 6.6 through 6.8.

When using an anesthetic, it is very important to emphasize differences in reactions among different individuals of the same and different species. Fluctuation in the reactions to anesthetics is common in small rodents. Intravenous delivery of anesthetics is difficult in small rodents thus, abdominal administration is used instead of i.v. injection. When intraperitoneal injection of anesthesia is performed, the anesthetic is like a disposable bolus quickly injected into the abdominal cavity and does not gradually enter into the body like vein titration; however, satisfactory anesthetic effects can be obtained. When anesthetizing rabbits and larger laboratory animals, the venous approach can be used to administer anesthetic. When the injected anesthetic dose reaches approximately 50% of the expected value, the

Table 6.9   MAC Value (%) of Inhaled Anesthesia in Different Species

	Ether	Halothane	Enflurane	Isoflurane	Nitrous Oxide
Humans	1.92	0.75	1.68	1.15	105
Primates	–	1.15	1.84	1.28	200
Dogs	3.04	0.87	2.20	1.41	188
Pigs	–	–	–	1.45	–
Sheep	–	–	–	1.58	–
Cats	2.10	0.82	1.20	1.63	255
Rats	3.20	1.10	–	1.38	150
Mice	3.20	0.95	–	1.41	275

Note: MAC—minimum alveolar concentration.

remaining drug should be injected at a slower speed to achieve the most appropriate anesthesia depth. Administration through the tail vein in mice and rats can be performed if professional knowledge is adequate.

In the following, several commonly used anesthetics will be introduced respectively.

6.5.4.1 Barbiturates

The two, short-acting anesthetic barbiturates that are widely employed in laboratory animals are thiopentone and methohexitone, as well as the long-acting anesthetic, pentobarbital. All of these anesthetics may produce hypnotic effects, but they lack inner pain-stopping effects. Therefore, only the higher doses that lead to cardiovascular and respiratory depression may obtain the anesthesia effects required for surgical operation. If barbiturates are injected intravenously, the dosage can be adjusted accurately and usage is relatively safe. It is difficult to predict the effects of short-acting barbituric salts through intraperitoneal injection. As such, the short-acting barbituric salts are not advocated for intraperitoneal injection. Thiopental sodium has a high pH, thus, intraperitoneal injection can produce serious stimulation reactions and is not recommended. Intraperitoneal injection of pentobarbital can produce the effects needed for surgical anesthesia, but the ideal range for anesthesia use is narrow, it is less secure with a high mortality, and is therefore not an ideal anesthesia method. In addition, many intraperitoneal injections of anesthetics may potentially harm animal organs and tissues.

6.5.4.2 Dissociation Anesthetics

Ketamine and tiletamine are the widely used dissociation anesthetics. For some larger laboratory animal species, especially nonhuman primates, dissociation anesthetics may obtain shallow surgical anesthesia effects, but the muscle relaxation effects are poor. In laboratory animals like small rodents, dissociation anesthetics have almost no effects unless a large dose, which is dangerous for the animals, is used. If dissociation anesthetics are mixed with sedatives, the anesthesia effects are improved significantly. Finished products of tiletamine include tiletamine and zolazepam. Ketamine can be combined with sedatives such as acepromazine, midazolam, and diazepam. Anesthetic effects will be improved if ketamine is combined with xylazine or medetomidine. The effects of these painkillers can be weakened by a special antagonist (adenosine triphosphate), and the recovery time is relatively shorter.

6.5.4.3 Neuroleptanalgesics

These are also known as nerve analgesics and are a strong and effective mixture of analgesic and antipsychotic medications. The widely used commercial preparations are fentanyl/fluanisone, fentanyl/droperidol, etorphine/methotrime prazine, and etorphine/acepromazine. When these drugs are used alone, marked analgesia can be produced, but the degree of muscle relaxation is lower and breathing is inhibited. For many animals, fentanyl/fluanisone and benzodiazepine can be used in combination. In this way, not only can surgical anesthesia effects be produced, but it can also make animal muscles relax and lower the degree of respiratory depression.

6.5.4.4 Steroid Anesthetics

Alphaxalone/alphadolone are effective anesthetics used for many laboratory animals. However, these anesthetics are not suitable for dogs and rabbits because the solvent used in these anesthetics will cause release of histamines in dogs; in rabbits, only doses that may cause respiratory arrest have the ability to produce surgical anesthesia effects. For other animals, if injected intravenously, alphaxalone/alphadolone will produce mild surgical anesthesia effects within 5–15 min. Repeated injection or continuous dosing can extend the lasing time of anesthesia, but animal recovery time will not be extended.

6.5.4.5 Benzodiazepines

Benzodiazepines, such as diazepam, midazolam, and olazepam, are anesthetic accessories for laboratory animals. They have effects on the cardiovascular and respiratory systems, and are used in combination with ketamine, opioids, and inhaled anesthetics.

6.5.4.6 Other Anesthetics

Sympathetic nerve-exciting drugs: adrenaline drugs, such as xylazine and medetomidine, may suppress the central nervous system and cause the central nervous system to induce muscle relaxation. They are often used as anesthetics for large ruminants and horses. Medetomidine is a new, strong, and effective alpha-2 agonist; it can produce excellent sedation and analgesia effects in dogs, cats, rabbits, and rodents. The use of medetomidine can also reduce the doses of other anesthetics.

Propofol may produce surgical anesthesia effects in primates, dogs, cats, goats, pigs, and most rodents via i.v. administration. Lasting time of anesthesia is short (<10 min), and recovery is quick. Repeated or continued administration can extend the anesthesia effects, and the animal recovery time will not be overextended. The anesthesia effects of propofol in rabbits is not adequate to perform surgery.

Tribromoethanol can be used for laboratory animals, such as most rodents, and produce surgical anesthesia effects and relax animal muscle. When intraperitoneally administered, decomposing tribromoethanol can irritate animals and lead to serious agitation or even death. Thus, researchers must be sure to use freshly prepared tribromoethanol solution. Even with freshly prepared tribromoethanol solution, when the second anesthetic is used moments later, it can also lead to gastrointestinal disorders and death of animals. A new study suggests that even fresh tribromoethanol may also be irritating and this preparation requires caution.

Alpha-chloralose and chloral hydrate are two basic soporifics that can slightly inhibit cardiovascular function. Intraperitoneal injection can stimulate the animal gastrointestinal tract and may cause intestinal obstruction. Alpha-chloralose combined with urethane has carcinogenic effects, and therefore, their combination can only be used for terminal experimentation.

6.5.5 Assessment of Anesthesia Depth

For all anesthetics, the most important thing is to detect the depth of anesthesia to ensure that animals do not feel pain due to weak anesthesia or die from too deep anesthesia. It is a continuous process from the conscious to completely unconscious state, rather than a series of unrelated steps. Reflection, change of position, breathing frequency and depth, change of heart rate, blood pressure, and other reactions to (pain) stimulation are used as the judgement indexes of anesthesia depth. However, the parameters of different anesthesia depths often change according to different laboratory animal species and anesthetics.

The anesthesia process is generally divided into four phases:

- **Induction phase**: In this period, animals are conscious, in a state of mild pain loss and are quiet with mildly delayed responses.
- **Excitement phase**: Animals gradually lose consciousness and reflex activity, and muscle movements increase. The pupils begin to expand, the secretion of tears and mucus increases, and the eye movements appear incongruous.
- **Operation phase**: Respiratory rate is decreasing and breathing depth is increasing. Eyelid and corneal reflection disappear, muscle tension, and reflex response decrease, with no response to surgery or other stimuli. This is the most suitable time for operation. It can be divided into mild, moderate, severe, and excessive stage.
- **Anoxic (toxicity) phase**: Also known as shock stage. Suppressed vital centers leads to the slowing or stop of breathing and heart rate. Mydriasis and light reactions disappear. The animal is likely to die within 1–5 min. Using excessive amounts of anesthetic will reach this stage when performing euthanasia.

Most animal experiments require anesthesia to reach operation phase. However, when used in combination with different anesthetics, the reactions described above may change significantly. The following reflex reactions can be used to evaluate and determine whether animals have sufficient anesthesia.

- **Righting reflex**: When animals are in a supine position, they always try to turn to the prone position. Animals remain in the supine position under the effects of anesthesia.
- **Eyelid reflex**: When the inside and outside canthus is touched, animals will blink. This reflection disappears in anesthesia phase.
- **Pedal reflex**: When the skin between the fingers or toes is pinched, the legs of animals will react by bending and straightening; however, this reflex disappears during anesthesia.
- **Swallowing reflex**: In the absence of anesthesia, pulling the tongue or squeezing the throat will cause the animals to swallow.

- **Tail pinch reflex**: When anesthesia depth is not very deep, nails or hemostatic clamp devices may be used to pinch the animal's tail, which will lead to flicking of the tail, and occasionally sound.
- **Ear pinch reflex**: When rabbits or guinea pigs are conscious, their ears may be pinched to produce a head-shaking reaction.

The disappearance of the righting reflex and pedal reflex is the most practical index to evaluate anesthesia effects for most mammals. In the process of anesthesia, these reflex reactions gradually disappear. For example, when pinching fingers (toes), some sound may be produced, which has already disappeared before the withdrawal reaction. The intensity of withdrawal reaction also gradually declines, leading to the possibility of performing surgery when moderate to deep anesthesia disappears completely. Do not misdiagnose animal reactions in the excitement period, such as shortness of breath, or abdominal breathing of the operation period (sudden contraction of the abdominal wall) as inadequate anesthesia, and then add additional anesthesia as this will lead to animal death.

In general, it is desired for the animal to enter the anesthesia period as soon as possible. This mild and moderate anesthesia is adequate for most of the surgery, and severe or excessive anesthesia leads to a high danger of death. At that moment, the operation must be suspended and first aid performed, otherwise, once animals enter the fourth period (shock period), more than 80% of the animals will die.

Physiological changes in different anesthesia periods are shown in Table 6.10.

6.5.6 Ventilation Methods and Devices

6.5.6.1 Artificial Ventilation

Artificial ventilation refers to the use of manual pressure devices or mechanical ventilation devices to forcibly put gas into the lungs. In animal thoracotomy or

Table 6.10   Physiological Changes in Different Anesthesia Phases

	Breath	Pupil	Eye Movement	Reflex	Muscle Tone	Pulse, Blood Pressure
Induction phase	Regular	Normal	Voluntary	Exist	Normal	Pulse ↑ Blood pressure ↑
Excitement phase	Irregular	Expand	Involuntary	Exist	Tension	Pulse ↑ Blood pressure ↑
Operation Phase						
Mild	Depth ↑ Speed ↑	Contract	Involuntary or restrained	Eye conjunctivas ↓	Slight relaxation	Normal
Moderate	Normal	Normal	Restrained	Throat ↓	Moderate relaxation	Normal
Severe	Depth ↓ Speed ↓	Slightly expand	Restrained	–	Extreme relaxation	Pulse irregular Blood pressure ↓
Excessive	Abdominal respiration	Moderate expand	Restrained	–	–	Pulse weak ↓ Blood pressure ↓

Table 6.11 Respiratory Parameters for Artificial Ventilation

Animal Population	Frequency (per min)	Tidal Volume (mL)	Inspiration/ Expiration Time (%)	Pressure (cm H₂O)
Mice	100–130	0.5–1	35/65	5–15
Rats	50–180	3–10	35/65	5–15
Guinea pigs	30–50	8–20	35/65	5–15
Rabbits	30–50	40–60	35/65	5–15
Birds	6–12	Variation with body size		5–15

administering medicine for muscle relaxation, artificial ventilation is required. If the operation time is more than 2 h, artificial ventilation can be used to ensure adequate gas exchange. In artificial ventilation, positive pressure ventilation is the process of inhalation, and exhalation is the passive process, which is due to the elasticity of the lungs. Increased lung pressure reduces the cardiac output when breathing. To minimize this effect, it is better to shorten the inspiratory time, which accounts for approximately 30% of a breathing cycle.

Although each animal has its own ventilation requirements, usually 10–15 times/ kg body weight of the tidal volume is able to maintain the normal respiratory function of animals. The breathing rate of different animals is not the same. Animals like dogs, sheep and pigs have a breathing rate of 10–15 times/min, while that of rabbits and rodents is 50–150 times/min (Table 6.11).

When using mechanical ventilation, air pressure must be monitored to avoid excessive charging and barotrauma. In general, for small animals, the inflatable excessive pressure should not exceed a 10 cm water column. In order to obtain good artificial ventilation, CO_2 content in the exhaled gas may be measured by a CO_2 analyzer, thus it is able to be kept within 4%–5% of the normal range. However, many CO_2 analyzers cannot accurately record CO_2 concentration breathed by small animals whose body weight is less 500 g because the amount of CO_2 is too little to be recorded.

When performing artificial ventilation, in order to make the basal metabolic rate of body fluid up to 10–15 mL/kg h^{-1}, an indwelling venous catheter must be inserted to input liquid to replenish the loss of blood.

6.5.6.2 Endotracheal Intubation

To insert one tube into the trachea, the animal must be first put under anesthesia. When animals rely on artificial ventilation to keep breathing, they must be endotracheally intubated. In addition, when the animal respiratory depression occurs, even if animals have autonomous breathing, endotracheal intubation is also a useful technique because it can keep the airway unobstructed and assist with ventilation.

For different animals, the implementation of endotracheal intubation is different, but the operation is simple to perform as long as a suitable laryngoscope is chosen. The endotracheal intubation technique is similar to that used for humans. Suitable corresponding endotracheal intubation techniques should be chosen for laboratory animals. Small laboratory animals, such as rodents, rarely receive endotracheal intubation.

Table 6.12 lists suitable specifications for endotracheal intubation in several animals.

All animals that need endotracheal intubation should be under full anesthesia. Open their mouth and pull the tongue forward to avoid the mastication and deglutition reflex. Then, the researcher can see the epiglottis and fold after inserting the laryngoscope. For cats, pigs, rabbits, and nonhuman primates, when performing endotracheal intubation, first spray local anesthetic into the animal's throat to reduce the risk of laryngospasm. Before inserting the tube into the airway, use a local anesthetic ointment on the tube. When performing endotracheal intubation in mice, rats, hamsters, and guinea pigs, a special laryngoscope is needed because the operation is more difficult, but can be performed with the aid of a 4-mm diameter otoscope. Animals are placed in the prone position, and local anesthetic on cotton swabs is spread on the mucus membranes of the mouth and throat to desensitize to stimuli. With the animal's head back and up, perpendicular to the desktop, use a cotton plug to push the tongue out of the mouth. Then, insert the otoscope into the oral cavity, at which the front side of the epiglottis can be seen. Put the stylet through the otoscope and bring the soft palate dorsally. The epiglottis will lower down and vocal folds will appear. Move the stylet through the vocal fold and carefully remove the otoscope. Then, insert the endotracheal tube along the stylet and remove the stylet after insertion. The tube cannot be inserted too deep to avoid one-lung ventilation.

For rabbits, a laryngoscope blade can be used to assist with endotracheal intubation, and the operation can be performed without seeing the throat. Rabbits should be placed in the prone position; the head is raised up to make the lower jaw perpendicular to the desktop. Put the thumb and index fingers between the upper and lower jaw to restrain the head, and make its mouth open to insert the tube via the soft palate. When the tube arrives at the throat and is going to pass through the vocal cords, the researcher should listen to whether breath sounds appear or not. A small mirror can be put at the end of the pipe or use a few hairs to check whether the intubation position is proper. When the animal exhales, the mirror will become blurred or hair will be blown away. As the larynx is vulnerable, be especially careful of damaging the throat during intubation. In-proper intubation will lead to airway obstruction due to laryngeal edema or bleeding, and eventually cause death. If intubation is difficult

Table 6.12 Diameter and Length of Endotracheal Tube for Different Animals

	Outer Diameter (mm)	Inner Diameter (mm)	Length (cm)
Mice	1.0	0.5	3
Rats	1.8	1.0	13
Hamsters	1.6	1.0	8
Guinea pigs	2.0	1.5	10
Rabbits	2.0–3.5	1.8–3.3	15

to implement, and rehabilitation after surgery is not required, tracheotomy can be performed.

For most birds and reptiles, it is simple to perform endotracheal intubation. After anesthesia induction, open the mouth and insert a tube into the trachea.

6.5.6.3 Anesthesia Respirator

For the animals' oxygen supply and inhaled anesthetics, different breathing machines can be used. The basic requirement of any breathing machine is that this machine can supply plenty of oxygen and discharge CO_2. The simplest technique to transfer oxygen and anesthetic gas is to use an airtight mask. Through mask or endotracheal tube, animals may be connected with a T tube. When using this system, in order to prevent used air from being sucked in again, the rate of fresh air that flows from the anesthesia machine will increase to 3 times/min for most animals' air capacity. Per minute air capacity is the amount of gas inhaled in 1 min, that is, respiratory capacity for 1 breath (tidal volume, about 15 mL/kg body weight) multiplied by the frequency of breathing. When anesthetizing large animals (body weight is more than 20 kg), the high quality and fresh gas flow rate may be not economical, thus a breathing machine with CO_2 absorber can be used. In addition, maintaining a sufficient and stable anesthesia depth needs a lot of experience.

6.5.6.4 Neuromuscular Blocking Agents

In order to perform artificial ventilation when using endotracheal intubation, animals can be given neuromuscular blockers (muscle relaxants) to relax the skeletal muscles. When using these drugs, the animal should be completely restrained, and even after recovering consciousness, it is not permitted to move in response to pain stimulation. Muscle relaxants should be used under strict conditions and administered by experienced professionals or technical personnel. When using muscle relaxants, monitor the animal's heart rate and blood pressure at all times.

The suitable and commonly used muscle relaxants and dosage for laboratory animals are shown in Table 6.13. Pancuronium, D-tubocurarine, and gallamine are common muscle relaxants. Naloxone is an antagonist. The characteristics of these drugs should be understood in detail before use.

6.5.7 Anesthesia Process Monitoring

In the whole process of anesthesia, it is necessary to observe and record some physiological indexes to ensure animal physiological activities remain within the normal range. The complexity of the observed animal physiology depends on the type and duration of animal experimentation. The assessment of an animal's basic situation can be done through simple observation. For example, the color of the mucous membrane, breathing pattern and frequency, heart rate and pulse, etc. The observation method of these indicators is simple, and is suitable for anesthesia evaluation in most animals. However, in the process of experimentation, it is very difficult

Table 6.13 Commonly Used Muscle Relaxants and Antagonists for Animals (mg/kg)

	Rats	Mice	Guinea Pigs	Rabbits	Cats	Dogs	Sheep	Goats	Pigs
Muscle Relaxant									
Pancuronium	–	2	0.06	0.1		0.06			
Alcuronium	–	–	–	–	0.1	–	–	–	–
Atracurium	–	–	–	–	0.5	–	–	–	–
Vecuronium	–	–	–	–	0.1	–	–	–	–
Gallamine	–	1	0.1–0.2		1			4	2
D-Tubocurarine	1	0.4	0.1–0.2		0.4			0.3	–
Antagonist									
Naloxone			0.1		0.05–0.1		0.1		

to observe repeatedly. Some indexes, such as blood oxygen, CO_2 concentration, and temperature, cannot be observed simply. Therefore, the use of electronic monitoring instruments is very useful. Monitoring equipment can make evaluation parameters more accurate. ECG and heart rate, cardiac output, arteries, pulmonary arteries and central venous pressure, CO_2 waveform figures (exhaled CO_2 concentrations), arterial blood gas and acid base environment (PaO_2, $PaCO_2$, pH, excess alkali, bicarbonate, etc.), arterial blood oxygen saturation, respiratory volume and respiratory frequency, airway pressure, body temperature, and electroencephalogram (brain) are all parameters applicable to electronic monitoring during animal experimentation.

Electronic detection devices are also applicable to animal anesthesia monitoring. Especially in the extension phase of anesthesia, the amount of inhaled anesthetic and respiratory oxygen concentration can be reasonably calculated. The selected electronic detection devices should be able to work well in the process of monitoring small animals. For example, when the heart rate is over 250 times/min, many heart rate monitors cannot count heart rate normally, while the heart rate of rodents is often more than 250 times/min at rest.

When anesthetizing laboratory animals whose body weight is less than 10 kg, maintaining the animals' body temperature requires extra attention. After anesthetizing, the body temperature of small animals decreases quickly. This is an important cause of increasing mortality, therefore, some equipment, such as an electric lamp, should be used to maintain the normal temperature of anesthetized animals.

Intravenous transfusion is essential as it can supplement the loss of moisture in the respiratory tract and blood loss caused by the surgery. As basic reference indicators, when normal saline (0.9%) is infused through the vein, the infusion speed is approximately 10–15 mL/kg h^{-1}. For small animals, subcutaneous or intraperitoneal administration can be used, but the absorption with these methods is relatively slow and the effects for acute fluid loss are not obvious. The liquid temperature should be close to the animals' temperature in case the animal's temperature drops excessively due to infusion.

If the animal's eyes are still open in the process of anesthesia, eye ointment or artificial tears should be used to prevent cornea from drying, or pull the eyelids down to cover the eyes and restrain the eyelids.

In the process of anesthesia, preliminary judgements about animal anesthesia can be made through visual inspection.

- **Respiration**: Rodents will stop breathing for a few seconds, then begin deep breathing in the process of anesthesia, which means the anesthesia is excessive; in general, thoracic breathing is mild anesthesia. The deeper the anesthesia is, the closer it approaches to abdominal breathing. Irregular breathing indicates that the animal is going to recover or anesthesia is excessive.
- **Color of mucosal membranes**: Under normal anesthesia, the color of mucosa (oral, anus) is pink, which shows enough oxygen; if the color is purple, which is a cyanosis phenomenon, it indicates a lack of oxygen.
- **Microvascular hyperemia time**: Press down on the animal's gums and then let go again; the time needed for gums to return to normal pink again is the microvascular hyperemia time. The normal time is less than 2 s. More than 2 s indicates poor heart output function.
- **Pulse**: Monitor the femoral artery of the hind legs, maxillary artery (large animals) and heart rate (rodents).
- **Reflex**: Pain reflex between toes is regarded as an index of anesthesia, and the oropharynx reflex can be used as an index of inhaled anesthesia recovery.
- **Eye reflex**: Nystagmus indicates shallow anesthesia. In the early stage of anesthesia, pupils dilate when excited and then narrow with the deepening of anesthesia level; pupils will markedly expand with excessive anesthesia.

6.5.8 Anesthetic Complications

Most anesthetics have many pharmacological side-effects, and in rare cases, can also cause serious complications. This is because, in some extreme cases, the decline in function of important organs triggers the automatic death program, and thus leads to shock and death.

6.5.8.1 Respiratory Depression and Stop

When the anesthetic concentration is too high, spontaneous ventilation function recession will occur, which leads to hypoxia and hypercapnia, increase of the nonoxygenated hemoglobin concentration in blood and skin mucosa to become bluish-violet. If this condition continues to develop, animals will stop breathing and undergo cardiac arrest. When this happens, if using inhaled anesthetics, immediately cut off the sprayer to make anesthesia circulate with oxygen; if using continuous-dripping anesthetic, cut off the infusion pump. If oxygen can be supplied, the fresh air velocity should be 3 times the amount of exhaled gas per minute. For small animals that have a tube inserted, supply oxygen through a mask and use artificial chest compressions to assist ventilation. If animals exhibit respiratory depression during anesthesia, continue the artificial ventilation until the anesthetic concentration metabolism comes to a safe level. When CO_2 concentration declines due to artificial ventilation, stop the artificial ventilation. When the CO_2 concentration increases to a normal level, animals will regain autonomous breathing. Doxapram hydrochloride (5–10 mg/kg) can sometimes stimulate ventilation function.

When animals are placed in a nonphysical environment, the animals' pulmonary perfusion and ventilation will be affected, which leads to ventilation—perfusion disorder, decline of oxygenation in the arterial blood, and increase in CO_2 tension.

6.5.8.2 Heart Failure and Hypotension

Hypotension can be caused by vascular expansion or insufficient arcotic myocardial contraction by anesthetics, blood loss, or tissue ischemia during operation. The signals of hypotension include reduced capillary perfusion, reduced arterial pressure, tachycardia, and pale color of skin and mucosa. Intravenous fluid drip is usually used to correct hypotension, and a blood drip to correct excessive bleeding. Blood cross-matching in animals is difficult, thus a blood donor and receptor must be from the same species of animal with the same type of acid citrate dextrose. Blood should be dripped at a speed of 10% of the animal's entire blood volume every 30 min. If there are acute hemorrhages, accelerate the rate of blood transfusion accordingly. If there is no condition for blood transfusion, plasma volume expanders instead of blood transfusion may be used. For example, plasma substitutes, such as haemaccel, gelofusine, or normal saline solution, may be infused.

6.5.8.3 Arrhythmia and Asystole

If there is no ECG monitoring system used to detect the heart activity, arrhythmia is difficult to find and diagnose. Arrhythmia can cause severe circulatory disturbance, which needs necessary treatment. In the process of the deep anesthesia, hypotension and vagus nerve stimulation can cause bradycardia. Tachycardia may be caused by pain, hypoxemia, hypercapnia, or blood volume deficiency. If animals enter cardiac arrest, cardiac compression can be used *in vitro*. For dogs, it is 70–80 times/min, and it should be supplemented with the above-described methods of correcting breathing difficulty. As a first aid measure, adrenaline (0.1–0.2 g/kg) may be administered through deep tracheal and intracardiac injection.

6.5.8.4 Regurgitation

Dogs, cats, and primates may vomit during the process of anesthesia, which may lead to asphyxia if vomit was intaken into the lungs. As mentioned earlier, fasting for 12–16 h can reduce the risk of vomiting in the process of animal anesthesia. Ruminants will exhibit food reflux during the anesthesia procedure. For these animals, it is necessary to use an endotracheal tube with a sleeve. Allowing these animals to fast for 16–24 h may be conducive for reducing gas accumulation in ruminants.

6.5.8.5 Hypothermia

Under anesthesia, the animal's body temperature may be too low. Artificial ventilation, infusion control, and vasodilation may not only cause disorders in body temperature regulation, but also increase the surface heat loss. Therefore, during anesthesia, the animal's body temperature must always be monitored and provide heat when necessary.

6.5.9 Postanesthesia Nursing

Postoperative monitoring is necessary to determine whether animals can recover quickly or not. After anesthetizing, the temperature regulating center of the animal is temporarily out of control. For small animals, incubators, light bulbs, heating pads, towels, and other measures are used to maintain the animal's body temperature. For young newborn animals, keep the environmental temperature at 35–37°C, and for large animals, 25–30°C is acceptable, and heating pad and heating lamp should also be provided. Check body temperature regularly to ensure that the method is fully effective. For bedding, it should be comfortable, and provide insulation and thermal insulation. Sawdust or small wood chips are not ideal bedding because the animals can breathe in powder, and sawdust can stick in the animal's eyes, mouth, nose, etc.

After surgery, the animals should be put in cages where they can be easily observed, and kept clean and quiet. When placing animals, keep the neck unbent, lie it on its side, and make the airway open. After 4 h, move the animals to lie on the other side to avoid lung congestion and gravity pneumonia. Tubes used in surgery should be removed after the recovery of the swallowing reflex, then continue to observe the respiratory condition until the animals wake up. If the animals vomit, put its head in a position below the neck and abdomen to avoid suffocation or aspiration pneumonia. Normal animals need approximately 40–80 mL/kg liquid per day. After recovery from anesthesia, some animals are often unable to eat or drink; at this time they must be fed orally manually or through intraperitoneal injection to supplement body fluid. The wounds of laboratory animals are easily polluted by dung, urine, and bedding. When necessary give antibiotics to prevent infection. Observe the animals' surgical wounds after operation to prevent animals from biting, licking, grasping, or tearing at the wound.

Pain can cause animals to produce a series of physiological reactions, and the appropriate way to reduce the pain after operation should be studied. To provide appropriate analgesics at the appropriate stage, technical personnel should know the extent of the animal's pain. It can be difficult to evaluate the extent of the animal pain as we need to know the normal and abnormal behavior changes. The important signs of abnormal behavior are posture change, reduced food and water intake, and weight loss. As rodents are less active during the day, it may be more accurate to estimate the health status by observing animals in a dark state.

Some analgesics can be used to treat animals, such as morphia, buprenorphine, and nalbuphine, or some nonsteroidal antiinflammatory drugs (NSAIDs), such as flunixin and carprofen. The implementation of local anesthesia can also reduce post-operative pain. In general, it is necessary to use opioids to control postoperative pain. Of the existing drugs for many kinds of laboratory animals, buprenorphine exhibits a long lasting time (6–12 h), and it is able to relieve pain safely and effectively. The effects of NSAIDs are poor, but fluorine, carprofen, and some newly developed analgesics demonstrated similar efficacy with opioids. In many cases, within 24 h after the first injection of opioids, NSAIDs can then be used for 24 h. In this way, post-operative pain can be alleviated effectively. In general, animal postoperative pain rarely lasts more than 72 h.

Common analgesics, route of administration and dosage are summarized in Table 6.14.

Analgesics have some side effects, which may disturb certain original experimental records. Opioid agents can cause respiratory depression, hypotension, and constipation, but these effects have almost no clinical significance in animals. NSAIDs can reduce the synthesis of prostaglandins and affect wound healing. These analgesics may obstruct blood coagulation and affect kidney function.

The pain tolerance of animals is different from that of human beings and as analgesics also have side effects, there are still reservations about whether to use analgesics after operation. However, to protect animal welfare and under the premise that it does not affect the result of the experimentation, depending on the condition of animals, analgesic injection within 24–72 h after surgery should be used to alleviate the pain of animals. If systemic analgesics are prohibited, animals exhibiting surgical trauma can be given a local analgesic, such as bupivacaine infiltration, which may provide 4–6 h of short-term pain suppression.

6.6 EUTHANASIA

After animal experimentation, if the pain and suffering of animals cannot be avoided, as well as, from the perspective of experimental study, there is no need for these animals to survive, then there is truly no need for these animals to live. In animal experimentation, sometimes in order to retrieve the animal tissues and organs, the animals must be put to death. If animals have an incurable disease, exhibit pain, or other special circumstances, animals must be euthanized. In scientific research or other cases when animals need to be executed, euthanasia must be performed.

There are some special requirements that need to be considered when performing euthanasia. The most important thing is that the method of euthanasia must be humane. When the aim of scientific research is to obtain animal tissues and organs, the method of euthanasia should have no effect on the animal tissues and organs. In addition, the method of euthanasia should be reliable, effective, economic, and easy to implement, as well as safe for the personnel.

Before implementing any euthanasia, research personnel should receive proper training.

The chosen euthanasia method must be humane. The disposal method for animals should be gentle and minimize the suffering of animals at the end of life.

Table 6.14 Analgesics and the Corresponding Dose Used to Alleviate Pain Postoperatively (mg/kg)

	Rats	Mice	Guinea Pigs	Rabbits	Dogs	Cats	Primates	Pigs	Goats/Sheep
Aspirin	120 per os, 4 h	100 per os, 4 h	85 per os, 4 h	100 per os, 4 h	10 per os, 6 h	!	20 per os, 6–8 h	—	—
Buprenorphine	0.05–0.1 s.c., 12 h	0.05–0.1 s.c., i.v., 8–12 h	0.05 s.c., 8–12 h	0.01–0.05 s.c., i.v., 8–12 h	0.01–0.02 s.c., i.v., i.m., 8–12 h	0.005–0.01 s.c., i.v., 8–12 h	0.01 i.v., i.m., 8–12 h	0.01–0.05 i.m., 8–12 h	0.005–0.01 i.m., 4–6 h
Butorphanol	1–5 s.c., 4 h	2 s.c., 4 h	—	0.1–0.5 i.v., 4 h	0.4 s.c., i.m., 3–4 h	0.4 s.c., 3–4 h	—	—	—
Codeine	60–90, per os, or 20 s.c., 4 h	60 s.c., 4 h	—	—	0.25–0.5 plus paracetamol, per os, 6 h	—	—	—	—
Flunixin	2.5 s.c., i.m., 12 h	2.5 s.c., i.m., 12 h	—	1.1 s.c., i.m., 12 h	1 per os, 24 h	1 s.c., 1–5d	2.5–10 i.m., 24 h	—	1 s.c., 24 h
Ibuprofen	—	—	10 i.m., 4 h	10 i.v., 4 h	5–10 per os, 1–2d	—	—	—	—
Morphine	2.5 s.c., 2–4 h	2.5 s.c., 2–4 h	2–5 s.c., i.m., 4 h	2–5 s.c., i.m., 2–4 h	0.5–5 s.c., i.m., 4 h	0.1 s.c., 4 h	1–2 s.c., 4 h	Total dose 20 mg i.m., 4 h	Total dose 10 mg i.m., s.c., 4 h
Nalbuphine	4–8 i.m., 4 h	1–2 i.m., 3 h	—	1–2 i.v., 4–5 h	0.5–2 s.c., i.m., 3–8 h	1.5–3 i.v., 3–8 h	—	—	—
Paracetamol	300 per os, 4 h	100–300 per os, 4 h	—	—	10–20 plus codeine per os, 6 h	!	—	—	—
Pentazocine	10 s.c., 3–4 h	10 s.c., 4 h	—	5 i.v., 2–4 h	2 i.m., 4 h	8 i.p., 4–6 h	2–5 i.m., 4 h	2 i.m., 4 h	—
Phenacetin	200 per os, 4 h	100 per os, 4 h	—	—	—	—	—	—	—
Pethidine	10–20 s.c., i.m., 2–3 h		—	10 s.c., i.m., 2–3 h	10 i.m., 2–3 h	10 i.m., s.c., 2–3 h	2–4 i.m., 3–4 h	2 i.m., 4 h	Total dose 200 mg, i.m., 4 h

Note: i.p., intraperitoneal injection; i.v., intravenous injection; i.m., intramuscular injection; s.c., subcutaneous injection; per os, oral administration.

Pheromones released by frightened animals may arouse anxiety and pain in other animals. Therefore, under no circumstance should animals be put to death in front of other animals.

Commonly used methods of euthanasia can be divided into two categories: medicinal chemical methods and the physical methods (Table 6.15).

6.6.1 Medicinal Chemistry

This refers to use of a drug or other compounds to execute animals. The most commonly used method is to use excessive general anesthetic to make the animal enter cardiac arrest, respiratory failure, and death. Pentobarbital is a commonly used drug for euthanasia due to relative rapid onset, simple use, and price. It is administrated via abdominal or i.v. injection (100–150 mg/kg). Barbiturates can cause vasodilation and organ passive congestion, which may affect histologic study.

Inhaled anesthetics used for euthanasia are ether, halothane, isoflurane, and halothane. CO_2 is also often used as an euthanasia inhalant. If animals are placed in an environment containing 100% CO_2, many animals may die due to dyspnea. When

Table 6.15   Commonly Used Methods to Euthanize Laboratory Animals

Drugs/ Methods	Species	Position and Mechanism	Safety	Application	Induction Rate
Ether	Rodents, cats, dogs, birds	Deactivates brain cortical, subcortical, medulla oblongata, and other vital centers	Explosive, flammable	Easy to implement in glass box	Slow
Fluothane	Rodents, cats, dogs, birds	Similar to ether	Chronic inhalation is harmful	Easy to perform	Fast
CO_2	Rodents, cats, dogs, birds	Similar to ether, Myocardial depression	Harmless	Sealed box	Very fast
Barbiturates	All animals	Similar to ether	Safe	Animals need to be restrained	Fast
Cervical dislocation	Body weight <200 g	Directly causes brain death	Safe	Needs training	Very fast
Decapitation	Rodents, small rabbits	Directly causes brain death	Basic safety	Easy to perform	Fast
Microwave	Mice, rats, and animals with same weight	Needs special instruments	No histological change in the animal's brain	Easy to implement if animal body position is correct	Fast for small animals
Freezing	Body weight <20 g	Inactivates enzymes in animals	Safe	Needs liquid nitrogen	Fast

using this method, as the animals are conscious and able to feel pain, it is not an ideal method. If the animal is placed in an environment that has a certain humidity, CO_2, O_2 in a ratio of 6:4 mixed gases, after losing consciousness, raise the CO_2 concentration to 100%, and continue this condition for at least 10 min in order to ensure that animals die. At this moment, the animal is entirely in an unconscious state. Before its death, it does not feel any pain. Newborn animals have a certain resistance to CO_2, and need to be exposed to the CO_2 environment for 30–60 min. If the time is too long, it is not a preferred method. Inhalants, such as CO_2, can induce pulmonary edema, which may affect subsequent studies that need animal tissues and organs.

6.6.2 Mechanical and Physical Methods

There is a possibility of potential interference with some experimental results for most medicinal chemistry methods, therefore, mechanical and physical methods of euthanasia may be more suitable for animals for the use of certain tissues and organs for biochemical and histological examination. If possible, before using mechanical and physical methods to sacrifice animals, sedatives or narcotics should be used. All mechanical and physical methods may cause pain, and this should be considered when choosing a method of euthanasia. Mechanical and physical methods include decapitation and dislocation of cervical vertebra. Decollation is to cut the head from the neck directly by using scissors or cutter. Cervical dislocation is to fracture the spine and block the important organ (e.g., heart and lungs) nerve impulses by stretching and rotating the animal's neck, which eventually cause death. Decollation and cervical dislocation techniques are suitable for mice, rats, hamsters, gerbils, dogs, cats, and birds; however, it is not suitable for big animals. If the mechanical and physical methods can be done quickly and professionally, it will not bring pain to animals. After the implementation of the mechanical and physical methods, the animal should be bled or have its brain damaged to ensure that the animal has already died.

For newborn animals and animals whose body weights are less than 20 g, perform euthanasia by immersing them in liquid nitrogen so that they can be frozen quickly. Whether this method can make animals lose consciousness rapidly has been questioned. Another alternative is to use microwave irradiation on the central nervous system to make the animals die immediately, leaving the biochemical properties of animal tissues and organs unchanged. If using microwave irradiation, the corresponding equipment is required.

For large laboratory animals, such as large dogs, pigs, ruminants, and horses, they can be first anesthetized to lose consciousness, and then cut the carotid artery to make the animals bleed and die immediately.

6.7 PATHOLOGICAL ANATOMY AND SAMPLE COLLECTION

Pathological anatomy is an important part of animal experimental study. By employing pathology knowledge, animal pathological changes can be checked to study disease occurrence and development.

In clinical practice through autopsy, we can on the one hand check whether the diagnosis of animal disease is correct, and then timely sum up the experience, as well as improve the quality of diagnosis and treatment. On the other hand, for the outbreak of diseases, such as infectious diseases and parasitic diseases, we can make an early diagnosis and take effective measures to prevent and control disease development. In addition, the accumulation of pathological anatomy materials provides important data for comprehensive research on animal diseases.

Before anatomy, basic information about animals and animal diseases should be first learned, including clinical tests, physical examination, and clinical diagnosis. In addition, we should also carefully check the body surface characteristics of animals and natural holes, mucous membranes, coat and skin, which must be noted by autopsy personnel.

Autopsy records are the important basis of an autopsy report as well as the original scientific data for comprehensive analysis. The content of autopsy records should be complete and detailed to truly reflect pathological changes. Moreover, the observation process should also be recorded in detail. Do not recall from memory to avoid omissions or errors. Record order should conform to the autopsy order.

A full autopsy record should include the system changes of organs as these changes are linked to each other. Sometimes to the naked eye, some changes are not obvious; however, presumed unimportant changes may be important clues for diagnosis. If omitted, it will be difficult to diagnose. The whole picture of certain diseases can be summed up by detailed materials. In addition, in order to identify the pathological changes, a clear understanding of the normal state of each organ is needed because the typical pathological changes between sick animals and the animals in its normal state cannot be simply reviewed. For necropsy, the needed equipment includes scissors, tweezers, and alcohol cotton, and sampling equipment includes sterilized scissors, tweezers, glass plates, sterilized saline or phosphate-buffered saline (PBS), straw, small tubes, formalin, and specimen bottles, all of which should be prepared in advance.

Here is an example of mouse autopsy.

- **Observe the animal's clinical symptoms**, including appearance (whether abnormal mental state), respiratory system (twang, nostril dirt), digestive system (attached dirt on anus), surface (presence of dirty hair, scabs, coat gloss, tumor, etc.), and measure body weight.
- **Anesthesia**: Put the mouse in an anesthesia bottle. After euthanasia with ether, re-restrain.
- **Blood collection**: Use alcohol cotton to disinfect all body surfaces, and then use tweezers to pick up the right fore armpit skin. Next, use scissors to cut along the body axis for 2 cm. Finally, scissors are used to cut off the alar arteriovenous and the blood is collected through a straw.
- **Take a trachea swab**: Use sterilized scissors to cut the midline of the neck skin and expose the trachea. Open a small hole in the sterile trachea, and swab a wet cotton swab with sterilized saline or PBS into the airway lumen.
- **Incision of chest and abdomen**: Use sterilized scissors to cut along the line from the neck to the lower limbs, then peel in the direction of limbs, and restrain and pin

the skin. Cut from the rib cartilage to get the sternum, and then expose the chest. The abdominal cavity is exposed along the midline incision.

- **Observe the viscera**: Record abnormal form, and if necessary, take some materials for bacteriology and histological examination.
- **Bacteriological tests**: Use the new sterilized scissors and tweezers to retrieve internal organs and put them in the sterilization plate to cultivate as soon as possible. They can also be stored in a refrigerator for a short time.
- **Selection of histological materials**: Select organs or viscera that have lesions or are going to be observed, and then soak in 10% formaldehyde solution. Sampling should be comprehensive and representative to show the development of disease. A tissue should include the lesion and also surrounding normal tissues; important structural parts of organs should be included as well. For example, for the stomach, intestinal samples should include from the serosal to mucosal layers of tissues to see the intestinal lymph follicles; kidney should include the cortex, medulla, and renal pelvis; heart should include the atrium, ventricle, and valve parts. In a large and important lesion, more than two tissues from different parts should be taken to represent the pathological morphological changes at each stage.

The size of the tissue should be 1–1.5 cm wide, and approximately 0.2 cm thick. When necessary, the size of the tissue block can be increased to 1.5–3 cm, but it should not be more than 0.5 cm thick for it to be easily restrained.

When restraining tissue blocks, the case number should be written on a little piece of paper by pencil, and put into the stationary liquid with the tissue to restrain. Stationary liquid, numbers of tissues, number, and restrain time should be written on the bottle.

Commercially available formalin solution is 37%–40%, and can be applied after diluting 4–5 times for stationary liquid.

BIBLIOGRAPHY

Liu E, Yin H, Gu W. *Medical Laboratory Animals.* Science Press, Beijing, China, 2008.

National Research Council. *Guide for the Care and Use of Laboratory Animal*, Eighth Edition. National Academy Press, Washington, 2011.

Pearson T, Greiner DL, Shultz LD et al. Humanized SCID mouse models for biomedical research. *Current Topics in Microbiology and Immunology* 2008;324:25–5.

Perry AC, Wakayama T, Kishikawa H et al. Mammalian transgenesis by intracytoplasmic sperm injection. *Science* 1999;284:1180–3.

Porteus MH, Carroll D. Gene targeting using zinc finger nucleases. *Nature Biotechnology* 2005;23:967–73.

Qin C, Wei H. *Laboratory Animal Science*, Second Edition. People's Medical Publishing House, Beijing, China, 2015.

Reagan-Shaw S, Nihal M, Ahmad N. Dose translation from animal to human studies revisited. *FASEB Journal* 2008;22:659–61.

Van Zutphen LFM, Baumans V, Beynen AC. *Principles of Laboratory Animal Science*, Second Edition. Elsevier Science Publishers, Amsterdam, Netherlands, 2001.

CHAPTER 7

Animal Experimentation Design

Changqing Gao

CONTENTS

7.1 VARIATION CONTROL

An important factor to be considered in the design of animal experiments is the variation unrelated to the research purpose in the experiment. Laboratory animals are living creatures and even a minute difference in the feeding conditions may cause a difference in their reactions, which may lead us to a wrong conclusion. Therefore, in the design of animal experiments, the influence of these factors must be taken seriously and appropriate strategies should be adopted to control them.

In animal experiments there are three main sources of variation: (1) variations between researchers; (2) inherent variations in the animals; (3) variations between the animals and the environment.

In this section, we will discuss these three variations and provide some suggestions on how to control them.

7.1.1 Variations between Researchers

There are two main variations: first, the implementation of the animal experimentation process (such as injection, oral drug administration, surgical intervention, etc.) is not standardized; the second is the lack of accurate measurement methods. Together, these two factors are often the key factors that cause the variations in the experimental results.

7.1.1.1 Variations in Implementation of Experimentation

The control of the experimental process should be considered first to ensure the consistency, accuracy, and integrity of the experiment. Even for a simple operation in the animal, the researcher should be appropriately trained to improve the quality of the operation. For example, it is necessary to use a new small volume syringe and needle to administer injections in an animal study. Animal study staff should be trained to be meticulous about all the details in the operation such as avoiding drug liquid spills and inconsistent operations in animal studies. In animal studies, performing operations is often necessary. Standardized procedures are the way to minimize variation, and surgical procedures in animals should be the same as in humans so that the residual variation is measurable. When measuring the blood loss during an operation, the swabs may need to be weighed. The weight of the resected tissue should be estimated each time, and the exact location of the intervention and its relationship with the anatomic marker should be the same each time. As even very small variation in the operation, for example, an inconsistency in blocking blood flow to the ligation, will affect postoperative recovery and tissue blood flow, thereby causing unexpected variation to the experimental results. Similarly, in order to reduce animal tension during the experimental process, procedures should be taken to avoid irritating the animals as it will not only affect the metabolism of the animal, but also change the rate of local blood flow in the animals and affect the drug absorption.

7.1.1.2 Inaccurate Measurements

Although the measurement techniques and tools are becoming increasingly advanced, many measurement methods are inherently inaccurate. For example, the migration rate, density, and size of gel points and bands are influenced by many subtle factors, which are not easily controlled. The inaccuracy of these methods and techniques will markedly reduce the significance of statistical analysis in the study. In addition to the application of appropriate measurement techniques and correct operation, a number of observations for the same measurement must be made to control the variation in the measurements.

In animal experiments, there are many measurements related to the natural behavior of animals that are more difficult to measure than many other physiological parameters and hence, more difficult to control their variation. For example, it is not easy to give a quantitative measurement for the expression of the pain of an animal. To accurately measure such behaviors, researchers have tried several methods. Decomposing a behavior into different components may be a useful method to improve accuracy and constancy. To this end, Keating et al., Sotocinal et al., and Langford et al. developed methods to measure an animal's grimace to reflect the intensity of pain in rabbits, rats, and mice by decomposing the animal's facial expression into several parts, that is, the intensity of orbital tightening, cheek flattening, change of nose shape, whisker position, and ear position. When the observers are well trained to follow the protocol, this method may achieve a high degree of consistency between different observers, and the complex behavioral elements can be used for statistical tests. Nevertheless, the results obtained by different observers as well as those obtained at different times by the same observer should be often checked. Training and accreditation are the most efficient ways to reduce variance among observers and among the results obtained from the same observer.

In many animal studies, it is difficult to find the most appropriate method to measure animal behavior on the spot. Scientists have been inventing instruments and tools to record animal behavior in a way that can be easily analyzed at a later stage. For example, in order to collect the parameters of a behavior, researchers are using an automatic recording instrument invented for recording animal activity that can be attached to the animal body. The use of these devices has enabled scientists to develop a digital data technology that is conducive to statistical analysis. Another strategy is to put animals in a situation in which the target behavior of animals in the given scene is certain arousal, thereby accurately recording the animal's response. The water maze, the conditioned chamber and different learning instruments are examples of this type of measuring device. Using such instruments, animals are usually first trained for a period of time, and then their performance is recorded and analyzed.

Another way to ensure the quality of the data from animal experimentation is to apply the good laboratory practice (GLP) concept in the study, which is that each step of the animal experiment must be conducted in accordance with the known standard operating procedures (SOP), and the validation of this process is through the signature of both the researchers performing data collection and an independent quality assurance personnel. Application of GLP greatly reduces the variation caused by the measurement and observation methods in scientific research, suggesting that controlled operation can greatly improve the accuracy, which greatly reduces the number of animals used worldwide in the pharmaceutical industry and research laboratories by reducing background noise.

7.1.2 Inherent Variations in Animals

In addition to the variation in genetic background among animals, a number of other effects, which must be considered in the design of experiments, can also increase the differences among animals.

7.1.2.1 Different Animal Sources

Many animal studies have demonstrated that the sensitive time for behavioral development is in the neonatal period, which decides the subsequent adult behavior pattern. Thus, the neonatal environment is critical. However, as far as we know, there are no two suppliers using exactly the same animal husbandry management procedures or exactly the same environment to raise animals such as the same light cycle, with the same microbial background.

Therefore, different sources of animals imply the existence of potential differences between animals. Indeed, environmental factors lie behind much of the irreproducible experimentation. In addition, due to genetic drift, animals from different suppliers may have variations in their genetic composition, and combined with the environmental factors, may become a complex manifestation.

7.1.2.2 Differences in Animal Feeding and Management

Animals may feel the feeders through visual, auditory, or olfactory senses. Observations have demonstrated that simply the replacement of the feeding technician while other conditions are fixed can make the reproductive performance of mice decrease significantly for a few weeks; whereas, if the original breeders leave for a short period of time and come back to work, this performance decline is limited. Thus, small changes in animal husbandry management may have a huge impact on animal behavior and other physiological conditions. Therefore, when conducting animal experiments, it is better to put the animals in a new environment or replace the feeding staff at least a week before starting the experiment to give the animals a period of time to adapt themselves to the changes.

7.1.2.3 Different Animal Housing

In an animal facility, the temperature of the air near the exhaust port space is 3–4°C higher than the air near the inlet space, and the relative humidity is 5%–10% higher. In the same room, there are different chemical and microbial compositions in different areas of the room. The place of a cage in the rack relative to the intake and exhaust position, and its vertical position in the rack, whether there are inclusions (including bedding and the amount of feed in the box) in the cage may affect the animal's survival microenvironment. If a cage is placed in an upper position of the rack, animals inside of the cage will feel greater light intensity than those in the lower places. For example, albino animals do not have the ability to regulate the light that reaches its retinas, which can lead to retinal damage, which may in turn change the function of the pineal gland and may even affect the reaction of the animal to treatment. The range and frequency of the hearing of a mouse and many other animals are much higher than those of humans. The range of sound that affects rodents is 20–90 kHz and the devices and cages crashing against each other and landing, the sound of the air supply system, the air conditioner, and electronic equipment can all cause such noises. However, the measurement of the noise has not been well noted because human ears cannot hear these noises. Characteristics of the bedding material will also affect animal metabolism, including effects on animal responses to experimental treatment. For example, bedding made of pine and eucalyptus wood contains substances that can induce animal liver enzyme activity and cytotoxic effects, directly affecting the characteristics of drug metabolism of the animal.

Animals are sensitive to these environmental factors, and even a small change in the environment can affect their behavior and physiology. The consequences of these affecting factors may include an increased degree of nervousness in the animals, reproductive performance changes such as delayed mating, their offspring becoming smaller or even no offspring, mortality rate of the offspring during lactation, or growth rate of the offspring. Therefore, it is important to keep temperature, humidity, illumination, and noise level controlled during an experiment.

7.1.2.4 Animal Health Status

Researchers do not usually use sick animals for experiments. However, animals with subclinical diseases are not easy to identify, which may result in serious outcomes. This may happen by first affecting specific organs, inhibiting their function, and changing the animals' physiological and biochemical status. When a certain disease affects the function of a specific organ or a specific biochemical metabolic pathway, and the experimental process involves the function or the pathway, then the experimental results are likely to be affected. Or second, a disease may have a broad

impact on the animal's health, making it sensitive to changes in experimental manipulations in the study. Usually, if a group of animals are infected, some of them are
seriously affected while others may not be significantly affected, and the majority is
in an intermediate state between the abovementioned extreme conditions. The biggest issue is the variance in infection impact between individual animals. Therefore,
specific germ-free (SPF) animals that exclude most of the infectious factors in natural conditions are now required in animal study. This requirement guarantees the
minimization of the individual differences in infection.

7.1.3 Variations Resulting from the Interaction between Animals and the Environment

In conducting an experiment, in addition to the factors discussed above, such as
the inherent variations between animals, variations produced during operations and
those related to the way of raising animals, consideration and control of more complex variations, which may arise from the physiological and behavioral responses
and biochemical reactions of animals to environmental stresses or stressors, must
be taken. The reactions may interfere with each other or work together, resulting
in greater variations than expected. For example, most rodents are gregarious, and
living alone is a kind of stress. In such a case, if the experimental design requires
separate test animals, some kind of nonexperimental "partner," such as covering,
play items, or other companion or enrichment, should be provided to minimize the
stress caused by isolation. For these species, some are gregarious while others are
different. The hamster, for example, is a solitary animal and should not be raised
together even after grouping for experimentation. Thus, whether animals should be
raised together or not should be addressed according to the nature of the animals in
the study.

Indeed, sociability has a significant impact on the characteristics of animals.
There is evidence showing that the difference in body weight among each individual
for mice raised with 2, 4, or 8 animals per cage is smaller than that among mice
raised alone. For the majority of species of animals living together, relatively clear
dominance hierarchies will be established. In such a case, the dominant animal in
the environment is an absolute master, playing the main dominant role, and the other
animals are in a subordinate position, which leads to differences in many aspects
among animals in the same group. Animals at the bottom of the hierarchy often
spend a lot of time and energy to avoid conflict with other animals. Especially during
the process of the establishment of social hierarchy, the animals are quite aggressive. Some strains of male mice, such as SJL, BALB/c, and C57BL/6, show a strong
nature to attack in social groups, resulting in animals continuously suffering from
stress, which in turn affects their behavior, physiological, and biochemical characteristics. The attacks between animals may last a few hours to a few weeks before
the hierarchy is reached, after that the stress and corresponding hormonal adrenal
glucocorticoid levels will decline. Therefore, experiments should not be conducted
immediately after grouping, instead allow a period of time to let the animals establish a stable social hierarchy, then perform the study. As the hierarchy is relatively

stable in small groups compared with in a large herd, the number of animals in a cage should not exceed five.

It is suggested that during the experimental process the composition of the social groups of animals should not be changed, for if an animal is taken away from the group, competition will increase until a new social structure is created. Accordingly, operations on animals or if an animal in a group becomes ill may trigger a new attack.

Researchers often find that animals are hurt in the morning when they are examining them. Attacks often occur a few hours after dark while the staff are not present. However, sometimes the differences in the research results apparently caused by stress are obvious even if there is no obvious trauma or behavioral change in animals.

Environmental enrichment is a complex issue. It has been demonstrated that 3–5 rats per cage easily formed a stable hierarchical society and their aggressive behavior was the lowest. However, in such situations, adding objects for environmental enrichment will significantly increase continuous aggression. Hence, adding enrichment objects in studies may or may not improve, or even worsen the research results according to different situations. In the application of environmental enrichment strategies, the effects on the social hierarchy of the animals should be taken into account because the use of enrichment objects will increase the complexity of the environment and animals will fight over preferences of regional environment. For example, in a group of rats, the dominant rat often occupies the preferred position for a very long time, while other individuals cannot enjoy the enrichment objects, but are vulnerable to the attack of ruler. Therefore, environmental enrichment strategies are to be adopted to ensure that there are enough "preferable" positions in the social space. The setting of the environmental enrichment objects should be simple, uniform, attractive, and effective, without increasing the competition among the animals in the cage. Equally important, do not change the environment settings when the cage is replaced, so as to avoid the occurrence of new problems.

The experimental procedures and operator may impact the animals. Therefore, the same operation or procedures should be conducted by the same personnel to ensure that the treatment is the same. This is because observation has demonstrated that animals show different responses to different experimental operators and breeders, suggesting some of them caused animal stress or discomfort during operation. Therefore, training of the staff is important. In addition, if the operation or administration needs to be done at consecutive time points, it should be done by the same technical personnel.

There are many other factors that are often neglected. For example, changing environmental conditions during experimentation, changing the conducting time in the day, which will affect the rhythm of the animals, or not paying attention to the relationship between the timing of the experimental work and the feeding process.

Handling animals is critical. When animals feel stressed, they may emit ultrasonic voices, a discomfort signal to the wait-to-treat animals. If an experimental operation causes an animal discomfort, the untreated animals may become restless and have a disturbed endocrine status, thus, the results from these animals are affected and cannot be compared with those that are not stressed. One way to over-come such problems is to train the researching team members before experiments start. It is also necessary to allow the animals to adapt to operations before experiments. To give the operators more chance to handle the animals, the handlers may give some rewards, such as food, to the animals after handling to reduce stress.

Accordingly, the order of treating animals is also an influencing factor. If the operation of animals is always carried out in a specific order, the results will likely be deviated. For example, if the same animal is always administered a drug first, the wait-to-treat animals will sense it and be stressed, and therefore, the animal's behavior and physiology will be different. The test results and those from its corresponding control group cannot be compared. However, if the order of drug administration is randomized by groups, the group differences will increase, making the number of animals used for statistical processing increase. In such a case, a randomized block design is suggested to reduce the effects. For example, if there are four groups in a study, it is appropriate to set four animals as a module. The first module of the four animals is randomly assigned to each of the four groups, then the next four animals compose another module.

One of the most effective ways to reduce the variations that are caused by non-professional operations is to establish a specialized laboratory animal center where there are professional operation experts to help researchers with performing operations. This is the case now for almost all universities in Western countries and China. For example, if we would like to produce monoclonal antibodies, in theory, because these animals are of an inbred strain and are in the same conditions, the reactions between individual animals should be very consistent. However, the results obtained in the laboratory vary; some animals produce relatively high levels of antibodies, while others are very weak in response to the antigen. With such results, an animal research center should be established with standardized technology, especially a skilled operation technical staff who may not cause stress to the animals. After years of study, a strong immune response from each animal may be achieved.

7.2 PROCEDURES OF ANIMAL EXPERIMENTAL DESIGN

For all research, reproducible and effective data are obtained through a correct scientific design and standard operation of the experiments. Experimental design is the key to success, which is directly related to the quality of experimental data and determines where the data will be published. If the experiment is not well designed, the results will certainly be a waste of time and resources, and animal experimentation is even more so.

7.2.1 Initial Steps in the Design of Experiments

When doing research, the following route is usually followed: we first observe a phenomenon we are interested in and put forward a question. Then we try to find an answer to this question by reviewing the literature. If there is no clear answer, we may, based on our knowledge and the facts, put forward a hypothesis (educated guess) to explain the phenomenon. Next, we setup a condition and test the hypothesis. This whole process is called experimental design.

7.2.1.1 Hypothesis

To clarify the research objectives and put forward a hypothesis is the key premise of experimental design. We have to know the meaning of the experiment, be it for improvement of human or animal health, or to increase the understanding of disease processes.

A hypothesis is a proposed explanation for a phenomenon. For a hypothesis to be a scientific hypothesis, the scientific method requires that one can test it. For each experimentation, we may set a null hypothesis, meaning there is no causal relationship between the study parameters; a preparation hypothesis is one in which there is a causal relationship; and a nontestable hypothesis, meaning no clear answer can be obtained from the results. For example,

Question: Does drug A relieve an animal's pain?
Null hypothesis: Drug A does not relieve an animal's pain.
Preparation hypothesis: Drug A does relieve an animal's pain.
Non-testable hypothesis: The data obtained do not allow a clear conclusion to be made on whether drug A relieves an animal's pain.

Even as the test is going on, these hypotheses can be modified. However, to clearly put forward a question and hypothesis is the foundation and premise of an experimental design.

During this process, we may need to review the literature. Now many resources, such as PubMed, Medline, NCBI, Google Scholar, Scopus, etc., can be very helpful to help us know what has been done in the field, what has yet to be done, what kinds of animal models have been established, what kind of discomfort and pain will be experienced by the animals, and the procedures to relieve them.

A hypothesis is a suggested explanation for a phenomenon that cannot be satisfactorily explained with the available scientific theories. In research, a hypothesis can generally be transferred into a comparison between data from two or more groups. Then the sample size of a certain effect (or expected difference between groups) can be calculated at a certain probability level. It is true that the smaller the difference between the groups (effect) or the larger the group variability, the larger the sample size needed to observe the differences.

7.2.1.2 Animal Models

In reviewing the literature, you may find that someone has done similar work with an animal model; this is a good point for you to start. If there are several models being used in the publications, you may choose one according to your research purpose and conditions after comparison of the advantages and disadvantages of all the models. For example, you may need to use primates as a model to study human reproduction; however, you find it is too expensive or not realistic under the current conditions. You may like to use a smaller animal, but not rats or mice. You then review the literature and find that several aspects of guinea pig reproductive functions are more similar to that of primates when compared with rats and mice, for example, a true luteal phase as a component of its relatively long reproductive cycle and more similar distribution of hypothalamic gonadotropin-releasing hormone neurons as compared with rats and mice. You may choose the guinea pig as an alternative model to study. Indeed, studies using guinea pigs as an animal model to study neuroendocrine regulation of male reproduction did produce interesting results.

The following suggestions may help with selecting an animal model:

- Review literature thoroughly, discuss with colleagues in the same research field, and contact the supplier or the animal model resource database to determine the source channel of the animal model.
- Use animals at the lowest phylogenic level, which may have less ability to feel pain and discomfort, if they have to experience pain during the experimentation in line with the 3Rs (replacement, refinement, and reduction) principle.
- Check that the species or strains of the animal to be used in research have the specific features or characteristics necessary to the study, just as mentioned above in the case of guinea pigs.
- Consider the maintenance conditions of the animal model during the experiment.
- Consult veterinarians before making the final decision.

7.2.1.3 Collaborators

In a study, it should be clear what kind of technical operation will be used and who has the corresponding operational expertise. Most of the time, you may need a collaborator or even multiple collaborators to finish the study. In such a case, it is necessary for the partner to understand the whole design of the experiment, the calculations and the conditions for collection of the samples. A large animal facility should be able to provide technical operations related to animal experimentation and training, as well as expensive facilities, equipment, and other services.

7.2.2 Procedures of Animal Experimental Design

7.2.2.1 Details in Operation

When testing the hypothesis, we need to know what parameters to observe, which should be able to accurately reflect the effects of the treatment, the method to produce data, the way to obtain samples, the list of equipment and/or software to use, the statistical method to analyze the data, investigators to participate and the time they can invest. More importantly, we need to provide a detailed scheme of the process of the operation.

From the beginning, the researchers should know how long the experimentation will last and how long the animal model can survive, the expected disease process in the model (to determine the most appropriate time for the measurement of the disease), the time of the personnel participating in the project, and the cost of the study. If animals are treated with chemical or biological substances, it is necessary to determine the most appropriate method of administration (e.g., oral administration of soluble substances or injection). All experimental procedures shall be specified in detail in SOP to meet the requirements of the GLP standard. Finally, the method of data analysis should be clear; an appropriate statistical test should be specified in the design stage.

The observation unit should be specified at this stage. The unit is used for data collection, which can be an individual animal or group of animals. For example, when testing for drug treatment or surgery, an individual animal can be an observation unit, but when testing for determining the teratogenesis of environmental factors, a whole nest of animals is a unit.

7.2.2.2 *Number of Animals to Use*

The number of animals in each group should be set at the design stage. Although in general statistics textbooks, we can find a formula to calculate the number of animals, the particularity of the laboratory animals should be considered in the experimental design, so as to ensure that statistical significance will be achieved. The number of animals per experimental group is generally determined by differences expected in the results among the groups and the statistical test methods used. For details, readers may find answers in statistics books. In practice, researchers now can get help from the website of Experimental Design Assistant in the National Centre for the Replacement, Refinement and Reduction of Animals in Research (https://www.nc3rs.org.uk/experimental-design), which provides tailored, study-specific advice and feedback on experimental plans and statistical analysis. It includes tools for randomization, blinding, and sample-size calculations, which help to address the most common flaws found in animal studies.

7.2.2.3 *Setting Controls*

Different variables (such as genetic, environmental, infectious factors, etc.) may impact the results of animal experiments. To eliminate the impact of these external variables or that from unknown variables, it is necessary to set up controls. In doing so, it is necessary to make a direct correspondence between the control group and the experimental group. Usually, the control types include positive control, negative control, blank control and media control, as well as analogy control.

- **Positive control:** The researcher will receive a treatment or test with a known result. This result is usually what researchers expect from the treatment, so it gives them something to compare. For a good animal experimental design, it is necessary to have a positive control, which can be used as a parameter to compare the response of the unknown treatment.
- **Negative control:** One in which the response expected in the positive group is not expected when treated with a known treatment.

- **Blank control:** The process of simulating the treatment group, but in fact does not give the animals a treatment. For example, if castration was performed in the treatment group, the procedure of castration should also be performed, but the testis should not actually be removed in the blank control group.
- **Media control:** When the test compounds have to be dissolved in a medium, such as saline, dimethyl sulfoxide (DMSO), or paraffin oil, before they are given to the animals, a media control group has to be setup. Animals in the media control group receive the solvent only; other operations should be the same as done in the treatment group. When compared with the control group, the effects of the solvent can be determined.
- **Analogy control:** Similar to the positive control, the analogy control is to compare the effects of an unknown treatment to that of a well-known treatment. For example, to test whether a new antibiotic is effective in an animal model, we may choose an antibiotic whose effect is well established in the clinic as an analogy control.

7.2.2.4 Blinding

The double-blind method is routinely used in clinical trials. Increasing evidence shows that blinding is also necessary in animal study. Vesterinen et al. reported that nonblinded animal study may significantly increase the treatment efficacy in multiple sclerosis studies.

7.2.2.5 Randomization

In an animal study, even though animals are similar in their biological characteristics, nonrandomized design may result in significant bias. Observations demonstrated that studies not randomized may overestimate treatment efficacy. Vesterinen et al. reported that nonrandomized animal study may double the treatment efficacy in multiple sclerosis studies.

In order to make sure to avoid bias, animals must be randomly assigned to the experimental groups. To achieve this, it is necessary to use animals of the same inbred strain with the same biological characteristics (such as age, sex, weight, etc.). The methods for randomization used in practice are as follows:

- Number all the animals and then pick the numbers from a "black box" or random number table. The animals corresponding to the randomly drawn numbers or the ones selected from the table are assigned to different study groups. For example, the first animal is assigned into group one, the second into group two, third to group one, fourth to group two, and so on.
- If there are multiple study groups, a random number table or computer program will be very helpful.

7.2.2.6 Other Considerations

It is best to conduct further confirmation and review after the completion of the experimental design; a pilot study is useful to test the rationality of the experimental design and provide basis for further improvement.

- **Approval from relevant agency:** Currently, the use of laboratory animals needs to be approved by the Institutional Animal Care and Use Committee (IACUC) in all developed countries and in many developing countries. This confirmation process is crucial to animal experimentation. They may check whether the use of animals is in accordance with relevant laws (please see Chapters 1 and 8 for details) and the welfare of the animals. In addition, many funding organizations ask for the approval of the IACUC in the submission of the project application if the project involves animal experimentation. If the study also involves materials that are hazardous, the program must be recognized by other committees or departments. For example, if an infectious factor or recombinant DNA is used in the experiment, it must be confirmed by the biological safety committee. If radioactive isotopes or radioactive materials are used in the experiment, it is necessary to get confirmation from the radiological safety committee.
- **Data input and analysis:** The input and analysis of data are the responsibilities of researchers. It is not unusual to make mistakes in the collection of data. For example, the identity card of an animal may be misplaced. Therefore, data verification procedures that can be conducted through a computer data input system are mandatory to check extreme figures automatically.
- **Training:** It is necessary to train all the personnel who participate in the research project. Skilled staff will carry out the experiment more smoothly.
- **Consultation:** Published literature is a good resource in helping design an animal experiment. However, discussion with your peers and with researchers from all the related fields will never be overemphasized. Opinions from experts, such as the IACUC members, can be very beneficial and can be incorporated in the study design. As the quality of data determines whether the data are worth publishing and where to be published, it is worthwhile to discuss with your peers about how to improve the quality of your data and the experimental design of your study.

7.3 FINAL STEPS IN ANIMAL EXPERIMENTAL DESIGN

As the problem to solve is different, the animal species used in the experiment will not be the same as will the size of the sample.

7.3.1 Pilot Study

A pilot study is a smaller version of a proposed research study, conducted to refine the methodology of the later one. It should be as similar to the proposed study as possible, using similar subjects, the same setting, and the same techniques of data collection and analysis. A pilot study may be useful to obtain the predictive data, by which the operation and technical procedures are tested and improved. It can be used to explore the differences between individuals in the experimental system, to provide a basis for the formal experiment to determine the number of laboratory animals needed. In addition, it may also help to determine whether the variables in different experimental conditions can be measured with sufficient accuracy. Meanwhile, the necessary conditions to carry out the study can be checked. For example, if an investigator wants to measure the effects of a drug on the growth of a tumor in an animal model and the laboratory has the method to measure the size of tumor, what is needed to measure the differences of the size of the tumor in different groups can be confirmed. In such a case, the investigator can use 10 mice to determine the size of the tumor before and after administration of the drug. In such a kind of pilot study, as there is no available data to be used as reference, the number of animals needed in the study can only be calculated based on experience and speculation. The results of this pilot experiment can provide a rough estimate of the standard deviation and the rate of success in implanting tumors in the animals, and provide a basis for the calculation of the sample size. If the standard deviation of the size of the tumor measured in these 10 animals is quite small compared with the size of the tumor, and if the change of the size of the tumor after treatment is obvious, the chance of success for this study is great and the possibility to detect the effects of the drug is large. The researchers can further study the whole process of the responses of the animals to the treatment and also the responses in different combinations with other drugs.

Data obtained in the pilot study are useful in the application of new funds. It is worth reminding that all the pilot studies should also be approved by IACUC.

7.3.2 Sample Size Determination

The following four factors affect the estimation of sample size: (1) the smaller the chance of type I error (α), that is, rejection of the null hypothesis when it is in fact true, the larger the sample size needed; (2) the smaller the chance of type II error (β), that is, acceptance of the null hypothesis when it is in fact false, the larger the sample size needed; (3) the smaller the admissible error, the smaller the sample size needed; (4) the larger the standard error (SE) of the population, the larger the sample size needed.

According to different types of data, the calculation of sample size varies. Here we do not discuss in detail about the calculation of the sample size, for which we suggest the readers to consult statistic books, as well as the related website, where the experiment design assistant can help you to calculate the sample size you need as shown previously.

7.3.3 Other Specifics to Consider in Animal Studies

In animal studies, many other parameters, such as age, body weight, food and water consumption, and auxiliary variables, should be recorded in addition to the specific responses. These auxiliary variables have a special and important role in the design of animal experiments and interpretation of their results. Usually they should be incorporated into the study design and analysis, which can increase the accuracy and strength of the experiment, and even obtain additional scientific information. All the information we get from animal experimentation should be properly used and reported so that the readers can interpret the experimental results themselves.

In animal studies, there are some auxiliary variables, which are different from those in statistics referring to the variables other than those elicited by the target reactions. The majority of the time, they are not necessarily directly related to objective response variables. They may be those that are not used in the design of experiments and that are not involved in the formal analysis of the main experimental responses. However, skilled animal experimentation investigators may take full advantages of these variables to use them as part of the experimental process and use them in the statistical analysis.

Many variables can play a role in different exploratory analyses; auxiliary variables may be useful in the analysis of auxiliary function. Some factors, such as gender, body weight, and age, are common variables and are considered to be universal, and are the main factors to be considered when comparing different experiments. Some of the universal variables, such as body weight and age, have to be considered in experimental design as a factor in the module, and some can be introduced into the statistical analysis as concomitant variables and covariates, supplementary information, conditional variables, and auxiliary variables.

The auxiliary variables have recently become increasingly used in animal experiments using specific statistical methods. Data collected from each individual animal should be kept well as an important part of the experimental records as these auxiliary variables may be very useful in comprehensive consideration when comparing different experiments.

7.3.4 Application of Auxiliary Variables in Animal Study

The auxiliary variables, such as body weight, can be very useful in the validation of the value of treatment responses. For example, when extreme data are found in an animal, we may go back to check the body weight of the animal. If it is found that the body weight of the animal has also decreased, then we might suspect that the animal had an infection, and thus data from this animal should be removed from analysis.

After assignment but before starting experimentation, comparison of the body weight among animals of different groups may be useful to check whether randomization is achieved. If there is significant difference in their body weight between different groups, then it is very likely that randomization has not been reached. In the process of performing experiments or at the end of experimentation, verifying body weight is still useful in checking randomization if we know that treatment should not affect the animals' body weight. If at the beginning there is no significant difference in body weight between different groups and if we found that a significant difference appears during the process of the experiments, then we have to suspect that other factors, such as the placement of the cages of different groups in the rack, which should have been randomly placed to avoid affecting the strength of light exposed to each cage, may have impacted the animals' food intake and behavior, and thus, affected the results.

7.3.5 Experimental Design

One of the main purposes of experimental design is to effectively assign treatments to different groups so that the differences observed have resulted from the treatments rather than other causes. Furthermore, the observed responses will be more accurate results of the treatments by controlling the known variables. Here we discuss some animal experimental designs.

7.3.5.1 Completely Randomized Design

A completely randomized design is one in which each animal has the same probability of being assigned to any treatment group in the study. For example, if there are four treatments, A, B, C, and D, being studied, then each animal should have a 1/4 probability of being assigned to any one of the four treatment groups. To achieve this, animals should be the same in their biological characteristics, and they should be numbered first, then be assigned to each study group using a table of random numbers or a computer program.

7.3.5.2 Randomized Block Design

A randomized block design is one in which the study animals are first divided into nonoverlapping "blocks" and then randomly assigned to the treatment groups separately within each block. For example, an experiment in which a group of animals is separated by gender to form two blocks. The male animals are randomly assigned to all the study groups using a completely randomized design approach, then the same for the female animals. In this way, if there is an effect from gender, it will be "blocked out," as animals from both genders are equally distributed in all the groups.

In most of the experiments in biomedical research, the animals used are from similar homogeneous groups. The age, body weight being limited to a certain range, and other factors may become an important packet when considering the module. For example, as each cage can only hold five mice, cages became an important auxiliary variable and the similarity in feeding between individual mice in a cage is greater than that in other cages. Cages become a grouping factor when considering the module. In addition, it is impossible to deal with all the animals or measuring responses at the same time on the same day. In this case, the time to process or measure is an important variable that may affect the outcome; grouping is an important module factor.

7.3.5.3 Split-Plot Design

The split-plot design involves randomly assigning animals with a common nature ("whole plots") to all the study groups, then randomly assigning subunits of each animal to a second treatment group. For example, animals with hypertension were randomly assigned to different treatment groups and then those with high blood glucose may be randomly assigned to different treatment groups.

7.3.5.4 Cross-Over Design

The cross-over design is used in the animal study in an effort to use the animal as its own control. In this design, animals are randomly assigned to one treatment for one period of time, then switched to another treatment for a second period. For example, if two drugs, A and B, are being injected into the animals, one group of animals would be assigned to receive drug A, another group to receive drug B. The next time, the animals that have received drug A will receive drug B, while those that have received drug B before will receive drug A. In such a way, both drugs A and B will be tested in the same animal and their effects can be compared in the same animal. In this design, residual or carry-over effects of the treatment given in the first period may affect the results found in the second period. To avoid this, a wash-out period between the two treatments is often required. If there are no carry-over effects or the carry-over effects can be washed away, or if the carry-over effects are the same when drug A is followed by drug B as when drug A is followed by drug B and the investigator is blinded, there is no problem in using all the data to estimate the true difference between the effects of the treatments.

7.3.5.5 Paired Design

Animals with the same nature and other characteristics, such as being rearing in the same cage, are paired and then randomly assigned to different groups. The factors to consider in pairing are often the nontreatment ones.

7.3.5.6 Factorial Design

Factorial design is the one in which the effects of two or more treatments are investigated at different levels in all combinations. For example, if we would like to know the effects of different doses of drug A, say A1, A2, and A3, and the effects of different doses of drug B, say B1 and B2. In addition, we also would like to know the effects of their combination. We could answer these questions using separate experiments. However, the more economical way to do so is to study them simultaneously in a single experiment. In this study design, we have two or more factors to study (in the above example, drug A and drug B), and we also have two or more levels (three levels for drug A and two for drug B) with different combinations, so we now have $3 \times 2 = 6$ different treatments, that is A1B1, A1B2, A2B1, A2B2, A3B1, and A3B2.

There are many things in common between animal study design and other biomedical studies. However, the design of an animal study is also unique compared with other study designs due to the special factors related to the animals. These factors may become variables that lead to biases if not well controlled; however, some of them may be used by experienced researchers to gather more information if properly considered.

BIBLIOGRAPHY

Gao CQ, Fraeyman N, Eertmans F et al. Further evaluation of the biological activity of the unique gonadotropin-releasing hormone peptide in the guinea pig brain. *Neuroscience Letters* 2011;487:246–9.

Gao CQ, Kaufman JM, Eertmans F et al. Difference in receptor-binding contributes to difference in biological activity between the unique guinea pig GnRH and mammalian GnRH. *Neuroscience Letters* 2012;507:124–6.

Keating SC, Thomas AA, Flecknell PA et al. Evaluation of EMLA cream for preventing pain during tattooing of rabbits: Changes in physiological, behavioural and facial expression responses. *PLoS One* 2012;7(9):e44437.

Langford DJ, Bailey AL, Chanda ML et al. Coding of facial expressions of pain in the laboratory mouse. *Nature Methods* 2010;7(6):447–9.

Liu E. *Animal Models of Human Diseases*, Second Edition. People's Health Publishing House, Beijing, China, 2014.

Liu E, Yin H, Gu W. *Medical Laboratory Animals*. Science Press, Beijing, China, 2008.

National Research Council. *Guide for the Care and Use of Laboratory Animal*, Eighth Edition. National Academy Press, Washington, 2011.

Qin C, Wei H. *Laboratory Animal Science*, 2nd edition. People's Medical Publishing House, Beijing, China, 2015.

Silverman AJ. Distribution of luteinizing hormone-releasing hormone (LHRH) in the Guinea pig brain. *Endocrinology* 1975;99:30–41.

Sotocinal SG, Sorge RE, Zaloum A et al. The Rat Grimace Scale: A partially automated method for quantifying pain in the laboratory rat via facial expressions. *Molecular Pain* 2011;7:5.

Van Zutphen LFM, Baumans V, Beynen AC. *Principles of Laboratory Animal Science*, Second Edition. Elsevier Science Publishers, Amsterdam, Netherlands, 2001.

Vesterinen HM, Sena ES, ffrench-Constant C et al. Improving the translational hit of experimental treatments in multiple sclerosis. *Multiple Sclerosis* 2010;16(9):1044–55.

CHAPTER 8

Organization and Management of Animal Experiments

Enqi Liu and Jianglin Fan

CONTENTS

8.1 OBSERVATIONAL STUDY AND EXPERIMENTAL STUDY

In general, new scientific knowledge is obtained either through observational study or experimental study. An observational study is a type of study in which individuals are observed or certain outcomes are measured. No attempt is made to affect the outcome (e.g., no treatment is given), meaning that the researcher simply observes behavior in a systematic manner without influencing or interfering with the behavior. It is well known that observational study allows us to clarify the relations of things, for example, the estimation of the average age of people in a country, or the approximate number of deaths from coronary heart disease, etc. By observational study, we can also learn about the relations of things. For example, tobacco smoking and lung cancer are closely correlated. However, we cannot conclude that tobacco smoking would definitely cause lung cancer; to prove it, we need to conduct large numbers of experimental studies.

Experimental study is a procedure carried out to verify, refute, or validate a hypothesis. Experimental study can provide insight into cause and effect by demonstrating what outcome occurs when a particular factor is manipulated.

The approaches to experimental study are different than to observational studies, each of which has different steps. The two most important steps are experimental design and analysis of the data obtained in each group during the process. By analyzing the data, the researchers can easily come to conclusions. The experimental study is based on an observational study or some simple experiments. These experiments are designed on one hand, to reveal the biological or life phenomena, and on the other hand, to abide by biological or life properties. For example, before a drug is put on the market, its side effects and toxic effects must be examined by conducting many experiments to decide its daily dosage.

Philosophers have different opinions on how to obtain new knowledge. However, they agree that experimental study is based on early observation and explanatory theories are formed by different observations, in which logic and intuition play an important role. According to the inductive–hypothetic–deductive philosophical principle, this period applies the method of inductive reasoning. For example, a group of similar chemical agents cannot induce cancer in different laboratory animals; we can get a general assumption by inductive reasoning: these similar chemical agents cannot induce cancer in all animals (including human beings). However, it is possible that some animals are special cases or new chemicals do not follow the rule. On the basis of general assumptions, a specific and effective hypothesis is produced. For example, a new chemical will neither lead to cancer in human beings nor is it a carcinogen in rats. This hypothesis (including reasoning and prediction) allows people or rats take the compound without being affected by cancer. To reason the unknown from a hypothesis is called deduction.

Now, hypotheses can be proven by experimental study. When a hypothesis is not denied by experimental results, it will be accepted. Experimental results (including some accidents) may affect a new inductive–hypothetic–deductive–experimental– concluding philosophical principle (a philosophical point of view), or overthrow the current hypothesis and form a new one.

The purpose of animal experiments is to demonstrate a certain hypothesis. There is a not well-understood concept in animal experiments, which is "to reason from animals to human beings." Some scientists believe that we should conduct experiments on animals first and then the results can be analogized to human beings, but the details of how to carry out experiments precisely are seldom described. However, the potential philosophy comes from inductive reasoning, or from a series of the taken observations (e.g., a certain drug is safe for mice and dogs) to undone observations (e.g., the drug is also safe for human beings). However, the inductive–hypothetic– deductive–philosophical principle considers the aforementioned process a violation of logic. Many opponents of animal experiments also believe it is Illogical, and list some special cases to prove that some animal experimental results had not been confirmed in humans (e.g., penicillin can kill guinea pigs). Based on this, *in vitro* study may be increasingly difficult to be proven reasonable. It is obvious that if it is difficult to explain how to extrapolate from mice to human beings, then it will be harder to prove the analogy from bacteria or cell lines to human beings as reasonable.

A toxicologist will make a series of assumptions and test them in experiments. The first assumption is likely to be that the compound has no toxicity. This hypothesis can be demonstrated by a cytotoxicity test and other *in vitro* techniques. Failure of the experiment can illustrate that the compound is a toxicant, and the hypothesis is temporarily denied. The experiment should, of course, be as scientific and strict as possible. Another hypothesis may appear based on the previous result: this compound is not a carcinogen in mammals. This hypothesis can be verified in rats or other laboratory animals. If proven that it is a carcinogen in rats, this hypothesis will be ruled out. The compound will be given up in most of the patients. However, another hypothesis may be put forward. Based on this, it is obvious that there is no process of extrapolation. When a hypothesis is given, homologous functions have been confirmed before the start of the experiment.

Designing animal experiments under the influence of philosophy is convenient. Extrapolation is not used, and *in vitro* study and the use of lower organisms may come to a conclusion, thus, the pressure to use primates will be greatly reduced. In order to obtain reliable experimental data to support and test the hypothesis, how to conduct animal experiments appropriately and rationally must be considered.

8.2 DEDUCTION FROM ANIMAL EXPERIMENTS

When conducting animal experiments, the possible results must be fully analyzed because the results of animal experiments will be finally applied to the study of human or other animal diseases, according to the corresponding principle. Here, the word "corresponding" means the similarity in evolutionary significance among different species (including animals and human beings) in morphological structure and physiological phenomena. The corresponding principle requires selecting the proper laboratory animals according to certain purposes and courses, and then extrapolating the study results to achieve the desired results.

There are two forms of deduction for animal experiment results, which refer to extrapolating from quality and quantity, respectively. Qualitative extrapolation means to deduce from the laboratory animals' process of dealing with pathological and physiological changes to other animals or human beings. Quantitative extrapolation means to evaluate certain doses of a compound's positive or negative response on other animals or human beings, based on the animal experiment. Quantitative characters of animals and qualitative characters' effects on physiological reactions are flexible and relative. For example, a mammal's digestive system and body surface area have certain proportion relations. The smaller the animal's organs are, the greater the relative surface area is, and thus, the higher the digestive level. If an animal's body size is small, its liver, kidneys, and heart share a large proportion of its weight. As a result, when studying dose–effect relationships, it is better to use digestion weight (the absolute weight$^{0.75}$) or body surface area instead of absolute body weight.

The discomfort or stress level of laboratory animals can also influence the deduction of animal experimental results. As shown in Table 8.1, laboratory animal models, target animals' similarity and discomfort level were combined to obtain general reference opinions for animal experimental result deduction. In terms of the same animal strain or species, the reliability of animal experimental result deduction is mainly decided by experimental design. Furthermore, deduction of animal experimental results is also influenced by animal's genotype, sex, age, and physical

Table 8.1 Deduction of Animal Experimental Results

Animal Model and Target Animal	Discomfort Level During the Process	Deduction Type	
		Quality Type	Quantity Type
Match	Light	+++	+++
	Serious	+++	++
Do not match	Light	++	+
	Serious	++	+

Note: "+" in the table means poor reliability of animal experimental results deduction, "+++" means good reliability.

condition. Special attention should be paid to deciding the influence of genes in different species.

In the production process for industry, agriculture, food, and drugs, the evaluation of product safety is often involved. In order to reduce the danger to human beings, animal experimentation is an effective method; however, the results of animal experiments cannot completely assure human safety. A safe result in an animal experiment may be disastrous in a different species. For example, according to the US Food and Drug Administration (FDA), two different kinds of laboratory animals are commonly used in toxicology tests to procure infallible data to the greatest extent.

The results of animal experiments cannot be absolutely extrapolated to humans because the results of animal experiments would be unlikely confirmed in a one-on one-human study. The results of animal experimentation are reached under certain special conditions. The extrapolation of animal experimental results is a kind of experiment study designed to decrease the risk of similar processing in humans.

8.3 ANIMAL EXPERIMENT PROCESS

8.3.1 Select a Study Topic

In general terms, the selection of a scientific study topic is decided by the development of society and people's living standards (e.g., the occurrence of cancer is associated with the environment). In narrow terms, the selection of the scientific study topic is determined by study goals and the ability of the institution, or determined by researchers' interests.

Animal experimentation is sometimes because of a need in modern society or laws and regulations by government. New chemicals (biological products, pharmaceuticals, etc.) must pass a series of animal experiments before being certified. As for the methods of animal experiments, corresponding regulations have been established in many countries.

8.3.2 Propose a Hypothesis

A hypothesis is a proposed explanation for a phenomenon. For a hypothesis to be a scientific hypothesis, the scientific method requires that one can test it.

To actually propose a complete and scientific hypothesis, three stages must be passed:

1. Observe and record all the relevant information
2. Analyze and classify the information
3. Propose hypothesis based on the information

In practice, it is difficult to judge which information is a relevant information. In the process of judgment, intuition plays a very important role. A large number of documents must be referenced, and computers and databases are effective.

By analyzing and classifying the information, a preliminary hypothesis is proposed.

General hypotheses include a kind of speculation about the relations between the phenomena observed and possible phenomena. Bold vision is critical, especially when the proposed hypothesis breaks traditional concepts. Only in this way can we make and confirm great discoveries. We all hope our own bold speculations to be fact, thus replacing existing theories, but demonstration by animal experiments is needed.

The methods of proving fruitful hypotheses are polytropic, and sometimes not scientific because a new hypothesis must be strictly tested to form a theory.

All initial hypotheses need to be tested by a large number of experiments, and this requires careful selection of the correct methods to carry out the experiments. The pre-experimental results must be seriously analyzed or the experiment results may instead reject the initial hypothesis.

For example, when judging a hypothesis, the experiment should be finished first, otherwise there is a tendency to draw a conclusion soon after the start of the experiment. This will make the experiment not proceed as planned. When the experiment is deliberately designed, a keen interest for science allows us to become surprised. However, in very rare cases, these findings will make the current hypothesis unacceptable, thus the identification of the current hypothesis is important.

We can reach a reasonable and scientific hypothesis through the following three points:

1. Professional conditions, which allows the test of hypothesis. For example, animal experiments that include the use of special strains of rats or special treatments are needed during the process. If there are errors in the experiments, sometimes the hypothesis will be disproved (e.g., to demonstrate a chemical compound is nontoxic to rats). Due to the wrong animal strain, the opposite conclusion may be obtained. Good experimental design can avoid these problems. By choosing different species, strains of laboratory animals, experimental processes, methods of administration, or feeding-and-management conditions, we can obtain concordant experimental results.

2. Experimental ability. A small or terrible experiment design only uses the obvious experimental materials, ignoring influences of minor factors. These influences may be very important to the biomedical study, and very appropriate to testing the hypothesis. The attitude toward hypothetical science not only has bold assumptions, but also strict and original thought to overthrow these assumptions. A good experimental design should have enough evidence for biomedical study to ensure the reliability of the experimental results.

3. Hypotheses can be better proven by experiments. If the experimental results cannot approve the current hypothesis, necessary adjustments to the hypothesis should be made to make the hypothesis more factual.

8.3.3 Select Experimental Materials

When selecting the appropriate experimental materials, many things must be considered, including animals and even human beings, for the use of animal models in the experiment. To obtain general experimental results, the selection of laboratory animals is of great significance because we are going to correlate the results to other animals or human beings.

8.3.4 Prepare a Study Scheme

In the process of relating precise hypotheses, full consideration of the possible experiments must be given in the experimental process. Many researchers feel free to choose their appropriate experimental methods. However, the determination of experimental sample size is affected by the experimental statistical level, and some effects are often caused by objective factors. For example, when selecting laboratory animals, selection depends on the experimental operation technology for existing animal species.

The list below includes some points that should be noted when designing a study scheme:

1. To deduce from facts and rationale to general hypothesis.
2. Summarize the hypothesis briefly.
3. Accurately debate and reason to test the hypotheses.
4. Schematic experimental design, including all processing in the experimental process.
5. Explanation of experimental operations and description of the method of measurement.
6. Briefly state the discomfort that the animals may feel.
7. Demonstrate by applying animal numbers and statistical methods. Provide relevant statistical analysis and statistical methods used to discuss experimental feasibility.
8. Briefly state the experiment preparations including methods, processes, and data analysis.
9. The practical demands of the experiment such as personnel, name of the study director, and funds needed.

The correct establishment of a study scheme has four benefits:

1. To judge whether the experimental purpose will/will not be acceptable to the ethical committee or Institutional Animal Care and Use Committee (IACUC).
2. To allow peers to review the study results.
3. To apply for grants.
4. To prepare a detailed experimental scheme and evaluation, and report experimental results.

8.3.5 Formulate a Detailed Plan of Animal Experiments

A detailed schedule of all animal experiments is needed. The experimental process should be clearly given in the experimental scheme including the selection of animal species and strains, animal nutrition and feeding regimen, sample collection, and euthanasia of the animals. Detailed drafts should include several aspects: the researcher, laboratory animals, and experiment technology. The researcher should also be aware of the conditions that are used to test scientific hypotheses.

All the experimental data should be collected and analyzed using mathematical and statistical methods when an animal study is completed. Finally, we must judge whether the results proved the hypothesis.

8.3.6 Evaluate and Report (Writing a Paper)

Drawing a conclusion based on statistical analysis of animal experimental results may prove the original hypothesis. This may be difficult, however, for some special hypotheses.

Not all experiments can strictly test the hypothesis. During the animal experimental study, special animal strains, special management, special feed, and special devices are often selected. If the animal experiment proves that the previous hypothesis was mistaken, but the experiment itself is successful, researchers may doubt other conditions such as different strains of animals, feeding regimen, or other experimental equivalents. For general hypotheses, a wide range of conditions should be tested. For example, the results should be consistent even if using different species or strains of laboratory animals, or different breeding management. To prove a hypothesis like "a compound is non-carcinogenic to mammals (including human beings)," the results should be the same using different animals or different methods of administration. Hypotheses like this can be tested. If we select another strain of rats, and find the compound is carcinogenic, it means our hypothesis is disproved. However, for some hypotheses, they need to be tested many times in different species and strains of laboratory animals under different conditions. Such hypotheses are difficult to test and a series of experiments may be required.

The final period is to analyze the experimental data, draw a conclusion, and finally present the conclusion in a domestic or international meeting to submit to scientific journals. The target journal requires specific formatting of the paper. All formal study reports usually include the following aspects:

1. **The title and the researcher,** including all of the experimental personnel involved. All the participants should make his/her own contribution.
2. **Abstract,** briefly introduce the purpose of the study, experimental design, methods, and conclusion.
3. **Introduction,** simply introduce the related background knowledge, similar study situation domestically and abroad, and raise possible hypotheses. If the hypothesis is invalid, state the expected results.
4. **Materials and Methods,** the results of animal experiments should be repeatable, detailed descriptions of the laboratory animals and animal experiment conditions, so as to be repeated in the same conditions when submitting an experimental result.

Laboratory animals: Species, genetic background, strains (inbred strain, outbred strain, other genetic control), microbial control level, grouping, treatment, age, gender, body weight, etc.

Animal experiment conditions: Experimental facilities' type (barrier, isolation, laminar flow), temperature, humidity, air flow, illumination (photoperiod, intensity and spectrum of light, natural, or artificial), cage type (size, shape, material, filter cap's condition), the number of animals per cage, bedding (type and change times), transportation and feeding time, etc.

Nutrition: Feed (including manufacturer, composition, formulation, sterilization, quality control methods, etc.), feeding regimen (free or restricted intake), drinking (bottled, disinfected).

Experimental procedure: Chemical substances, drugs (dosage, manufacturer, purity, additives), experimental period, animals' estrous cycle, physiological status, administration route (oral or intravenous injection), sample collection (blood, urine, feces), anesthesia method (anesthesia procedures, type and dosage of anesthesia, drug delivery route), animal execution technology, euthanasia (methods), processing and preservation of animal tissue or organ samples, etc.

Experimental design: Formal design (completely randomized principle, randomized group, Latin design, etc.), assessment methods and statistical analysis methods (parametric, test and variance analysis, regression and correlation, nonparametric), etc., should be introduced clearly. In addition, it should be noted whether the project is approved by IACUC.

5. **Results,** display experimental results with graphs, tables, and text. The text should be brief and clear.
6. **Conclusion,** show conclusive results briefly, and clarify the conditions of testing the hypothesis and scientific study achievements to offer good suggestions for future study.
7. **Acknowledgements,** express thanks to people and firms that have offered help during the study.
8. **References,** list the cited documents' title, authors, and source.

A manuscript from submission to publication usually needs a certain period of time. In general, the editor will deliver the paper to experts to peer-review, and the experts will put forward some relevant suggestions or comments on the quality of experimental design and working hypothesis to give guidance to the experimental method and to affirm the results. In general cases, the editor will return the manuscript to its authors and let him/her make necessary changes or supplementary experiments based on experts' suggestions. Here, some authors revise their paper while others stick to their opinions.

Arbitrary institutions or a third-party laboratory often does some experiments to detect the reliability of the manuscript, and they have the author explain the rationale behind the experimental design and validity of the experimental results. By going through this process, the results are likely to be published in renowned journals.

Once the paper on the animal experiment is published in an international journal, especially international English journals, the experimental results will draw many people's interests, as well as lead to some new hypotheses, or make the original hypothesis' experimental design stricter.

8.4 ORGANIZATION AND MANAGEMENT OF ANIMAL EXPERIMENTS

A biomedical study project using laboratory animals cannot be performed by one person. It involves many people and complicated processes. If the organization and management of animal experiments is improper, wrong conclusions may be obtained and the hypothesis may not be proven.

A successful and professional expert who conducts an animal experimental study is like a manager; he/she has to have a profound understanding of the organization and management of animal experiments to use the advantageous factors and overcome the disadvantageous factors.

The workload of animal experiments requires that one researcher cannot manage the whole project alone, therefore the personnel should be clearly divided and cooperate with others. Researchers must design, organize and conduct the experiment, manage funds, and write reports; technicians, who are certified to conduct animal experiments, must carry out detailed operations based on the experimental design; the animal feeders feed and look after the animals; the IACUC confirms the welfare of the animals in the experiment and whether the experimental method is humane.

8.4.1 Special Conditions of Animal Experiments

The organization and structure of animal experiments undergo changes because of optimization by the researchers and utilization of material resources. The specificity of animal experiments exceeds the demand for personnel and material resources. When designing animal experiments, the researcher should not only consider scientific fact and effectiveness, but also restrict the number of animals that are used, and try to avoid causing pain and suffering to animals in experiments. Such additional requirements have been governed by laws in many countries. IACUC monitors the process, and puts moral constraints on researchers to make sure anytime animal experiments are conducted, the number of animals is limited to a minimum and the methods are as precise as possible. There are many ways to achieve this goal: to enhance the communication between study institutions; to use dead animals instead of live animals when training; to recruit experienced, skilled administrators and technicians. Duty records have been put in law in many countries to avoid unnecessary and uncontrollable animal experiments. This also limits the number of experimental subjects or animals, and evaluates the animal pain scale.

Many countries require study institutes to establish an IACUC or similar ethical committee whose task is to judge whether an animal experiment is necessary. If an animal experiment is necessary, it must abide by laws and regulations during the process. In these countries, an animal experiment project should be submitted to IACUC. Once the animal experiment is approved by IACUC, it is certified to be performed.

When designing an animal experiment, the project leader must understand the relevant laws and regulations, and give full consideration to the following points:

1. Whether the animal experiment project is in line with the national and local laws or regulations. It should be noted that, in many countries, not all animal experiment projects will be approved.
2. Whether the institute (or laboratory) is approved by the government and qualified to carry out animal experiments.

3. Whether the researcher is qualified to design or conduct animal experiments.
4. Whether there are enough reasons to explain the necessity of the animal experiment.
5. Whether there is enough professional staff (feeders, technicians, etc.) to make sure the experiment will run smoothly in the future.
6. Whether the source of the laboratory animals is legitimate.
7. Whether the animal facility meets the requirements.
8. Whether there is proper equipment and reagents to anesthetize or euthanize animals. Whether all the relevant personnel know how to use them.
9. Whether the local animal welfare organizations or IACUC know about the animal experiment, and whether they are informed before the start of experiment.
10. Whether the animal experiment is approved by the IACUC.
11. Whether the use of laboratory animals is recorded and the record is saved.

Promotion of animal welfare is increasing, which makes many countries' animal experiment regulations undergo revision many times. The main concern is that the qualification requirements for personnel (including the researchers of design, guidance and implementation, and the feeders) involved in animal experiments are increasing.

The above-mentioned factors may be strict and broad for organizers and managers of animal experiments. However, a good design scheme will not only lead to accurate results, but can also avoid using too many animals. It not only conforms to moral constraints, but saves time and energy. Thus, a strict experimental design is beneficial to the conduct of animal experiments.

8.4.2 Influencing Factors on Animal Experiments

The reproducibility and reliability of animal experiments is highly dependent on standardization. When designing an animal experimental project, the influencing factors on animal experiments, which are summarized in Table 8.2, must be considered. These requirements must be fulfilled to minimize the variation of experiments and animals. Standardized requirements for animal experiments are the following: to use the same animal supplier; use the same manager and technician; use the

Table 8.2 Influencing Factors on Animal Experimental Results

Influencing Factors	References
Genetic quality	Strains, reproductive system, suppliers
Biological status	Sex, age, body weight
Health status	Suppliers, health status, barrier system
Nutrition	Suppliers, nutrients, water
Breeding conditions (cage)	Feeding type, bedding, stock density
Breeding conditions (room)	Ventilation, temperature, relative humidity, illumination, noise, other animals
Transportation	Transport method, transport box, feed, and water
Animal feeding	Qualified breeder
Experimental technique	Qualified technician, standard technology, time management

same animal facility; and conduct the experiment at the same time every day. This requires intimate cooperation between researchers and auxiliary personnel (including manager and technicians).

8.4.3 GLP Standard and Animal Experiments

Animal experiments are a special field that use animals to conduct a biomedical study, in which organization and management play important roles. In an animal experiment, problems, such as animal management, chemical substances, drug utilization, discussion of medical experimental methods, diet, etc., are often encountered. Management procedures should follow the relevant state laws and regulations. Many countries follow the Good Laboratory Practice (GLP) requirements for animal experiments. GLP is the law for organization and management, project implementation, recording and reporting, and file maintenance during the process of evaluating drug safety. The purpose is to ensure integrity, reproducibility, and repeatability of drug safety evaluation.

The US FDA was the first to establish and implement GLP in December 1978. It indicated that any laboratory that is not in conformity with the GLP standard has no qualification to conduct animal experiments for toxicity to declare new drugs, and all the secure experimental data provided shall not be accepted by the FDA. The European Union enacted GLP in 1980 and implemented it in 1981, while Japan enacted GLP in 1982 and implemented it in 1983. China first issued GLP in December, 1993, and revised it for the first time in October 1999, later revising it again in 2003.

With the development of science and technology, and the public's attention to drug safety, the GLP clauses are gradually increased and improved, and are gradually recognized by many countries. The principles of GLP are almost the same in different countries, and the content is very similar as well. GLP has become the international standard of new drug development. The US FDA has built agreements among many countries including Britain, Germany, France, Switzerland, Italy, Japan, Sweden, and the Netherlands. The ability to mutually accept and mesh conditions allows animal experimental results in one country to be accepted and registered by another country, thereby reducing unnecessary repeated animal experiments.

The evaluation of new drug safety requires toxicological study on animals. GLP has detailed and specific requirements for the use and management of laboratory animals. For example, GLP demands that before the start of a new drug study, there must be a clear plan of animal experimental study. The plan should include the detailed experimental mission and scope, experimental procedure designed to achieve expected purposes, and name and qualifications of the researchers who are responsible for the experiment, and all observational records. All the procedures that are involved must be recorded in detail, and act according to work regulations and standard operating procedure (SOP). There must be standardized operating methods for animal feeding, health condition, experimental technique, observational method, animal inspection, data collection, and processing. GLP principles are also concerned with laboratory design and animal facility, as well as specific requirements for animal facility structure and environmental conditions. Moreover, the quality of laboratory animals must meet the requirements, and proof of all these conditions must be kept appropriately.

Although the primary purpose of GLP is to ensure accuracy of animal experiments in drug toxicity studies, the promotion of GLP principles to other studies involving animal experiments is helpful to make the study project standardized and increase the reproducibility of the results.

The GLP principles are, however, not fit for all animal experimental studies because every type of animal experiment has its own special level for standardization.

8.4.4 Centralization of Animal Experiments

In order to serve researchers, many institutions conduct animal experiments by centralizing management and utilization of spaces and facilities. There are three levels of arrangements in the centralization of animal experiments:

1. **First level:** supply room for laboratory animals. This facility is only for purchasing, producing, breeding, and quarantining laboratory animals.
2. **Second level:** "hotel" for laboratory animals. In addition to the first service, it is for placing or feeding animals that are not commonly used or require specialized placement, such as sheep, dogs, cats and primates, as well as all the animals used during the experiment. Even if a researcher does not have his/her own animal room, he/she can make use of this facility as a "hotel" for laboratory animals. However, if the animals are going to have surgery, radiation, x-ray, imaging, or be used in other projects, they need to be moved out from the "hotel" for laboratory animals.
3. **Third level:** animal experimental center, which is the highest form for centralization of animal experiments. It not only includes all the functions of the previous two arrangements, but also avoids moving the animals throughout the experiment. The affiliated animal facility in comprehensive universities or institutions is this kind of facility.

For researchers, centralization of animal experiments means the centralization of collaborators, animal rooms, and equipment that do not belong to the same department. The centralized management of animal experiments requires the staff to have clear goals and rich knowledge, and be able to actively engage in the experiments. An animal experiment center is quite different from other study departments, as they pay special attention to the quality, health, and welfare of laboratory animals. Centralization may give rise to some problems, such as there are more people going in and out using equipment, thus bringing problems to the health condition of animal rooms and equipment maintenance. Furthermore, it cannot be guaranteed that every researcher can use animal facility alone. They may have to share with others or apply to use the animal facility. The advantage is that experienced staff can be used to the greatest extent, thus leading to effective work assignments, growth of knowledge and skills in broader fields, more efficient use of animals, and opportunity for creating the best space and technical facilities. In addition, centralization of animal experiments can enhance the communication between researchers and institutions, and provides the opportunity to understand and exchange animal testing information.

8.4.5 Best Experimental Arrangements

When doing animal experiment plan arrangements, one of the key factors is to arrange time properly. All the animals should be arranged in the same period of time, in the same process, and every step of the experimental arrangement from the beginning to the end of experiment should be considered.

Excluding general factors that suit all experimental types, factors that are specific for animal experiments should also be taken into consideration:

1. Must set aside enough time for purchasing animals. The animals must fulfill experimental requirements (sex, age, weight, numbers, etc.), special attention should be paid when purchasing uncommon animals or strains, or ordering animals from noncommercial suppliers.
2. Must set aside the necessary quarantine period to ensure animals' health conditions. Animals that are newly bought or animals that come from the same laboratory animal center should be quarantined.
3. The experimental plan could be delayed by unexpected or natural death of animals.
4. Must set aside enough time to learn and master new technology.
5. Must set aside enough time to prepare specialized animal feed.
6. It is necessary to do small-scale pilot experiments when bringing in new methods or technology.
7. Must set aside enough time to be accepted by IACUC or other authorities.
8. Implementing logistical resources needed for the experiment also takes time. For example, animal facility, staff (breeders and technicians), purchasing animals, reagents, and special equipment needed for the experiment (isolator, laminar flow), etc.

Before the start of the experiment, a prospectus containing the cost estimation and expense budget is needed. If the funds are insufficient or the funds may be over budget, how enough funds to meet the budget will be raised is a critical point. Animal experiment technicians and experts are willing to offer help during the study, and we must consult with them.

8.4.6 Working with Others

Most animal experiments cannot be done without cooperation between the subjects and departments. The researcher must act as a commander and ensure all the necessary activities are done by appointed staff at the right moment. If you follow the following principles, you will have a greater chance of success:

1. To decide everyone's role after talking to all the staff involved in the experiment. When making the final decision, the point of the decision maker must be considered. All the staff involved must have the pre-knowledge of what must be done, how to do it, and when to do it.

2. To motivate partners. To tell them the purpose of the animal experimental study and let the partners know their important roles. Tell them the actual progress and results (including positive, negative, and setbacks), and explain why the plan has changed. You can also motivate your partners by asking for advice and discussing.

3. To record all the meetings, discussions, and suggestions, to sign an agreement and make daily records.

4. A brief record is needed for every counseling meeting, including how the consensus was reached or how opinions were kept; a copy of the written agreement is printed and the technicians involved should sign it. The contents include, the starting date of the relevant project, the number of animals required, and drug delivery approach and dosage. Relevant personnel should fill in the study data form every day, which include animal species and strains, sex, number, weight, health condition, size of tumor, death date, anesthesia, euthanasia, etc., and deliver it to relevant staff to copy study projects, study reports, and related information.

5. To communicate timely, and change the experiment plan as little as possible. Animal experiments can waste time due to inadequate preparation, and adjustments and changes of experimental plans are sometimes necessary. However, before the adjustment or change, it should be fully discussed and noted to make sure all the relevant staff know and understand the changes.

6. Follow the experimental protocol by yourself and make sure others do so as well.

8.4.7 Safety of Animal Experiments

Centralized utilization of animal experimental facilities have higher management requirements. The principal of the animal experimental center should be a veterinarian or relevant professional expert who is certified to manage. He/she should know all the workings within the animal experimental center to verify the safety of staff, prevent accidents, and animal infections. The design of animal experimental projects should not be limited to animals themselves, the researcher or staff may bring some harmful chemicals (to both human beings and animals), such as radioisotopes, carcinogens, cytotoxic substances, or other harmful substances, to the laboratory. As part of the experiment, we sometimes use these hazardous reagents to induce illness or vaccinate the animals with infectious agents to induce infection, thus making animal models for biomedical study. Hazardous reagents may do harm to workers' health and animal-borne disease can also infect researchers. Moreover, some animal-originated materials used for animals can sometimes be allergens to researchers.

When planning the location of an animal experimental project, the joint participation of both researchers and animal facility management personnel is required. Researchers often ignore the influence of integrated environment on study results. The most typical case is animals being infected by pathogens that affect the experiment results (such as viruses, bacteria, or parasites). Researchers are unable to control the microorganisms in animals nor conduct thorough veterinary study, which is where the staff in the animal experimental center play a leading role. Animal infectious disease is one of the most significant and serious incidents during an animal experiment. When designing an animal experiment, the help of professional animal experimental staff is a must.

Most commonly used animals are kept in special facilities and do not have infectious diseases that can affect the health of researchers or the experimental results. It is safer to use SPF animals. When using wild animals (such as primates) to conduct the study, it should be especially noted that some serious diseases (infectious diseases of animal origin) can be transmitted from wild animals to people, and some of them can even cause death. When designing such experiments, work closely with veterinary and medical experts to fully estimate the problems that may appear in the experiment such as what kind of cages for animals should be used. Seek opinions instead of doing it all alone.

Caution is needed in experimental studies that use contagious animals with incurable diseases. For example, when using HIV or viral hepatitis animals, researchers should be extremely cautious. Such animal experiments should be conducted in Biosafety level 3 facilities. The animals should be raised in totally sealed, negatively pressured housing to make sure the air within the facility will not leak outside and cause contamination. Such facilities are relatively safe for researchers.

8.5 STANDARDIZATION OF ANIMAL EXPERIMENTATION

The standardization of animal experiments refers to the stabilization of the quality of any given animal (or animal population) and their environment, which is the standardization of animal quality and experimental conditions. Standardized animal experiments can increase the reproducibility of experimental results and make the results within different laboratories be comparable. Standardization can also decrease the variation of quantification and measurements of the same animals within the given animals during the experiment. From a statistical standpoint, to reduce measurement variation between individual animals can reduce the number of animals needed in each experiment.

8.5.1 Variation Analysis

The standardization of animal experiments only involves the known possible variations. Measurement variation is derived from two cases: the first one is the known between-experiment variation or between-group variation, and the second one is within-experiment or interindividual variation, also called within-group variation. Both of the two variations come from the animals themselves or environmental influences during the experiment.

1. **Between-experiment variation.** When repeating a particular animal experiment, although the animal is from the same group, different measurements are often reached. Treatment changing the average measurements of the control group and experimental group is referred to as between-group variation. If there is interaction between treating factors and experimental conditions, there is an additional factor

that can change the treating process in each experiment, and this may lead to the wrong explanation of experimental results. In order to accurately evaluate the efficacy of the treatment, we must repeat the experiment. Treatment effects should be real effects, systemic error should not impair nor exaggerate.

Reducing the effects of between-experiment variation can reduce same experiment repeatability, thereby reducing the number of animals used. From a scientific point of view, experimental results should not rely on the time and place of the animal experiment, rather, it should be based on repeated experiments.

2. **Within-experiment variation.** In an experiment, quantified measurements between the same animals is intrinsic interindividual variation. This kind of within-experiment variation includes, the variation caused by experimental process, analysis variation, intraindividual variation, and intrinsic interindividual variation. Intrinsic interindividual variation originates from measurements of each individual animal. Each animal has its own intrinsic characteristics that are not dependent on the treatment. Intraindividual variation is a variation in animal body that cannot be standardized, and this kind of variation can make daily measurements fluctuate.

If statistical efficiency (the probability to measure the real value) is continuously increasing the interindividual variation in measurements (increasing standard deviation), the number of animals needed for each experiment will increase. It is feasible to decrease the number of animals by reducing the interindividual variation in the experimental results.

3. **Origination of between-experiment and within-experiment variation.** An important origination of animal experimental variation is the animal itself. The differences between animals in one group or differences between animals in two groups originate from the animal's age, weight, number of littermates, and other differences that may have existed before the experiment. These differences can increase the between-experiment and within-experiment measurement variation, and if this variation influences the variation between control group and experimental group, it will influence the experimental results. The genetic interindividual differences (including sex) can also increase the variation of measurements. Biological factors (the microorganisms that animals carry, estrous cycle), physical factors (illumination, temperature, etc.), chemical factors (nutrition, sleep, etc.), group factors (breeding density, the interaction between individuals), and environmental factors in the experiment can increase the intergroup and intragroup variation in experimental results.

The measurements acquired in animal experiments are basically determined by the interaction between genetics and environment, which show in different levels: the influence of environment on fertilization, embryo development, and sexual maturity stage is called primary milieu. The interaction between primary milieu and genotype decides the animal's phenotype. After the primary milieu, the preenvironment still influences the phenotype, thus forming the dramatype, and the preenvironment is called the secondary milieu. When animals are influenced by the experimental process and treatment, it is referred to as the tertiary milieu.

According to the type of measurements, the interaction between environment and genotype has different effects on between-experiment and within-experiment variation and treatment.

8.5.2 Standardization of Animals and the Environment

Theoretically, individual differences caused by genetics can be excluded by using the animals with the same genotype (isogenicity). Animals from an inbred strain or F1 hybrid share the same genotype, and the measurement of interindividual variation observed in the experiment using such animals is much lower than animals with different genotypes (e.g., outbred animal). However, genetics is one of the factors that can affect measurements. For example, for the reason of differences in body weight of same sex mice that are born on the same date, the environment before and after accounts for 20%, and analytical error and intangible error account for 30% of the total variation. If the same environment control is performed for each animal with the same genotype, an uncertain error, caused by the interaction between genetics and the environment in the different developmental stages, still exists.

In order to control genetic factors and reduce interindividual variations, the best method is to use animals of the same genetic background (such as inbred strains or F1 hybrid animals) in different experiments. However, not all animal experiments require using animals with the same genotype, and many experiments (such as toxicity tests) require animals with some degree of genetic variation (such as outbred strain animals).

The organisms carried by laboratory animals can influence the multivariation of measurements. Some animals with latent infection by organisms can markedly increase the variation of measurements between individuals in different groups and the same experiment. Such variation can be reduced by using animals that carry a specific microbial background (such as SPF animals).

In the series of purchasing, transporting, and experimenting, the experimental results may be altered by the living environment. All transportation will cause endocrine and metabolic reactions in the animals. This "transportation stress" can influence different individuals within limits and increase interindividual variation. After one week, the animal is likely to be in a new stable physiological state. Animals arriving in a new place is a change in the physical, chemical, and microbiological environment, and whether that environment is temporary or permanent, it will increase the measurement of interindividual variation. Therefore, it was suggested that, depending on the type of animal experiment studies, animals purchased from outside should have a three-week adaptation period prior to use in a new environment.

Although some countries have specifications on animal stocking density, in many barrier systems, if a top air supply and ventilating system that discharges air through the 4 bottom corners are applied, the 1.5 m point of the animal cage may have a higher temperature than the 0.5 m point. The feed intake of separately housed animals is higher than animals housed in a group. This is because the animals housed in groups will huddle together, thereby reducing the loss of body heat and reducing the energy demand. For example, the feed intake may be 7.6 g/male mouse/day when there is only one male mouse in one cage; however, the feed intake may be decreased to 4.2 g/male mouse/day when the stocking density is 8 mice/cage. Microenvironments inside the animal cage include animal populations, relative humidity, type of bedding, etc. All these factors that may cause measurement variation are factors that should be noted in animal experiments.

In the experimental process, animals with different feeding conditions should be equally divided into the control and experimental groups. In this way, the measurements between the control group and experimental group will not be biased due to the difference of feeding conditions. Local environmental conditions inside the animal facility should remain the same in different experiments. The ideal situation is all the environmental factors stay the same.

When using animals to conduct experiments, the standardization of the environment is necessary. An ideal, standardized animal experiment environment should consider animal welfare and engineering principles, although sometimes the variation of measurement in the experiment cannot be reduced.

Moreover, when standardizing animal experiments, standardization should not have any negative impacts on animal health. The animal's physiological and behavioral requirements should be considered, or standardization may not reach the expected goal. In general, researchers get animals from suppliers, thus it is impossible to decide or control standardization. However, researchers can ask suppliers for documents to prove genetic quality and microbial control level. This serves as a reference when designing animal experiments.

8.5.3 Standardization and Deducing Animal Experiment Results

In principle, when the design of an animal experiment is finished, the results depend on experimental conditions (animals, environmental factors). For common confirmatory experiments (such as vaccine activity testing), it is not a problem. However, for other experiments, the experimental data should be summarized to reach a conclusion. The standardization of experiments means a restriction in experimental conditions, which is not the same as reasoning of experimental results. Under what conditions can the animals used in the experiment represent all of a specific group of animals? Even if the animals used in the experiment can represent all of a specific group of animals, the experimental results are limited to this group of animals (the same strain, the same sex, and the same body weight) and limited conditions (biological, physical, chemical, and environmental features). This problem will be more complicated when the results are concluded. To confer animal experimental results to a general conclusion, especially from one species to another, due to the standardized experiment, experimental results can only be generalized in a limited way. Therefore, standardization is also important for deducing experimental results.

8.6 ANIMAL EXPERIMENTS IN TRANSLATIONAL MEDICINE

It is well known that almost all human disease diagnoses, treatments, and prevention methods were all initially developed and tested through animal studies and confirmed safe to perform. Animal experiments aiming at translational medicine (although not all animal experiments are for translational medicine study) have the following goals: (1) to provide basic theory on disease diagnosis, prevention, and treatment for humans; (2) to clarify pathogenesis in human diseases; (3) perform preliminary experiments before clinical tests of drugs and other treatments. However, when discussing translational study findings from bench to bedside, nearly 90% of study findings from animal experiments cannot be directly applied to humans. Therefore, well-designed and accomplished translational medicine animal experiments are of great importance to raise the credibility, accuracy, and clinical effectiveness of experiments tested with animals.

8.6.1 Translation of Animal Experiments

The translation of animal experiment study findings to clinical processes faces huge challenges. The following three renowned study reports or cases are about the necessity and limitations of animal experiments.

8.6.1.1 Findings from Animals Inconsistent with Those in Human Clinical Trials

According to a retrospective review involving 228 cases of animal experiments, released in the famous *British Medical Journal* in 2007, most of the study findings by animals are inconsistent with those in clinical trials. The author evaluated 6 interference measures proven to be beneficial or harmful for patients, retrieved the animal experiment reports published or not yet (the author does not know the outcome of experiments in advance), and assessed those outcomes.

The study outcome of six interference measures are as follows: (1) Corticosteroids were proven clinically useless in treating head injuries, and even raise the death rate of patients; however, they are beneficial in animal experiments. (2) Antifibrinolytics are effective in treating bleeding, but do not work in animals. (3) Thrombolysis can be used to treat acute ischemic stroke. Moreover, tissue plasminogen activators can improve neurosurgical function in animal models, the outcome of which is consistent with that in human beings. (4) Tirilazad was associated with a worse outcome in patients with ischemic stroke, while it reduced infarction volume and improved neurobehavioral scores in animal experiments. (5) Corticosteroids can clinically decrease the neonatal respiratory distress syndrome rate and death rate of newborns. However, in animal experiments, it only works to reduce the former, and is unclear on the latter. (6) Bisphosphonates can increase the bone mineral density of females with osteoporosis after menopause. It also works in animals with ovariectomy.

This review indicated that nearly all the outcomes of the 288 animal experiment findings demonstrated inconsistencies between animals and humans, except one treatment—thrombolysis treats acute ischemic stroke. Perel et al. suggested that the discordance between animal and human studies may be due to bias or to the failure of animal models to mimic clinical disease adequately.

Unfortunately, inconsistencies or contradictions are also found in other similar study fields. For example, over 500 neuroprotective animal experiment treatment plans have been created, but only aspirin and tissue plasminogen activator, used for thrombolysis, were proven to be effective. The study findings for tumors in animals and humans are different.

8.6.1.2 Only Few Animal Studies Can be Used in Human Clinical Practice

A report released in 2006 shows that only a small number of animal experiment findings that are published in world famous journals can be translated into clinical study. Drs. Daniel G. Hackam and Donald Redelmeier, from the Medical School of Toronto University in Canada, retrieved high-quality animal experiment study papers about disease prevention and treatment, which were published between 1980 and 2000 in seven top scientific journals (*Science, Nature, Cell, Nature Medicine, Nature Genetics, Nature Immunology,* and *Nature Biotechnology*), and were cited over 500 times and then tested in humans. Seventy-six articles were found to meet the above standards, of which the outcomes are all positive. Each article was cited 889 times, on an average. The animal experiment designs and methods of 38% of the 76 articles were reliable. In addition, most of those articles contained reliable gradients of medicine dose, relevant clinical outcomes, and long-term end outcomes. Most experiments are limited to random grouping, multi-hypothesis adjustment tests and blind evaluation of the outcomes; the quality of animal experiments did not significantly improve from 1980 to 2000.

Among the 76 relevant animal experiment studies, only 28 were translated into human clinical randomized trials, of which 14 studies demonstrated different outcomes from animal experiments to human clinical trials, 34 animal experiments have never been translated into human clinical process, and 8 findings from animals were finally used in treating human diseases (10.5%). It takes 7 years, on average, for findings from animal experiments to reach clinical translation.

We can infer that only 1/3 of high-quality animal experiment findings that were published in renowned journals, cited frequently, and had a huge impact can be translated into human clinical trials. Moreover, only 10% of animal experiment findings can transfer the tested process to human clinical practice. The results of animal experiments published in ordinary journals (not top journals) are of low quality and their possibility of clinical transfer will be very low.

The two cases above show us the great difficulty that translation medicine is facing. There are many factors that can explain the inconsistent outcomes: (1) Some clinical trials lack enough data to prove the effectiveness of treatment plans. For practical or commercial purposes, some clinical trials ignore the limitations of drugs found in animal experiments in their design (e.g., toxicity). (2) Some promising animal experiment projects fail to translate into clinical practice due to lack of data. Researchers may be inclined to choose positive data and ignore negative but effective data. (3) The animal models simplify human diseases, and those models cannot fully or correctly mimic the pathology and physiology of humans. Laboratory animals differ from human patients as they are usually young and have fewer complications. (4) Compared with clinical trials, animal studies proven negative or fruitless have no chance to be published, thereby giving an impression that animal experiment outcomes tend to be positive.

Here, we are faced with a drive to solve (2), and (3) relates to animal model problems and a need to improve the accuracy of the experimental animals, elimination of bias, and a need to improve the reliability of animal experiments to increase the efficiency by which animal experiments are translated to clinical practice.

The successful translation of basic animal study into clinical practice is not only the fundamental requirement of translation medicine, but is also the ultimate goal of scientific medical study.

8.6.1.3 Successful Cases of Translation from Animal Experiments to Clinical Treatment: The Development of Statins

In the latter half of the twentieth century, a milestone event in the developmental history of cardiovascular medicine was the statin. The use of statins changed life styles and the death rate of Americans with cardiovascular disease was decreased by 25%.

Before the mid-twentieth century, there was a lack of understanding of cardiovascular diseases as well as effective prevention and treatment. In 1948, the National Institutes of Health (NIH) launched a project in Framingham, Massachusetts, that later became the world-famous Framingham Heart Study Project. Its original purpose was to study coronary heart disease (CHD) performance and CHD risk factor in normal people, and to establish a new screening method.

The Framingham study initially consisted of 5209 males and females, aged 28–62, including 1644 couples and 596 families. After 10 years of follow-up visits, Framingham creatively came up with the concept of risk factors, and found hypercholesterolemia as a risk factor of cardiovascular disease, implying that reducing the cholesterol level in humans may reduce cardiovascular morbidity and mortality.

In the 1960s, through animal experiments and human clinical observation, scientists found that blood cholesterol is from food and the body's own synthesis. When the level of cholesterol from food is low, the liver can increase cholesterol synthesis to meet the needs of the body. Conversely, when the source of cholesterol is abundant, the synthesis and absorption of cholesterol in the liver and intestines will be inhibited. In 1966, scientists discovered that the synthesis of cholesterol in the liver is regulated by a liver enzyme particle called 3-hydroxy-3-methylglutaryl-coenzyme A (HMG-CoA) reductase. Later, they demonstrated that HMG-CoA reductase may lower blood cholesterol.

In 1971, a Japanese biochemist, Akira Endo, and his colleagues screened more than 6,000 microbial strains in Sankyo Pharmaceuticals Company, and finally extracted three compounds that could lower cholesterol levels from *Penicillium citrinum*, and one of them, mevastatin, can be specifically combined with HMG-CoA reductase to inhibit endogenous cholesterol synthesis, thereby lowering the blood cholesterol.

The safety and curative effects of mevastatin needed to be fully tested in animal experiments.

In 1974, Dr. Endo gave rats 20 mg/kg of mevastatin orally, and tested the plasma lipid level after 3–8 hours. He found that mevastatin could reduce the blood cholesterol in rates by 30%. However, this result was difficult to repeat. Dr. Endo was not certain whether this was due to experimental errors (such as technical problems) or that mevastatin was not effective. Subsequent animal experimental observations demonstrated that even adding 0.1% mevastatin into the rat diet for seven consecutive days did not change the blood cholesterol level. Moreover, increasing the dosage to 500 mg/kg for five consecutive weeks did not reduce the blood cholesterol level either. Mice also exhibited the same results.

Later, Dr. Endo continued to screen derivatives and analogues of mevastatin *in vitro*, but found that mevastatin was still the best. In 1977, he used rats to conduct a more detailed study. He demonstrated that 3–8 hours after the rats were given mevastatin, the synthesis of cholesterol in the liver was inhibited. This proved that mevastatin reacted very quickly in rats. However, as the rats were given a large dosage of mevastatin, compensatory HMG-CoA reductase in the liver increased by 3–10 times, masking the lipid-reducing effects of mevastatin. This initially accounted for the failure of the rat experiments. The nonionic detergent Triton WR-1339 can increase rat liver HMG-CoA reductase and synthesis of cholesterol, thereby generating a hypercholesteremia rat model. Giving hypercholesteremic rats 100 mg/kg of mevastatin reduced the plasma cholesterol by 21%. Although the experimental results gave hope, it was not enough to prove the efficacy of mevastatin.

There is 300 mg of cholesterol in an egg; two-thirds of which come from consumption, and one-third from autosynthesis. Due to the requirements for egg production, the level of cholesterol synthesis is higher in hens than in males. In 1978, Dr. Endo added 0.1% mevastatin into the hen diet for 30 consecutive days. As expected, hen plasma cholesterol was reduced by 50%, and it did not influence the animal's body weight, diet uptake, or egg production.

The success of the hen experiment gave Dr. Endo confidence, and he continued successful experiments on dogs and monkeys. He gave dogs and macaques 20 mg/kg of mevastatin, and plasma cholesterol was decreased by 30% and 21%, respectively. Mevastatin can significantly decrease "bad" cholesterol—low density lipoprotein (LDL) cholesterol, and slightly increase "good" cholesterol—high density lipoprotein (HDL) cholesterol. Later, the use of Watanabe heritable hyperlipidemic rabbits (LDL receptor deficient) also confirmed that mevastatin could reduce plasma cholesterol by 39%.

The above animal experiments demonstrate that mevastatin can significantly decrease plasma cholesterol in animal models, such as chickens, rabbits, dogs, and primates, but it is ineffective in traditional rodents. As different species of animals have different metabolic pathways for lipoprotein in their livers, the curative effects of mevastatin (reducing plasma cholesterol) are also different. After treating with mevastatin, synthesis of cholesterol in the liver decreases, chicken, dog, monkey, and rabbit plasma cholesterol consumption increases, and plasma cholesterol levels decreases. On the contrary, in rats and mice, their liver HMG-CoA reductase levels increase after mevastatin treatment, and synthesis of cholesterol in liver increases because rats cannot break down and use plasma lipoproteins. Furthermore, the main lipoprotein in rats and mice is HDL, not LDL, and mevastatin can also decrease rats' bile acid excretion. Due to the above reasons, mevastatin is not effective in rats and mice.

By repeated animal experiments and clinical trials, the lipid-lowering effects of statins were finally confirmed. In 1978, the FDA approved lovastatin by Merck to appear on the market, creating a human revolution for the treatment of hyperlipidemia, and its significance is comparable with penicillin. In the past 30 years, human heart disease and stroke mortality has fallen by 50%, and statins are the most important contributor. Fourteen international, large-scale clinical trials, involving 91,000 patients, confirmed that after statin treatment, the incidence of heart disease decreased by 30%. Currently, thousands of people worldwide use oral statins to prevent and treat cardiovascular diseases.

We can draw two conclusions from the statin study process. First, animal experiments are indispensible in medicinal development. Second, all scientific problems need to be explored from different sides and different levels to find a solution. Finally, selecting animal models in drug development is significant.

8.6.2 The Key of Translational Medicine is Animal Experiments

The US National Institute of Neurological Disease and Stroke (NINDS) discovered that many spinal cord injury animal experiments supported by NINDS could not be reproduced. This is mainly because the animal experimental design was incomplete or descriptions were not accurate. The problems were in how to randomly select animals, set groups, and define animal exclusion. By analyzing 100 animal experiment articles published in *Cancer Research* in 2010, scientists found that only 28% of the articles reported that the animals were randomized into groups, and only 2% of the articles used blinded treatment. Not one paper explained how to determine the number of animals in each group or how to avoid errors in results. In addition, hundreds of stroke, Parkinson's, and multiple sclerosis animal model experimental results demonstrated that if there are flaws in the clarifying key method parameter, the results may be biased. Two study reports mentioned in Section 6.1 of this chapter also explained that animal experiment design flaws and bias in results exist extensively in biomedical studies.

Here is a brief analysis of the reasons that may cause bias in animal experiment, so as to remind researchers to improve animal experiment design standards and data analysis ability, thereby improving the accuracy of the animal experiment and translating animal experiments to human clinical practice.

8.6.2.1 Internal Validity

Internal validity means to eliminate the possible bias by experimental design and implementation. The bias here means the estimated interventional effects are distorted to a true value, and it is due to the improper analysis of experimental design, implementation, or results.

Animal experiments having sufficient internal validity means the difference between different treatment groups of animals, in addition to the random error, is caused by different treatment. As shown in Table 8.3, four kinds of bias can cause different systemic errors among different treatment groups, lowering internal validity.

Table 8.3 Four Types of Bias That Influence Internal Validity

Bias Type	Definitions	Solutions
Selection bias	Bias allocation to treatment groups	Randomization, allocation concealment
Performance bias	Systemic differences in care between the treatment groups, apart from the intervention under study	Blinding
Detection bias	Systemic distortion of the results of a study that occurs when the person assessing the outcome has knowledge of treatment assignment	Blinding
Attrition bias	Unequal occurrence and handling deviation from protocol and loss to follow-up between treatment groups	Blinding, intention-to-treat analysis

Like any clinical trial, every animal experiment that has to test intervention effects should be established based on a detailed study plan including experimental design, implementation, results analysis, and experiment report.

8.6.2.1.1 Randomization

Randomization means to distribute the control group and experimental group randomly to ensure the forecast is impossible. In order to avoid selection bias, just like all methods used in human clinical trials, all animal experiments should be randomized. Randomization can prevent researchers from choosing a special individual animal for treatment, thus avoiding an individual animal that makes a certain effect especially obvious or especially inconspicuous as representative of the entire group's treatment effect. To group before treating may cause selective exclusion caused by prognostic factors of individual animals. The above problems occur in experiments where the researcher knows or can predict grouping, even if it is according to the prescribed rules, such as interval grouping or grouping according to time, or according to open randomized grouping rules. However, to randomly grab an animal in the animal cages may also consciously or unconsciously risk manipulation. For example, animals that run slowly may be caught first, thus forming a comparison of slow and fast, which cannot reflect true randomization.

If the animals already show features of a homogeneous group, such as inbred strain animals, randomization is less important. Mice, rats, or other rodents often have homogeneous characteristics. Factors that cause increases in variation are not animals themselves, but disease and infection. For example, most of the rat models of ischemic stroke have marked variations in the infarction area, not only because of individual anatomical differences of collateral circulation (the same inbred line is also different), but also because some individuals have better arteries than other individuals who easily get complications (e.g., periprocedural hypotension or hypoxemia) and affect the experimental results. Due to this kind of difference, it is necessary to do randomized group after injury or surgery.

Automated randomized techniques, such as random number generation, are widely used in clinical trials. However, some manual methods, such as tossing a coin or throwing dice, are also accepted because they cannot be manipulated.

8.6.2.1.2 Blinding

Blinding is also called masking. It means that researchers and staff who collect data and evaluate results do not know the control and treatment group. In a blinded animal experiment, researchers and other experimental personnel are not influenced by the experimental treatment group, thus avoiding attrition bias. Knowing the grouping may subconsciously affect additional treatment, the evaluation of results, and elimination of laboratory animals.

Triple blinding is the most commonly used method in clinical trials, where the patients, researchers, and treating personnel do not know the grouping. As patients do not know what kind of treatment will be given, their placebo effect should be the same as the control group. Animals are not sensitive to placebos, therefore animal experiments do not use the triple blinding method.

8.6.2.1.3 Sample Size Estimation

Sample size means the number of animals involved in the experiment. The choice of sample size is very critical to any control experiment design. The sample size must be large enough to detect the treatment effects on the scale of a given group more accurately, but at the same time, the sample size should be small to conform to ethical, legal, 3Rs, and actual requirements. Sample sizes needed for animal experiments should be confirmed before the start of the experiment by formal sample size calculation (for details, please see Chapter 7). In general, the number of noninbred animals is greater than inbred animals. Unfortunately, the estimate of the measured value of variation is based on incomplete data, and one minor mistake can lead to estimating the result as too low or too high. From an ethical perspective, effects of low estimation are not desirable because they may lead to the wrong conclusion that the intervention had no effect, and all the animal experiments are wasted. Overestimating the effects of the study is also immoral, but this happens less frequently.

8.6.2.1.4 Testing Physiological Parameters

Physiological variables (e.g., sex, age, breeding conditions change, pregnancy, etc.) can affect the final animal experimental results. If these physiological variables are not well controlled, the wrong conclusion may be obtained. As for whether the physiological parameters should be evaluated and how long to be evaluated depend on the design of the animal model and determined conditions.

8.6.2.1.5 Eligibility Criteria and Elimination

Eligibility criteria mean defining eligibility and elimination criteria. In other words, it is the criteria that designate which animals can be used in the experiment.

Due to the complexity of complications, many animal models are liable to complications; for example, during operation leading to blood transfusion causing brain or myocardial ischemia. Although it may not relate to animal experimental treatment, it has a large influence on the experimental results. If the eligibility criteria is set before the experiment instead of after experiment, and the staff who eliminate animals do not know the experimental process, it is possible to remove these animals that are infected or developed complications according to eligibility criteria.

In human clinical trials, eligibility and elimination criteria are usually used before the trials, but sometimes it is reasonable to use it after the start of animal experiments. However, when using the criteria, it should be limited to complications that clearly have no relation with experimental intervention, or it may lead to a bias. For example, a new treatment plan targeting colorectal cancer growth may make tumor-burdened animals become weak and increase their susceptibility to infection instead of inhibiting the development of tumors. Prematurely ruling out dying animals due to infectious diseases may lead to selective elimination of the largest tumor-burdened animals to cover the harmful effects of treatment.

8.6.2.1.6 Data Analysis

Many articles and books give detailed descriptions for data analysis of experimental results. However, although the data are sometimes simple and easy to analyze, improper methods are often used. Common mistakes are to do a t-test to data with no parametric data; calculate average and standard deviation from ordinal data; multiple observation from one animal as independent.

Analyzing all the data that are listed in the animal experiment according to the original intention of treatment, whether or not the intervention of these animals is complete, is intention-to-treat analysis. It is a kind of method analyzing randomized group test results, and it is welcomed in clinical trials because it can avoid the loss of associated nonrandom participant bias.

8.6.2.1.7 Study Behavior and Ethical Education

Many researchers and students heavily focus on the number of scientific papers published, the influence of the published article, such as impact factor of the journal, and have the concept that positive results are more likely to be published than negative results. Animal experiments should not only emphasize randomization, concealed grouping, and blinding, but also require a third party to fully monitor and audit the research work. Some academic groups do this to strengthen the management of animal experiments and increase transparency to ensure reliability of animal experiments.

8.6.2.1.8 Deviation in the Study of Animal Experiments

It was discussed previously that there is a wide deviation in acute ischemic stroke animal experiments, possibly because in this field, differences in animal experiments and human clinical trials are large and easy to recognize. When systemically evaluating the different experiments on acute ischemic stroke, other serious diseases, Parkinson's disease, multiple sclerosis, or amyotrophic lateral sclerosis, only one-third or less of the reported research have randomly assigned treatment groups, and the proportion with hidden therapy groups or blinded evaluation results are even less. Even if the study papers are published, randomization and blinding method used in the experiment are seldom written down. According to our observation, only 0%–3% of animal experimental studies reported the sample size calculation method. Almost all of the studies focused on whether the animals die during experiment, and 90% of the animals that died early were eliminated in the analysis. In a review of treatment of acute ischemic cerebral stroke studies, only one out of 45 articles mentioned set eligibility and elimination criteria in advance, and 12 studies mentioned the eliminated individual animals at the time of analysis. It is incredible that the other experiments designed by researchers ran smoothly. This phenomenon rarely exists in real animal experimental studies.

There are two factors that limit the interpretation of the data. First, the evaluation of confusion factors in systemic evaluation is based on article reports, and the contents of articles may be incomplete because the author may think these have nothing to do with the design of experiment and did not mention it. Next, the definition of randomization, concealed grouping, and blinding may be different. For example, to randomly pick up animals from cages can be defined as "randomization."

8.6.2.2 External Validity

External validity is the deduction of animal experimental results to humans. Even if the animal experiment design and implementation are very reasonable and can exclude bias, because there are major differences in detection among animal models of human disease, treatment strategies and real human clinical trials, the results of animal experiments may fail to translate to clinical medicine. Common reasons for this kind of low external validity are as follows: (1) Animal models used to induce disease in the experiment are young and healthy, while most patients exhibiting diseases are elderly. (2) The evaluation of a treatment is in a single animal (inbreeding or gene knockout), while human patients are diverse (including genetic diversity and environmental diversity). (3) Animal experiments often may only use single sex, male or female, while the emergence of human disease is independent of gender. (4) Single disease animal models lack similarity to multiple complications in human beings. (5) Extremely high or toxic doses can be used in animal experiments, while clinically it is not realistic or cannot be accepted by patients. (6) Time of evaluating experimental results is different in animal studies and clinical trials. Animal and human disease differences are not limited to pathophysiology including coexisting diseases, the application of combined treatment methods, treatment time, dosage, and the choice of testing results. Although internal validity may be based on the animals rather than certain diseases, external validity, to a large extent, is a factor decided by specific diseases.

8.6.2.2.1 Acute Disease Animal Models

As discussed previously, animal model studies targeting neurological disease have a low success rate for translating, especially ischemic stroke. With the increasing of age, the stroke incidence in humans increases continually. Stroke patients often have other health problems, and these problems can increase their risk of having a stroke. We know that in patients with acute cerebral infarction, occurrence of high blood pressure and hyperglycemia reached 75% and 68%, respectively. It is important to know whether candidate drugs can retain effectiveness with these complications. A large survey found that only 10% of the ischemic lesion studies used animals with high blood pressure, and less than 1% of animal studies used diabetic animals. Furthermore, almost all stroke models use young males; female animals are mostly ignored. More than 95% of the studies are done on rats and mice, and animals that are biologically closer to humans are seldom studied. In addition, most animal studies do not recognize that there is an inevitable delay between the possibility of onset of symptoms and treatment to the patients. In animal experiments, the average time is only 10 mins from the onset of ischemia to beginning of treatment, but it is infeasible in clinical trials. In most of the human clinical trials, functional results are the main measures for curative effects. However, animal studies usually depend on the infarction volume. Some studies demonstrated that the infarction volume and functional results are only medially correlated. Moreover, result evaluation for animal models is usually 1–3 days, a sharp contrast with the 3 months in human patients. For these reasons, except for thrombolysis, it is apparent why all treatment strategies that were proven effective in laboratory animals failed clinically.

The above-mentioned differences between animal models and clinical trials may have been the cause of failure in treatment for relieving lethal reperfusion injuries in patients with acute myocardial infarction. As serious complications in patients are ignored and there is inevitable initial treatment delay in clinical trials, the external validity of brain trauma animal model studies is limited.

8.6.2.2.2 Chronic Disease Animal Models

External validity of animal experimental models of chronic progressive disease is also challenged by other influencing factors. For the treatment of Parkinson's disease, researchers mainly rely on simulating damage, which lacks the substantia nigra striatum dopamine defects, to induce animal models instead of considering the late, progressive, and continuous degradation characteristics of human disease. In human clinical trials, interventions are conducted in the long process of chronic disease duration. Neuroprotective agents that are acknowledged in typical basic animal experiments are used before or at the same time in acute

Parkinson's injuries, which is an obvious difference with the human treatment for Parkinson's disease.

8.6.2.3 Publication Bias

The ideal evaluation for new clinical trial treatment strategies is based on previously published clinical studies. Systemic evaluation and comprehensive analysis on clinical data are done in order to choose the treatment strategies with the most potential. However, if the published study is based on a selective part of the results, even though meta-analysis is based on strict system analysis, it is misleading.

The bias in reporting of clinical trials has been widely studied. Strong evidence shows that only positive or significant studies are likely to be published. Conclusions having statistically significant differences are likely to be reported with all content instead of just the abstract. This is called a publication bias, and it can lead to overestimating the efficacy of the treatment and make readymade evidence unreliable when deciding.

Publication bias in animal experiments does not get enough attention. Reports on systemic analysis of intervention effectiveness studies for animal models of human disease demonstrated that four out of six articles have a publication bias. Another study conducted meta-analysis on 525 published articles. Edger regression analysis and Trim and Fill analysis results show that there is widespread publication bias. The study believed that publication bias may explain one-third of validity problems in the study of animal stroke.

Not publishing the results of animal experiments is unethical. It not only deprives researchers of their rights to share and use accurate data for clinical trial potential, but also because there was no contribution to the accumulation of knowledge and the animals were wasted. Furthermore, studies that exaggerate biochemical effects may lead to more unnecessary animal experiments to test invalid hypotheses.

8.6.2.4 Actual Strategy for Improvement

In many human clinical trials, many interventions that look promising have failed. This is due, in part, to a lack of preclinical testing with enough internal and external validity, that is, animal experiments, and animal experiment study articles tend to publish positive results. As a result, based on clinical trials and some preclinical testing, it is suggested that using standards similar to clinical trials when using animal models to determine the plan of disease treatment or writing reports ensures that the animal experimental study is based on high-quality data with no bias. For details, determining the sample size and eligibility criteria; deciding eligibility; reasonable experiment grouping and allocation concealment; blinding; paying attention to whether the physiological parameters of animals are monitored and controlled in a reasonable range; adopting accurate statistical and analytical methods are suggested.

Not only should human disease animal models' disease or injuries be as close as possible with humans, but age, sex, and complications should also be as similar as possible with the human disease. Researchers should prove the rationality of their methods in choosing animal models and measuring results. On the contrary, human clinical studies should also be designed to have reusable design conditions in which successful experiments can be as effective as possible. In order to fully explain the potential and limitations of a relatively new strategy, systematic review and meta-analysis should be performed on all evidence from animal studies before the start of human clinical trials. The history of the development of statin drugs adequately demonstrate that experimental evidence obtained from a single animal species, single animal model, or single animal type (strain) may not be enough. All animal models exhibit only a part of human diseases. Humans are not mice. Once one fully understands the basic knowledge of human disease and the pros and cons of animal experiments, animal experiments for translational medicine can be carried out to truly achieve medical conversion of animal experiment results.

BIBLIOGRAPHY

Endo A. The discovery and development of HMG-CoA reductase inhibitors. *Journal of Lipid Research* 1992;33:1569–82.

Hackam DG. Translating animal research into clinical benefit. *British Medical Journal* 2007;334:163.

Hackam DG, Redelmeier DA. Translation of research evidence from animals to humans. *The Journal of the American Medical Association* 2006;296:1731–2.

Kannel Wb, Dawber Tr, Kagan A et al. Factors of risk in the development of coronary heart disease—six-year follow-up experience. The Framingham Study. *Annals of Internal Medicine* 1961;55:33–50.

Landis SC, Amara SG, Asadullah K et al. A call for transparent reporting to optimize the predictive value of preclinical research. *Nature* 2012;490:87–91.

Liu E. *Animal Models of Human Diseases.* Second Edition. People's Health Publishing House, Beijing, China, 2014.

Liu E, Yin H, Gu W. *Medical Laboratory Animals.* Science Press, Beijing, China, 2008.

National Research Council. *Guide for the Care and Use of Laboratory Animal.* Eighth Edition, National Academy Press, USA, Washington, 2011.

Perel P, Roberts I, Sena E et al. Comparison of treatment effects between animal experiments and clinical trials: systematic review. *British Medical Journal* 2007;334:197–203.

Rea PA. Statins: From fungus to pharma. *American Journal of Science* 2008;96:408–5.

Russell WMS, Burch RL. *The Principles of Humane Experimental Technique.* Methuen, UK, London, 1959.

Singer P. *Animal Liberation: A New Ethics for Our Treatment of Animals.* Random House, New York, USA, 1975.

Stokes WS. Humane endpoints for laboratory animals used in regulatory testing. *Institute for Laboratory Animal Research Journal* 2002;43 Suppl:S31–8.

van der Worp HB, Howells DW, Sena ES et al. Can animal models of disease reliably inform human studies? *PLoS Medicine* 2010;7(3):e1000245.

Van Zutphen LFM, Baumans V, Beynen AC. *Principles of Laboratory Animal Science.* Second Edition. Elsevier Science Publishers, Amsterdam, Netherlands, 2001.

Vesterinen HM, Sena ES, ffrench-Constant C et al. Improving the translational hit of experimental treatments in multiple sclerosis. *Multiple Sclerosis* 2010;16:1044–55.

Zhao S, Liu E, Chu Y et al. Numbers of publications related to laboratory animals. *Scandinavian Journal of Laboratory Animal Science* 2007;34:81–6.

Index